ALLE ZEIT WACH 1842

Emil Fischer

AUS MEINEM LEBEN

With a Prologue and an Epilogue
by Bernhard Witkop

Reprint from
Emil Fischer, Gesammelte Werke
Edited by M. Bergmann

Reprint

Springer-Verlag
Berlin Heidelberg New York
London Paris Tokyo

Emil Fischer

ISBN-13: 978-3-540-07662-9 e-ISBN-13: 978-3-642-72796-2

DOI: 10.1007/978-3-642-72796-2

Softcover reprint of the hardcover 1st edition 1987

Printing: Mercedes-Druck, Berlin; Binding: Lüderitz & Bauer, Berlin
2152/3020-5 4 3 2 1 0

Emil Fischer's Living Legacy

A Prologue by Bernhard Witkop

From the point of view of the history of science, *transmission* is as essential as discovery (George Sarton).

Emil Fischer's autobiography was published posthumously three years after his death, July 15, 1919, by Max Bergmann in 1922 shortly after Kurt Hoesch's "Emil Fischer, sein Leben und sein Werk" had appeared. Other famous students of Fischer, Emil Abderhalden, Max Bergmann and Karl Freudenberg tried to act as chroniclers and custodians of Fischer's rich legacy.[1]

The Meaning of Memoirs – A Literary Excursion

Richard Willstätter who chose the same title "Aus meinem Leben" for his literate autobiography[2], contrasts Hoesch's „pathetischen Stil" (pompous style) with Fischers's refreshing informality. As a student and close friend of Adolf von Baeyer, Willstätter criticizes Hoesch's presentation for lack of depth and insufficient inside knowledge.

Willstätter's own characterization of Fischer has the ease and charm whose absence he criticizes in Hoesch's presentation: "Emil Fischer was always reserved, imposing and serious, but he had the capacity to unbutton and to relax into his charming and captivating self, though during the hectic years in Berlin his capacity for fun and relaxation, so liberally practiced during the years in Munich and Erlangen, was markedly reduced and finally suppressed. With some hesitation he displayed his Rhenish legacy of gaiety and *joie de vivre* and he did not easily adjust to the unbending sober ways of Berlin, counterpoint to Munich. What weighed sometimes heavily on his mind was the awareness of his growing mission which reinforced his latent drives for dedication, duty and perfection. This burden grew with the years and is to blame for the fits of melancholy in his later life which he was to end by his own hand. Only those who knew him better, comprehended what lifted and what aggravated this great chemist".

Fischer's authoritarian personality reacted sharply to dissenting opinions and he did not suffer fools lightly. Colorful gifted colleagues, such as his Division Head, Carl Harries, had difficulties in avoiding clashes. Willstätter, who visited Harries in his palatial mansion in Berlin-Charlottenburg, formerly belonging to Werner von Siemens, wondered about a huge painting in the study hanging with its face against the wall. When Willstätter questioned him, Harries turned it around with the comment: "This is Emil Fischer with whom I happen to be cross at the moment, so he hangs that way for punishment."

Max Bergmann, long-time friend and disciple, calls Fischer's autobiography "jenes eigenartigste schriftliche Dokument, das wir von Fischer besitzen" (that most singular written document, which we possess from Fischer's hand). Bergmann notices Fischer's restraint and reservation dictated by self-criticism in excess of normal bounds. These autobiographical notes were written to beguile the tedium of two convalescences, one in Locarno in the Spring of 1918 after the forth strenuous Winter of the First World War in which he suffered from bronchitis, gall bladder problems and pneumonia, leave alone the stress and depression caused by Fischer's losing battle with his many obligations to chemical education, research and war efforts.

Since 1895, when his wife Agnes, née Gerlach, had died of meningitis, after only seven years of married life, Fischer had been a widower. His son Walter Max (1891 - 1915) took his own life because of depression; his second son Alfred Leonhard Joseph (1894 - 1917) died, a military physician, of typhus in Roumania. Under these gathering dark clouds he continued his autobiography in the following summer, when in Karlsbad he took the famous waters to alleviate his biliary troubles.

The reader of the 200 pages of – incomplete – autobiographical notes would not suspect them to be transcriptions of stenographical records resulting from comparatively few sessions with the recording secretary,

Fräulein Margarethe Barth. In a style almost too relaxed, Fischer resuscitates all the encounters of persons, members of his family, colleagues, friends and notable contemporaries who crossed his way. In limning their character traits in loving detail he is full of charity toward everybody and has malice toward none. But what is missing are any reflections on Fischer's own inner life.[3] In the words of Max Bergmann:

> „Den Schleier vor dem Geheimnis des eignen Ichs zu lüften, hindert Fischer in seiner Selbstbiographie dieselbe Scheu, die ihn völlig verschlossen in allen Fragen des eigenen gefühlsmäßigen Empfindens gegen seine Mitmenschen und selbst seine besten Freunde machte. Nur jener Untertitel seiner Erinnerungsblätter, ‚Geschrieben in dem Unglücksjahr 1918' deutet, wenn auch in der zurückhaltendsten Form, die tiefe Depression an, die sich bei Fischer seit Beginn des Weltkrieges langsam vorbereitete und ihn nach dem Verlust zweier geliebter Söhne und nach dem für Deutschland verlorenen Krieg mehr und mehr niederwarf."

> "Fischer was kept from lifting the veil covering the secrets of his own person by the same shyness that prevented free communication of his innermost feelings with fellow human beings, even with his best friends. Perhaps the subtitle of his autobiographical notes 'written in the year of disaster 1918' may be an indication, if subdued, for the deep depression that had its beginning at the outbreak of the World War and that, after he was bereaved of his wife, of two beloved sons and after the war was lost for Germany, grew more and more and finally defeated him."

In personal memoirs the writer must decide what he will and what he will not tell, and in so doing makes crucial decisions about what is and is not important in his life – about what really matters to him, and about what he thinks will be most interesting and instructive to his readers. Memoirs can be made to serve many different causes. For instance *"The Autobiography of Benjamin Franklin"* is a classic, but much less an autobiography than an artful fabrication. The Franklin who emerges from these pages is not the real man as seen by his contemporaries and as revealed in many of his other writings, but a carefully contrived idealization of the real man. The autobiography gives us the legendary Franklin who was early to bed and early to rise, not the actual Franklin who kissed the girls and made them cry. What Franklin wanted to do was locate himself in American history and provide a model for youths to emulate. In the form of the memoir he found a convenient way to do so, just as in *"The Education of Henry Adams"*, its author found a convenient form for educating the public in his own, somewhat eccentric, theories of history.

Roots of Biochemistry in Chemistry

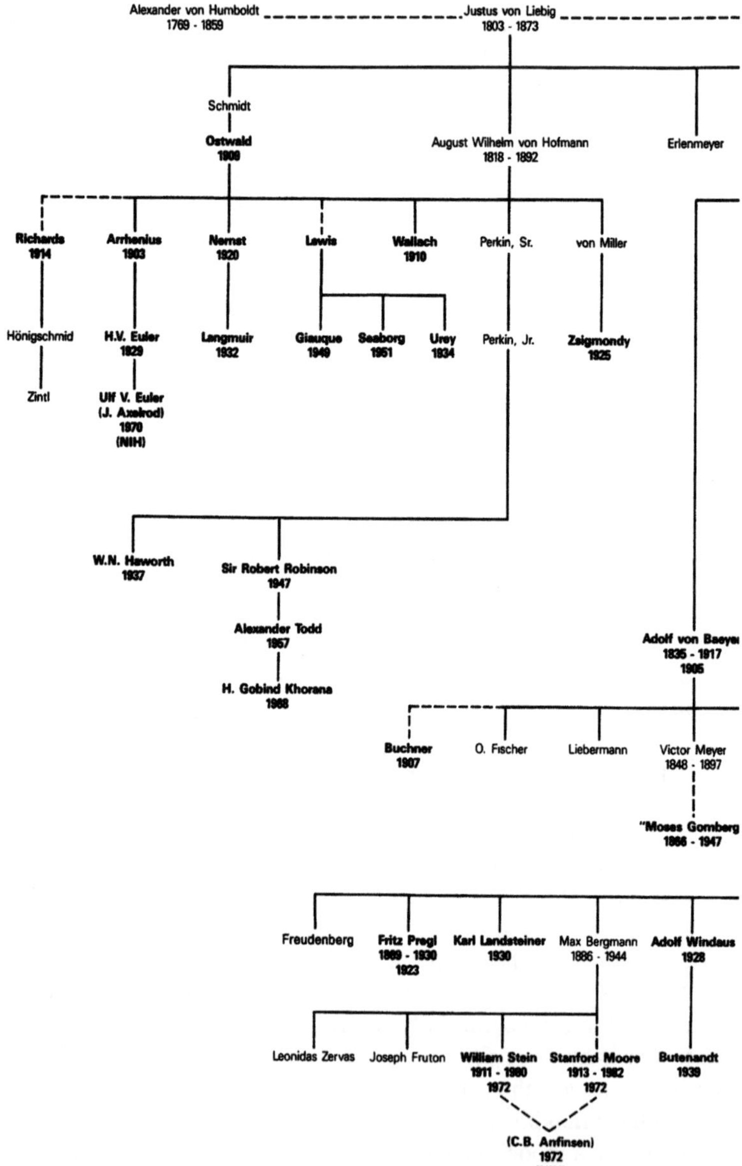

Table 1.
Emil Fischer (1852 - 1919) at the center of a great scientific tradition and surrounded by a "pride" of Nobel Laureates (in bold face with year of award) in CHEMISTRY, PHYSICS, PHYSIOLOGY and MEDICINE. Fischer's example

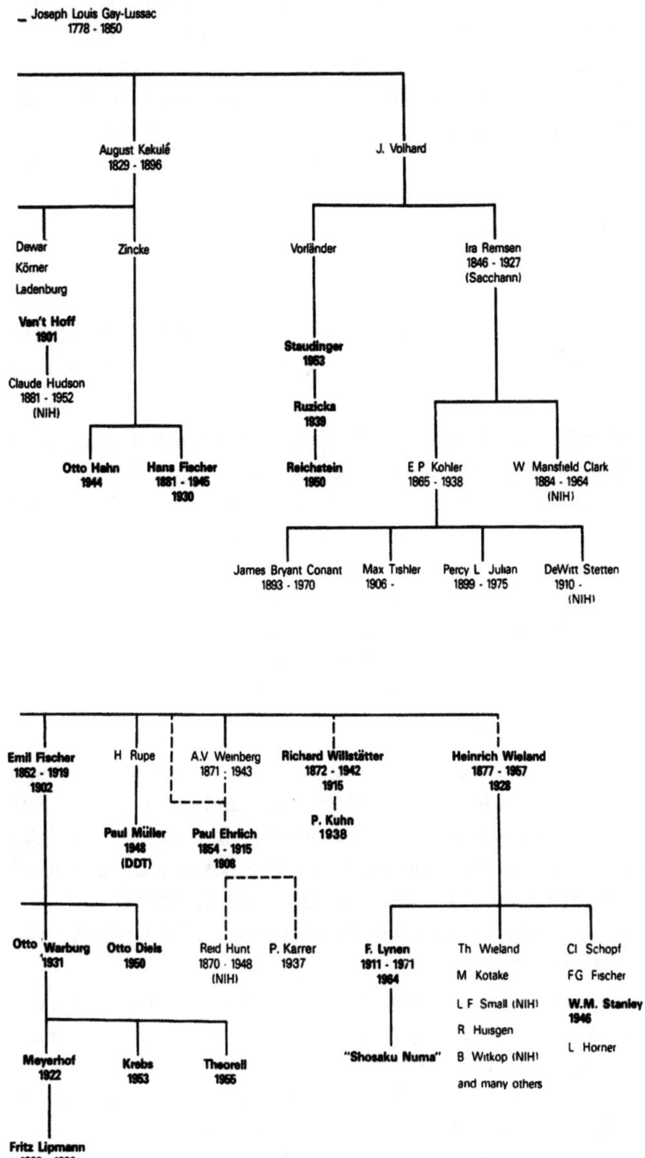

and importance still continue to show organic chemistry to be at the basis of biological, medicinal and clinical chemistry. Unlike pharmacology, biology, immunology and other fields of the life- and neuro-sciences organic chemistry never adopted nor required the modish epithet "MOLECULAR".

The many shades of memoirs in this context are further exemplified by H. L. Mencken's "Happy Days" in which he warns the reader: "I have made a reasonably honest effort to stick to the cardinal facts, however disgraceful to either the quick or the dead, but no one is better aware than I am of the fallibility of human recollection." If we combine this fallibility with the memoirist's desire to entertain and instruct, we wind up with something that is not exactly fiction, but certainly is no longer mere fact, though this kind of reporting may come a lot closer to the truth than a mere recitation of the facts ever could."

In this more literary than scientific analysis Fischer emerges as a writer who does not try to dazzle his readers with famous persons he has met, but attempts to remember what it was like when he and all the already famous persons were young and what the circumstances were in which he and they forged fame and career.

Like August Wilhelm von Hofmann (1818 - 1892) Fischer belonged to the now extinct breed of encyclopedic scientists who were able to comprehend and approach science in the way Alexander von Humboldt exemplified. The fragmentation into research islets and speciality journals came much later. A. W. von Hofmann filled three volumes with his "Erinnerungen an vorangegangene Freunde" (Remembrances of Friends Departed), truly a masterwork in which he portrayed the great classics, his teacher Justus von Liebig, Friedrich Wöhler, Thomas Graham, Jean-Baptiste Dumas, Quintino Sella, Adolph Wurtz *et al.* He also edited the letters of Liebig and Wöhler in two volumes (Braunschweig 1888).

Compared with these historical masterworks the reader of Fischer's autobiography must dampen his expectations, must understand the choice of the memoirist and follow his intent to find out with him what did and did not matter in his life, his science and his time. The passage of time itself sets accents that no writer can foresee. Like Banquo, the writer faces the same dilemma: "if you can look into the seeds of time and say which grain will grow and which will not . . ."

Clearly, Fischer's continuing merit and legacy is his innovative incorporation of biochemistry, enzymology, natural products, low- and high-molecular, into his type of comprehensive organic chemistry. We must see Fischer in this historical context (*Table 1*) in order to understand fully the extent of the tradition he represents and the legacy he left. He demonstrated that tradition is progress preserved and that progress is tradition carried forward. All the founders of modern biochemistry were his pupils and heirs, the Nobel Laureates Windaus, Butenandt, Warburg, Meyerhof, Krebs, Theorell and Lipmann. We are dealing here with an isolated phenomenon as important and unique as that of French Impressionism. Like post-impressionists there are many scientists in the old and New Worlds that owe their choice of topics and methods of experimentation to the inspiring example of Emil Fischer (see Epilog).

The Sensitivity of the Nose – a Forgotten Classic (pars pro toto)

In his autobiography Fischer refers briefly to his attempt to determine the sensitivity of the nose to odors. These experiments were done in Erlangen 1884 together with his colleague Franz Penzoldt. They were published in Proceedings of the Erlangen Medical Society 1886[4] and in the Annalen 1887.[5] Their centennial and the prophetic nature of these experiments merits their presentation in this preface. Fischer's official professional mansion in Erlangen happened to contain a large reception or banquet room with the impressive capacity of 230 m^3. The stone floor and the white-washed walls promised a minimum retention of odoriferous molecules. Here is the first English translation of the original procedure.[4]

> "Of the substance to be tested exactly one gram was weighed and dissolved in one liter of pure alcohol, and 5 ml of this solution was brought to the desired dilution, again with alcohol. The final solution, 1 - 3 ml, was transferred into a gas wash bottle with two curved glass tubes through a cork with two borings.
>
> Weighing, dilution and determination of the volume of the odoriforous solution, as well as the removal of all traces on the exterior of the gas wash bottle, all these operations were carried out by an independent helper not participating in the experiment and far away from the test room.
>
> For the test one of us (Fischer) sprayed the contents of the bottle into all directions of the closed room with the aid of a rubber bulb pressure device. The evaporation of the alcoholic solvent required 5 - 10 minutes. Then the air in the room was stirred by waving a large flag for about 10 minutes.
>
> When a special sign was given, the other experimenter (Penzoldt) entered the room in order to test the odor.
>
> The result was always controlled by a third unbiased and educated person.

In this fashion, after an initial exploratory preselection, two compounds "MERCAPTAN" (which in those days was short for ethyl mercaptan, b. p. 158°[6]) and chlorophenol (presumably o-chlorophenol m. p. 7°, b. p. 175°[6]) were tested as follows."

There follows the protocol which shows that Fischer greatly underrated the sensitivity of the nose which responded to as little as 0.01 mg of mercaptan in 230 m^3 *i.e.*, a dilution ratio of 1:60 billion. Fischer's later threshold value for ethane thiol[5] 7.1×10^{-13}M, is lower than new determinations, namely 5.6×10^{-11}, whereas his value for 2-chlorophenol of 3.4×10^{-11} compares favorably with the latest 8.1×10^{-11}M.[6] How instructive to compare two experiments done in the "sealing-wax-and-string" era of laboratory work with the modern age of electrophysiology and gas-liquid-chromatography-computer-oscillograph technology. Fischer's "equipment" cost practically nothing, while the contemporary grantee would submit a request for $ 150,000 to acquire all the state-of-the-art sophistry.

Fischer and The Receptor Concept

Fischer was well aware of the problem of sensitivity. He felt that he had to determine the volume of air, laden with the odoriforous molecules whose number or weight he wanted to know, that passed the *regio olfactoria* of the nose to be noticeable. He speaks of "Geruchsperception", percepton of odor, which is close to our common concept of *RECEPTOR*.[7] However, the passage or jump from perception to receptor was not made by Fischer in 1887, nor by John Newport Langley (1852 - 1925)[8] in 1878, but by Paul Ehrlich (1854 - 1915) in 1878 when he submitted his thesis to the University of Leipzig and postulated the "side-chain theory of cellular action" and the concept of selective affinity[9] that later on (1906) became his receptor theory.[10] It was not until 1894 that Fischer, impressed by the stereospecificity of the cleavage of glycosides by enzymes, coined the immortal "*lock-and-key*" analogy.[11] He skirted again the receptor concept in trying to correlate rotatory power (stereoconfiguration) with taste. In 1887 he did not try to speculate on a connection between the *regio olfactoria*, the odoriferous agent and the process of olfactory perception. However, he comments on sensitivity as follows:

"The best conditions for smelling tests are met by sniffing, *i. e.*, short inspirations through the nose which, because of the construction of nasal passages creates a strong turbulence of the air influx.

Penzoldt was able, after a deep exhalation, to repeat the inhalations about 20 times until his lungs were filled with an air volume of 5000 ml. In a single inhalation about 250 ml passed the nasal cavity.

According to Henle (Handbuch der Anatomie, 1873 II p. 855, 861) the volume capacity of the *regio olfactoria* in relation to the entire *cavum nasi* is about 1:5. This means that about 50 ml of air are necessary to come to an olfactory perception."

As impressive as Fischer's simple experimental setup is his circumspect approach to the problem of olfactory sensitivity. How do his pioneering thoughts compare with modern concepts?

Odorous intensities of a given substance can be expressed as a function of one physical variable, the concentration, *but only if the experiment is conducted under the almost natural conditions of sniffing.*[12, 13, 14] In addition there are three biological parameters, the threshold, suprathreshold intensity slope and saturation point. Current notions of olfactory perception suggest that the receptor neurons are rate-sensitive and that the idea of a threshold for sensation is imprecise and that a sensory system is better characterized in terms of a signal-to-noise-ratio.[15, 16, 17]

A leap of one hundred years has not invalidated Fischer's experiments, but has brought them into sharper focus. It may be difficult to assess a lower limit of sensitivity of the nose in terms of picograms per nose, for the mechanism of olfaction, but Fischer would be pleased, could he observe the growth of the seeds he sewed. The highly potent (tritiated) bell pepper odorant, 2-isobutyl-3-[^{3}H]methoxy-pyrazine, binds specifically and saturably to nasal-epithelium.[18] An odorant-binding protein localized in nasal glands has been isolated and may be a transport rather than a receptor protein.[19] Fischer's title: "*Über die Empfindlichkeit des Geruchssinnes*" is still a problem, more timely and significant than hundred years ago!

A Prophecy of Fischer's Practical Mind

In the same publication Fischer states:

"This extraordinary sensitivity of the nose toward mercaptan suggests the idea to use it in tests on air currents, diffusion of gases, tightness of ventilating systems of discovery of rifts, crevaces, passages or aquaeducts in mines, mountains and other terrain."

It was not until odoriferous city gas from cokeries was replaced by odorless or natural gas, that the need arose to deter customers from breathing such gas or to prevent leakage and explosions. Many city gas works introduced trace amounts of butylmercaptan as a tracer. Probably nobody then took the time or the trouble to recall or reread Fischer's pioneering paper. The writer of this preface shares the reader's joy in seeing Fischer's autobiography resuscitated for such new insights gained after so much water has flowed down the various rivers that served Fischer's abodes, starting with the Rhine (Euskirchen and Bonn), Isar (Munich), Regnitz (Erlangen), Main (Würzburg), Spree (Berlin) and finally the Styx.

The Meeting in Karlsbad (1902) and The Structure of Proteins

A hundred years ago scientific congresses were few and far between. The chemists usually participated in the annual event called "Tagung der deutschen Naturforscher und Ärzte" which Fischer attended for the first time in Hamburg 1876, in the company of his cousin Otto and his friend Wilhelm Königs (1851 - 1906), a jolly trio who fitted the Hamburgers' pun of being "*forsche Naturen*" (brisk blokes) rather than *Naturforscher* (explorers of nature).

One generation later on the afternoon of September 22, 1902, at the 74th Annual Meeting of the "Gesellschaft deutscher Naturforscher und Ärzte" in Karlsbad Fischer read a paper, abstracted in the Chemiker-Zeitung,[20] in which he summarized the isolation of amino acids and "*peptides*" from protein hydrolyzates. In the same way as he designated carbohydrates as disaccharides, trisaccharides, polysaccharides, he proposed that a substance of the type of glycyl-glycine be named a depeptide, a higher order tripeptide and so on to polypeptides or proteins. Fischer wrote in the auto-abstract:

> "Finally the speaker discussed the coupling of the amino acids in proteins. The idea that acid-amide-like groups play the principal role most readily comes to mind (liegt am nächsten), as Hofmeister also assumed in his general lecture this morning. The same conviction led him [Fischer] more than 1 - 1/2 years ago to initiate experiments to effect the synthetic linkage of amino acids."

We owe it to the noted biochemist and historian of science, Prof. Joseph S. Fruton, Yale University, that these historical events were recently detailed and fully clarified.[21] Here is Fruton's admirable account:

"The Karlsbad meeting that links Hofmeister and Fischer thus provides still another apparent instance of simultaneous scientific discovery. For the purpose of this essay, however, their encounter is of greater interest in offering a contrast in the style of two scientists with different backgrounds and personalities, who worked in different institutional settings, and who had met in the consideration of the same scientific problem. The biologist Hofmeister had already done extensive and important work on proteins, and held a large view of the role of chemistry in the study of physiological problems. The chemist Fischer had just entered the arena of protein research, and was emboldened by his notable success in the synthesis of sugars and purines to believe that synthetic organic chemistry could also solve the problem of protein structure. In scientific stature, as measured by public esteem, Fischer vastly overshadowed Hofmeister. At the time of the Karlsbad meeting Fischer already held the title *Geheimer Rat*, and a few months later he was awarded a Nobel Prize in Chemistry, whereas Hofmeister received no honors of this sort."

In the epilogue we shall reflect on the seeds of time and on how many of Fischer's grains grew to mighty trees in the hands of able disciples. Fischer's synthetic peptides, especially his largest, an octadeca-peptide with 15 glycine and 3 leucine units, may not have stood the test of today's exacting criteria of purity, but they were milestones notwithstanding on the glorious road traveled by his able scientific children and grandchildren, Bergmann, Fruton, Zervas, Stein, Moore *et al.* Times have to be ripe for certain ideas to come to fruition. Fischer had enough self-criticism to feel uneasy about the homogeneity of his higher polypeptides so that after 1910 there are no longer any publications from his laboratory on synthetic peptides which, however, still gushed forth, in questionable purity, from the laboratory of Emil Abderhalden, his pupil in Halle.

Half a century later 34 synthetic peptides made by Emil Fischer were made available by his son Hermann O. L. Fischer at Berkeley to test their homogeneity. C. A. Knight examined these venerable samples and, to everybody's surprise, found that all of them but three gave single spots upon paper chromatography. The contaminants of the three exceptions were traces of the constituent amino acids.[21a] Elvin A. Kabat to whom I am indebted for this recollection, reflects on this event in his historical perspectives and prolog to the first volume of Annual Review of Immunology (1983).[21b]

Peptides and Proteins: They may not crawl, but they refold.

"I have recently made the first decapeptide and will now try to reach an eikosapeptide whereupon one should be midway in the protein group." Fischer's views on proteins, their true molecular weights and their properties were as naive as optimistic. They resemble the prediction of the physiologist Du Bois-Reymond who encouraged Fischer: "If you synthesize proteins large enough, they will start crawling!" Not quite so! But in Fischer's optimism, is there not a premonition, not of the problem of synthetic life, but of the raison d'être of large biomolecules? It was Erwin Schrödinger in his essay "what is Life?" who first surmised that the function of large bio-molecules is that they carry messages.[21c] Fischer would have been the first to expect and accept the principle that the proper sequence of amino acids in an extended peptide specifies a message and governs refolding and homing-in to the original native protein. His methods still were too imperfect to accomplish the synthesis of a peptide with 124 amino acids with which C. B. Anfinsen (Nobel Prize 1972) demonstrated successfully the principle of reductive randomization and slow oxidative reactivation of ribonuclease.

Theme and Variations on Phenylhydrazine

Phenylhydrazine, a product discovered by chance in 1875[21d] became the key theme of Fischer's "Habilitationsschrift" in 1878 in Munich at a time when hydrazine, a name Fischer coined, had not yet been made.[22] This compound was his boon and his bane: it opened up a new route to indoles[23] blazed new trails in carbohydrate chemistry,[24] made possible Ludwig Knorr's (1859 - 1921) preparation of antipyrin in 1883, led Heinrich Wieland to the discovery of tetraphenylhydrazines and their dissociation into radicals.[25] But phenylhydrazine was also his bane, a fact that Fischer learned only 15 years later after he had been afflicted by respiratory and intestinal problems.[26]

Phenylhydrazones and osazones helped to correlate hexoses. Emil Fischer's elucidation of the configuration of glucose[27] and of all the other carbohydrates set precedents of experimental brilliance and elegance[28] that did not fail to impress the Nobel Committee which awarded him the Nobel

Prize for his pioneering achievements in the chemistry of sugars and purines in 1902. His legacy was carried on by Claude Hudson (1881 - 1952) and, by the British School, Thomas Purdie (1843 - 1916), J. C. Irvine, W. N. Haworth and E. L. Hirst.

The Fischer Convention

It was not until the early fifties when the ingenious method of Johannes M. Bijvoet (1892 - 1980), an epigone of *Jacobus Hendricus van't Hoff* (1852 - 1911), established and confirmed the absolute configurations of important reference compounds, such as 2R. 3R-tartaric acid, D(−)-isoleucine and other chiral building stones. There was no need to change Fischer's original assignments and his ingenious but arbitrary "convention".[28a]

As of today Fischer's undisputed "convention" serves as the point of reference for carbohydrates, mucopolysaccharides[28b], antibiotics and a host of other natural products whose principles of biosynthesis have fallen into place like a Chinese jigsaw puzzle in ivory.

The knowledge of the absolute configuration brings in another dividend: it allows the construction of substrates or inhibitors of enzymes of known architecture on the principle of Fischer's "lock and key" idea. Much of what today is called "rational drug design" rests on this approach. A timely topic are antiviral agents based on "fraudulent nucleosides" as therapeutic skeleton keys opening or blocking locks of strategic importance.

Another recent application of Fischer's convention is the utilization of chiral building blocks of "synthons" for the stereospecific construction of complicated natural products: Both enantiomers of one of the many pheromones of the olive fly, *Daucus oleae*, namely S- and R-1, 7-dioxa-spiro | 5,5 | undecane recently became accessible in remarkable purity by starting with D-glucose.[28c] This is but one exemple out of many.

On the occasion of the van't Hoff Centennial Vladimir Prelog elaborated on Emil Fischer's original ideas on asymmetry and expanded them to the modern and wider concepts of *pseudoasymmetry, prochirality* and *pro-pseudoasymmetry* in two- and three-dimensional space.[28d]

Dynamic Aspects of Carbohydrates

In his lock-and-key approach to the selective action of enzymes on α- and β-methylglucosides Fischer's interest in geometric configuration and molecular asymmetry anticipates recent research with carbohydrates as polyhydroxy-aldehydes and ketones that form essential structural elements in nature. We now know they confer individuality to biological surfaces, and are vital intermediates in the production and storage of energy in biological systems. They also provide a complete set of diastereomers for the study of structure – function relationships in chemistry and biochemistry, and many exhibit complex solution behavior. Monosaccharides, for example, exist in solution as interconverting linear and cyclic forms, additionally complicated by the potential of each to adopt various conformations. Chemical and biological reactions of these "simple" molecules often involve only one of the equilibrating forms, and mechanistic interpretations must take account of the dynamics of tautomer interconversion.

Cyclic forms of monosaccharides occur as glycosides in oligo- and polysaccharides, glycoproteins and glycolipids, nucleotides, and polynucleotides. In all, conformation is a critical determinant of biological behavior.

For the past 10 years, the solution conformations of oligosaccharides of the type 2 blood group, using synthetic oligosaccharides enriched at specific sites with stable isotopes, particularly ^{13}C, have been studied as tools for elucidating 0-glycosidic bond conformations. Stable isotope enrichment combined with NMR spectroscopy (^{1}H, ^{13}C, ^{2}H) and mass spectrometry provided opportunities to monitor carbohydrate reactions in solution and to elucidate their mechanisms.[29]

Methylated Purines Hundred Years Later

In 1882 Fischer ventured structural formulas for uric acid, caffeine, theobromine, xanthine and guanine. The correct formula for uric acid had been ingeniously deduced by Ludwig Medicus in 1875.[29a] He synthesized theophylline and caffeine in 1895 and uric acid in 1897. In 1897 he revised his formulas, when he recognized uric acid and related compounds to be "oxides" of a hypothetical base that he called *purum uricum* or purine. He met the challenge of the synthesis of purine in 1898 by reductive elimination of three chlorine atoms from trichloropurine, a difficult approach that was simplified two generations later.[30]

In 1903 he synthetized 5,5-diethylbarbituric acid which, under the name of barbital, veronal or dorminal, became one of the most popular hypnotics, not to be surpassed until 1912 by phenobarbital or luminal.

Willstätter reports an appropriate anecdote out of this time: When Fischer was vacationing in Sankt Blasien (Black Forest), the writer Hermann Sudermann recognized Emil Fischer and told him: "Oh Excellency, how happy I am to express my thanks for your wonderful Veronal sleeping pills. You have saved me. I don't even have to take them; it's enough that they lie on my bedside table!" "What a strange coincidence," was Fischer's amicable response, "Whenever I have trouble falling asleep, I pick up one of your novels. This method never fails and is even effective when I see the beautiful opus lying on my bedside table!" "Well", added Emil Fischer, laughingly, "Thereupon the poor guy just bolted!"

Chemical methylations and metabolic demethylations of caffeine, theobromine, paraxanthine and heteroxanthine and Fischer's comprehensive investigations on the building stones of nucleic acids, in the words of Kurt Hoesch, "may unhesitatingly be classified as his most perfect achievement . . . deeply pleasing to both scientists and aestheticians." The presence in nature of methylated purines and pyrimidines has been known for a long time, but it is only in the last thirty years that the discovery of the widespread occurrence of methylated purines and pyrimidines in nucleic acids has generated renewed interest in the biochemistry and biological significance of these compounds. The "molecular" biology of the *in vivo* methylation of DNA, RNA and t-RNA deserves a somewhat more thorough treatment in order to show how the passage of time deals with *cause* and *effect.*

Methylation for Gene Expression and Differentiation

As a rule, a methylation site consists of a cytosine or an adenine residue encoded in a short palindromic sequence. Both DNA strands are methylated in a sequential stepwise fashion[31, 32] and the rate constant of methylation of hemimethylated DNA is much larger than that of unmethylated DNA. As a result of these kinetic characteristics only unmethylated or fully methylated DNA is found in cells and hemimethylated DNA is absent. The methyl groups are located in the major groove of the DNA helix and are thus ideally placed to modify the interaction between DNA and proteins and their recognition, and perhaps in shifting the B-Z equilibrium in favor of Z-, or left-handed, DNA.[33]

There is a great difference between the amount and variety of the methylated bases found, respectively, in DNA and in RNA. DNA contains only one, or in bacteria two, methylated bases and these are found in amounts not exceeding 2 % of all nucleotides. The presence of more than 50 different methylated plurimodified nucleotides has been recognized in RNA and in some cases as many as 10 % of the bases of an RNA molecule may consist of modified bases.[34] The other factor that has hindered our ability to formulate a coherent hypothesis for the biological significance of RNA methylation is that each cell contains several different molecular forms of RNA, and that in different species the biological consequences of the methylation of analogous forms of RNAs, such as rRNA or tRNA, seems to be different. Finally, we are not yet able to manipulate RNA *in vivo*, modify it *in vitro* and/or reintroduce it in cells as has become almost routine with DNA. Hence the problems of identifying the specific role of a given base methylation in RNA has so far depended on the fortuitous isolation of mutants lacking a specific methylase or on the use of methylation inhibitors most of which are rather unspecific.[35]

The presence of methyl groups in various positions of the purine bases will affect both the ability to form hydrogen bonds with complementary bases in other nucleic acids and protein recognition. The great variety of methylated nucleosides in RNA suggests that the secondary and tertiary structures of the various species of RNA can be tailored to fit a specific function. In this connection, it is of interest that both the variety and the average number of methyl groups present in a tRNA molecule in prokaryotes is smaller than in higher eukaryotes. This observation correlates well with the greater complexity of physiological functions attributed to tRNA in higher eukaryotes as compared to primitive bacteria. Additional support for the hypothesis that there is a correlation between the number and variety of methylated bases and the biological versatility of different species of RNA comes from a comparison of the "methylation status" of m-RNA and t-RNA. m-RNAs are relatively simple macromolecules: they are made in the nucleus as a complementary copy of gene exons and then transported to the cytoplasm to be read there by the protein-synthesizing machinery. These two functions find an almost exact counterpart in the two types of posttranscriptional modifications that are found in mRNAs. The function of the plurimethylated cap structure at the 5′ end[36] is to regulate the translation and stability of the mRNA molecule, whereas the internal m^6N-A serves some kind of nuclear role in precursor processing and/or transport.[37] By contrast, tRNAs are the most complex of all biological macromolecules: they are a family of molecular species that differs in their base sequences and each member of which may interact either with only some or with all of a large number of different proteins and nucleic acids (amino acid activating enzymes, tRNA nucleotidyl transferases, initiation factors, elongation factors, GTP-dependent translocases, transformylases, etc). In addition, the metabolic control of gene expression involves tRNA modifications which serve as a means to adapt cells to environmental stress.[38] This multiplicity of biological roles correlates well with the variety and diversity of posttranslational modifications (more than 50 at the last count) that comprise simple methylations and complex plurimodifications as seen in Wybutosine, or 2-methylthio-N^6-isopentenyladenosine.[34] At the present state of our knowledge we are unable to assign any particular function to any specific modification, and much further work will be needed before we can dissect the relative contribution of a specific modification of a purine or a pyrimidine to the biological function of a single tRNA.

The reader may wonder why he is plunged headlong into the "discussion" that nucleic acids conduct with amino acids and proteins in the cell. Well, Fischer provided the theme and close to hundred years later molecular biologists study the variations, almost comparable to Bach's Art of the Fugue.

Fischer's prediction of "genetic Engineering"

Gene expression and differentiation were terms that were not known at Fischer's time. But his uncanny "sixth sense" for dormant importance and latent developments did not fail him in the field of nucleotides and nucleic acids, as the following passage, written in 1914, shows:

"With the synthetic approaches to this group we now are capable of obtaining numerous compounds that resemble, more or less, natural nucleic acids. How will they affect various living organisms? Will they be rejected or metabolized or will they participate in the construction of the cell nucleus? Only the experiment will give us the answer. I am bold enough to hope that, given the right conditions, the latter may happen and that artificial nucleic acids may be assimilated without degradation of the molecule. Such incorporation should lead to profound changes of the organism, resembling perhaps permanent changes or mutations as they have been observed before in nature."[34a]

This ingenious prophetic insight prompts us to correct Thomas Alva Edison's quip, "Genius is one percent inspiration and ninety-nine percent perspiration." In Fischer's case the balance is toward inspiration.

A Legacy of Lipids

Fischer's last work was on the esterification of glycerol by fatty acids. The aim of all his investigations was to apply the methods of organic chemistry to the biosynthesis and metabolism of substances in living systems, a program that was continued and completed by his only surviving son Hermann Otto Laurenz Fischer (1888 - 1960), who in 1932 synthesized the calcium salt of D, L-glyceraldehyde-3-phosphoric acid which Warburg, Embden, and Meyerhof utilized in developing their schemes of alcoholic fermentation and glycolysis.[39]

In 1932 Fischer's son accepted the invitation of Sir Frederick Banting, who with Charles Best had discovered insulin, to join the staff at the Banting Institute of the University of Toronto. He arrived there in the fall of 1937 with his wife, two sons and daughter, Erich Baer und other assistants, his private laboratory, including the *9,000 reference compounds representative of all of his father's work, and his father's library.* The historical collection of Fischer's original samples has recently been transferred from Berkeley to Munich where they are on display at the Deutsches Museum. H.O.L. Fischer continued work on trioses, glycerol derivatives, and glycerides. In his laboratory were Baer, J. M. Grosheintz, Leon Rubin, J. C. Sowden, and Henry Lardy, working on the synthesis of selachyl alcohol, the oxidation of glycols, the condensation of aldoses, the cyclization of glucose into myoinositol, and the synthesis of biologically interesting organic phosphates.

In 1948 H. O. L. Fischer became a member of the new department of biochemistry at the University of California at Berkeley. On 9 October 1952 he dedicated the *Emil Fischer Library*, containing his father's books and reference compounds, to the university. Fischer became chairman of the biochemistry department in 1953, holding that position and lecturing on carbohydrates and lipids until his retirement in 1956. Among those who worked with him at Berkeley were D. L. MacDonald, C. E. Ballou, E. A. Kabat, and S. J. Angyal, who investigated sugar disulfones, sugar dialdehydes, inositol derivatives, tetrose phosphates, hexose dialdehydes, D-erythrose-4-phosphate, galactitol, and the conformational analysis of sugars and inositols. On his retirement H. O. L. Fischer was given space in the Biochemistry and Virus Laboratory, where he conducted research on carbohydrates, assisted by Hans Helmut Baer, from the Max Planck Institute for Medical Research in Heidelberg. Starting from Fischer's observation that a sugar dialdehyde of the pentose series forms, with nitromethane, nitroinositols, Baer worked on the syntheses of 3-amino-3-deoxy-D-ribose and 3-amino-deoxy-D-mannose. Fischer was a member of many professional societies, recipient of the Sugar Research Award of the American Chemical Society in 1949 and of the Adolf von Baeyer Memorial Gold Medal from the Society of German Chemists in 1955, and was named professor emeritus by the University of California in 1956.

Umbrella in a High Wind or the Walden Inversion

Emil Fischer was an experimentalist. His legendary ability was summed up by A. W. Hofmann: "Er bringt selbst einen Limburger zum Kristallisieren!" (He can even crystallize a Limburger cheese). There are many methods bearing his name, starting with the Fischer esterification,[40] the Fischer Indole synthesis,[41] Fischer oxazole synthesis,[42] Fischer peptide synthesis,[43] Fischer phenylhydrazine,[44] phenylhydrazone and osazone reactions,[45] but we have only one theory that might be connected with Fischer's name, that of the Walden inversion, a phenomenon that he re-encountered – and intuitively tried to avoid – in certain replacement reactions both in his carbohydrate as well as peptide research.[46]

Fischer thought that substitution reactions occurring at asymmetric carbon atoms, in the sense of Kekulé, begin with an addition of the reactants. Whether inversion occurs or does not occur depends on which group leaves from the intermediate complex. For his stereochemical studies the Walden inversion emerged as a "bête noir", a Schreckgespenst or a monster rearing its ugly head.

A more complete and accurate explanation was advanced in 1923 in his classical monograph[47] by Gilbert Newton Lewis (1875 - 1946), an American student of Wilhelm Ostwald (1853 - 1932).

Fischer's student E. Bergmann with M. Polanyi[48] contributed to the mechanism of the Walden Inversion by postulating that the most economical path energetically was for a displacing group to start forming a bond at the same time that the leaving group was breaking its bond. In the transition state the asymmetric carbon was planar, and the reaction was finished by complete bond formation for the entering group and departure of the leaving group. The analogy commonly used to describe inversion by this mechanism was that the bonds around the asymmetric carbon had turned inside out *like an umbrella in a high wind.*

The seeds of Walden's and Fischer's many experiments rendered rich harvests and many new discoveries in the laboratories of C. K. Ingold, L. P. Hammett, Saul Winstein, Paul D. Bartlett and many other investigators, who studied *solvolytic* or *polymolecular reactions*, later extended by others to the stereochemical course of enzymatic reactions.

References

1. Abderhalden: Die Bedeutung von Emil Fischers Lebenswerk für die gesamte Medizin (Naturwissensch. 7 [1919], 860 - 68); Gedächtnisrede von Beckmann: Sitzungsber. Kgl. Preuss. Akad. Wiss. 1920, S. 698 - 703; Nachr. von A. Binz: Chem.-techn. Wochenschr. 1919, S. 212 - 13; Paul Diergart: Nachrufe auf Emil Fischer und seine Wertschätzung der Geschichte der Chemie (Mitt. Gesch. Med. Naturw. 19 [1920], 121 - 23); C. Duisberg: Emil Fischer und die Industrie (Ber. Dtsch. chem. Ges. 52 [1919], A, 149 bis 65); Martin Onslow Forster: Nachruf (J. Chem. Soc. London 117 [1920], 1157 - 1201); M. O. F.: Nachruf in: Proceed. Roy. Soc. London 98 [1921], A, L - LVII; E. Freudenberg: Emil Fischer in seiner Bedeutung für die deutsche Wissenschaftspflege (Naturwissensch. 7 [1919], 873 - 78); Gustav Haas: Emil Fischer zum Gedächtnis (Prometheus 31, S. 329 - 30); Otto von Fürth, Wiener Klinische Wochenschrift, *32*, 828 - 829 [1919]; C. Harries: Nachruf in: Umschau 23 [1919], 609; C. Harries: Emil Fischers wissenschaftliche Arbeiten (Naturwissensch. 7 [1919], 843 - 60). – Kurt Hoesch: Emil Fischer. Sein Leben und sein Werk (Ber. Dtsch. chem. Ges. 54 [1921], Sonderheft; K. Hoesch: Gedächtnisrede (Z. angew. Chem. 36 [1923], 47 - 49). – Über die Gedächtnisfeier in Euskirchen (16. 7. 1922) siehe Z. angew. Chem. 36

[1923], 40; P. Jacobson: Nachruf (Chem.-Ztg. 43 [1919], 565 - 69; Ludwig Knorr: Über die wissenschaftlichen Arbeiten und die Persönlichkeit Emil Fischers (Ber. Dtsch. chem. Ges. 52 [1919], 132 - 49); Nachruf v. A. Kossel: Z. physiol. Chem. 108 [1919], 1 - 2; L. Krauss: Nachruf (Südd. Apoth.-Ztg. 59 [1919], 504); Lockemann: Nachruf (Mitt. Gesch. Med. Naturw. 18 [1919], 266); J. Pollak: Nachruf (Osterr. Chem. Ztg. 21 [1918]. 172 - 75, 179 - 82); Karl W. Rosenmund: Nachruf (Ber. Dtsch. pharm. Ges. 24, S. 533 - 44); Th. Sabalitschka: Nachr. (Pharm. Ztg. 64, S. 427 - 28); A. von Weinberg: Emil Fischers Tätigkeit während des Krieges (Naturwissensch. 7 [1919], 868 - 73); H. Wichelhaus: Emil Fischers Verdienste um die Deutsche chemische Gesellschaft (Ber. Dtsch. chem. Ges. 52 [1919], 129 - 32); Nachruf v. A. Wohl: Chem. Ind. 42 [1919], 269 - 76). – M. O. Forster, "Emil Fischer Memorial Lecture", Transactions of the Chemical Society, 117, 1157 - 1201 (1920); G. Hilgetag and P. Paul, "Zur wissenschaftlichen Leistung Emil Fischers", Zeitschrift für Chemie *10*, 281 - 289 (1970).

2. R. Willstätter, Aus meinem Leben, Verlag Chemie, Weinheim, 1949; From my Life, translated by L. S. Horning, New York 1965.

3. F. Herneck, "Emil Fischer als Mensch und Forscher", Zeitschrift für Chemie 10 (1970): 41 - 48.

4. E. Fischer and F. Penzoldt, Sitzungsberichte der phys.-medizinischen Sozietät zu Erlangen, *18*, 7 - 10 (1886).

5. Justus Liebigs Annalen der Chemie *239*, 131 - 136 (1887).

6. In "Selected and Standardized Values of Suprathreshhold Odor Intensities for *110* Substances" by F. Patte, M. Etcheto and P. Laffort, Chemical Senses and Flavor, *1*, 283 - 305 (1975), methanethiol, Fischer's "Mercaptan" is No 66 and 2-chlorophenol, Fischer's "Chlorphenol", is No 37 in Table II containing 110 substances.

7. J. Parascandola, The Development of Receptor Theory, in Discoveries in Pharmacology, Vol. 3, Pharmacological Methods, Receptors and Chemotherapy, M. J. Parnham J. Bruinvels, eds., Elsevier, 1986, pp. 129 - 156).

8. Atropine and pilocarpine are mutually exclusive in their actions on secretion of saliva from the submaxillary gland, an observation that prompted J. N. Langley to postulate affinity of these agonists or antagonists to special substances in the nerve endings of the glands [J. Physiol. *1*, 339 - 369 (1878)], a conclusion which he refined and improved in the next 30 years.

9. P. Ehrlich, Contributions to the Theory and Practice of Histological Staining. M. D. Dissertation, University of Leipzig in F. Himmelweit (Ed.), The Collected Papers of Paul Ehrlich, 3 Vols., Pergamon Press, London, 1956 - 1960.

10. P. Ehrlich, Address Delivered at the Dedication of the Georg-Speyer Haus, in Himmelweit (ref. 7), Vol. 3 pp. 53 - 63 (1960).

11. E. Fischer, Einfluß der Konfiguration auf die Wirkung der Enzyme. I. Ber. 27, 2985 - 2993 (1894).

12. M. Stuiver, Biophysics of the Sense of Smell, Thesis, Groningen, Netherlands, 99 pages, 1958.

13. D. Tucker, Physical Variables in the Olfactory Stimulation Process, J. Gen. Physiol. *46*, 453 - 459 (1963).

14. Le Magnen, Etude sur l'excitation olfactive, Compt. Rend. Acad. Sci. 225, 1378 - 1380 (1947).

15. A Gas-Chromatographic (GLPC) Model for the sense of Smell. Variations of Olfactory Sensitivity with Conditions of Stimulation, by R. B. Nachbar and Th. H. Morton, J. Theor. Biol. *89*, 387 - 407 (1981).

16. Selective Olfactory Deficits in Case H. M., by H. Eichenbaum, T. H. Morton, H. Potter and S. Corkin, Brain 106, 459 - 472 (1983).

17. J. R. Mason, L. Clark and T. H. Morton, Selective Deficits in the Sense of Smell caused by Chemical Modification of the Olfactory Epithelium, Science *226*, 1092 - 1094 (1984).

18. J. Pevsner, R. R. Trifiletti, S. M. Strittmatter, and S. H. Snyder, Isolation and characterization of an olfactory receptor protein for odorant pyrazines. Proc. Nat. Acad. Sci. *82*, pp. 3050 - 3054, 1985.

19. J. Pevsner, P. B. Sklar, and S. H. Snyder, Odorant-binding protein: Localization to nasal glands and secretions. Proc. Nat. Acad. Sci. U. S. A., *83*, pp. 4942 - 4946, 1986.

20. E. Fischer, Über die Hydrolyse der Proteinstoffe, Chemiker-Zeitung *26*, 939 - 940 (1902).

21. Contrasts in Scientific Style. Emil Fischer and Franz Hofmeister: Their Research Groups and Their Theory of Protein Structure, by Joseph S. Fruton, Proceedings of the American Philosophical Society, *129*, 313 - 370 (1985). In Footnote 15 Fruton writes: There have been numerous treatments of this subject before and after the famous article by R. K. Merton, "Singletons and Multiples in Scientific Discovery", Proceedings of the American Philosophical Society 105 (1961): 470 - 486. For example see H. U. Baumann, Über Mehrfachentdeckungen (Münster 1972); the Fischer-Hofmeister peptide theory of protein structure is listed on p. 66. One view of the priority in this case was offered by Hofmeister's former assistant Otto von Fürth in his obituary notice for Fischer in Wiener klinische Wochenschrift 32 (1919): 828 - 829: "In 1902, on the occasion of the Karlsbad meeting, Franz Hofmeister first formulated scientifically the assumption of an amide-like linkage of the individual amino acids within proteins. E. Fischer was however the one who attacked the problem systematically".

21a C. A. Knight, Paper chromatography of some lower peptides, J. Biol Chem. *190*, 753 - 762 (1951).

21b E. A. Kabat, Ann. Rev. of Immunology, p. 7 (1983).

21c Cf. G. B. Kauffman and R. P. Ciula, Emil Fischer's Discovery of Phenylhydrazine, J. Chem. Educ. *54*, 295 (1977).

21d M. F. Perutz, Physics and the riddle of Life, Nature *326*, 555 - 558 (1987).

22. H. Wieland, Die Hydrazine, Stuttgart 1913.

23. Cf. P. A. Roussel, The Fisher Indole Synthesis, J. Chemical Education *30*, 122-125 (1953).

24. K. Freudenberg, Emil Fischer and his Contribution to Carbohydrate Chemistry, Advances in Carbohydrate Chemistry, *21*, 1-38 (1966).

25. H. Wieland, Ber. 48, 1078-1095 (1915).

26. L. Lewin, "Eine toxikologische Erinnerung an Emil Fischer", Naturwissenschaften 7, 878-892 (1919).

27. C. S. Hudson, "Emil Fischer's Discovery of the Configuration of Glucose,, J. of Chemical Education *18*, 353-357 (1941).

28. C. S. Hudson, "Historical Aspects of Emil Fischer's Fundamental Conventions for Writing Stereo-Formulas in a Plane", Advances in Carbohydrate Chemistry 3, 1-22 (1948).

28a R. Parthasarathy, The Determination of Relative and Absolute Configurations of Organic Molecules by X-ray Diffraction Methods, Vol. I, 185, Thieme, Stuttgart 1977.

28b Horst Kunz, Synthesis of Glycopeptides, Partial Structures of Biological Recognition Compounds, Angew. Chemie Internat. Ed. *26*, 294-308 (1987).

28c Cf. H. Redlich and W. Franke, Angewandte Chemie, 96, 506-507 (1984).

28d V. Prelog, Koninkl. Nederl. Akad. Wetenschap. Proc. 71B, 108 (1968); Helv. Chim. Acta 55, 2581-2598 (1972).

29. R. Barker and A. S. Serianni. Carbohydrates in Solution: Studies with Stable Isotopes. Acc. Chem. Res. *19*, 307-313, 1986.

29a L. Medicus, Ann. 175, 230 (1875); Cf. R. Huisgen, Angew. Chemie, *98*, 297-311 (1986).

30. H. Bredereck et al., Liebigs Annalen 659, 133- (1962).

31. Modrich, P., Structures of DNA restriction and modification enzymes, Quart. Rev. of Biophys. *12*, 315-369 (1979).

32. Yuan, R., Structure and mechanism of multifunctional restriction endonucleases, Ann. Rev. of Biochem. *50*, 285-316 (1981).

33. Behe, M. and Felsenfeld, G., Effects of methylation on a synthetic polynucleotide: The B-Z transition in poly (dG-m^5 dC). poly (dG-m^5 dC). Proc. Nat. Acad. Sci. U.S.A. *78*, 1619-1623 (1981).

34. Kersten, H., The role of coencymes and tRNA modifications in metabolic control of gene expression. In Borchardt, R. T., Creveling, C. R. and Ueland, P. M. (eds.): *The Biochemistry of S-adenosylmethionine as a basis* for drug design symposim, Bergen, Norway, pp. 163-174 (1986).

34a E. Fischer, Ber. 47, 3196 (1914). – Prof. Vladimir Prelog kindly directed my attention to this momentous passage.

35. Backlund, P. S., Jr., Carotti, D. and Cantoni, G. L., Effects of the S-adenosylhomocysteine hydrolase inhibitors 3-deazaadenosine and 3-deaza-ari-

steromycin on RNA methylation and synthesis. Eur. J. Biochem. *160*, 245-251 (1986).

36. Shatkin, A. J., Cell *9*, 645-653 (1976).

37. Rottmann, F., Narayan, P., Goodwin, R., Camper, S., Yao, Y., Horowitz, S., Albers, R., Ayers, D., Maroney, P. and Nilsen, T., Distribution of m^6A in mRNA and its possible biological role. In Borchardt, R. T., Creveling, C. R. and Ueland, P. M. (eds.): *The Biochemistry of S-adenosylmethionine as* a basis for drug design, symposium, Bergen, Norway, pp. 189-200 (1986).

38. Buck, M., and Ames, B. N., Cell *36*, 523-531 (1984).

39. For a short autobiography with portrait, see Hermann O. L. Fischer, "Fifty Years 'Synthetiker' in the Service of Biochemistry", in Annual Review of Biochemistry, *29* (1960), 1-14. Early papers on carbomethoxy derivatives of acids written with Emil Fischer can be found in Berichte der Deutschen chemischen Gesellschaft, *46* (1913), 1138-1148, 2659-2664; and *47* (1914), 505-512, 768-780. Fischer and Dangschat's work on quinic acid, shikimic acid, and their derivatives is reported in papers found in Naturwissenschaften, *19* (1931), 310-311; Berichte der Deutschen chemischen Gesellschaft, 65B (1932), 1009-1031, 1037-1040; and Helvetica Chimica Acta, 17 (1934), 1196-1200, 1200-1207; and 18 (1935), 1204-1206, 1206-1213. Some of the papers written in collaboration with Erich Baer and others on the trioses, dihydroxyacetone and glyceraldehyde, and their derivatives can be found in Berichte der Deutschen chemischen Gesellschaft, 60B (1927), 479-485; 63B (1930), 1732-1744, 1744-1748, 1749-1753; 65B (1932), 337-345, 345-352; Helvetica chimica acta, *16* (1933), 534-547; *17* (1934), 622-632; *18* (1935), 514-521, 1079-1087, *19* (1936), 519-532; *20* (1937), 1213-1226, 1226-1236; Naturwissenschaften, *25* (1937), 588-589, 589; and Chemical Reviews, 29 (1941), 287-316. H. O. L. Fischer published a review of his investigations of chemical and biological relationships between hexoses and inositols in Harvey Lectures, *40* (1944-1945), 156-178; and a review of his chemical synthesis of intermediate products of sugar metabolism in Angewandte Chemie, *69* (1957), 413-419.

40. Fischer, E. and Speier, A., Ber. *28*, 3252 (1895); Fischer, E., Z. Physiol. Chemie, *33*, 151 (1901), describes the application of his method to amino acids.

41. Ishii, H., Tetrahedron *29*, 1991-2003 (1973) presents a recent review.

42. Fischer, E., Ber. *29*, 205 (1898).

43. Fischer, E., Ber. *36*, 2982 (1903).

44. Fischer, E., Ber. *8*, 589 (1875).

45. Fischer, E., Ber. *17*, 579 (1884).

46. Fischer, E., Ann. *381*, 123 (1911).

47. Lewis, G. N., Valence and the Structure of Atoms and Molecules, New York, 1923, p. 113.

48. Bergmann, E., Polanyi, M. and Szabo, A., Z. physik Chemie, B20, 161 (1933); A. R. Olsen, J. Chem. Physics *1*, 418 (1933).

Emil Fischer

EMIL FISCHER
GESAMMELTE WERKE

HERAUSGEGEBEN VON M. BERGMANN

AUS MEINEM LEBEN

MIT DREI BILDNISSEN

BERLIN
VERLAG VON JULIUS SPRINGER
1922

AUS MEINEM LEBEN

VON

EMIL FISCHER

BERLIN
VERLAG VON JULIUS SPRINGER
1922

Vorwort des Herausgebers.

Der Unterzeichnete steht im Begriff, die wissenschaftlichen Schriften Emil Fischers zu einer Gesamtausgabe zusammenzufassen. Ihr mögen gleichsam als Einleitung einige persönlich gehaltene Worte des Forschers vorausgeschickt werden, seine Lebenserinnerungen. Das Manuskript verdanke ich der Freundlichkeit seines Sohnes, Dr. Hermann Fischer. Die vorliegende Fassung ist im Jahre 1918 während zweier Genesungsreisen nach Locarno und Karlsbad unmittelbar niedergeschrieben worden mit der Absicht, weitere Arbeitspausen zur Fortsetzung zu benutzen. Dazu ist es aber nicht mehr gekommen.

So mag das vorliegende Bändchen den Freunden Emil Fischers als letzter Gruß des Scheidenden gelten. Ich bin gewiß, es wird im Verein mit dem Lebensbild, das Professor K. Hoesch's Künstlerhand soeben gezeichnet hat, dem verstorbenen Meister neue Freunde gewinnen.

Berlin, im November 1921.

M. Bergmann.

Geschrieben in dem Unglücksjahre 1918

Aus meinem Leben

Mußt Du Gram im Herzen tragen
Und des Alters schwere Last,
Rufe Dir aus jüngeren Tagen
Die Erinnerung zu Gast!
(Kußmaul)

In der fruchtbaren Ebene, die nach Süden vom Vorgebirge der Eifel und im Osten von dem Villegebirge begrenzt ist, liegt die kleine Kreisstadt Euskirchen, mein Geburtsort, nur wenig entfernt von dem Erftfluß, der in der Eifel entspringt und zu Neuß in den Rhein mündet.

Die Stadt wird durchflossen von mehreren Bächen, die den dort heimischen Industrien sehr zustatten kommen und uns Knaben reichlich Gelegenheit zur Übung von Wasserkünsten gaben.

Sie zählte vor 60 Jahren etwa 3500 Einwohner, von denen wohl die Hälfte vom Ackerbau lebte, der sich in der fruchtbaren Umgebung lohnte. Außerdem gab es eine alte Tuchindustrie, die vorzugsweise für den Heeresbedarf arbeitete und die in den Kriegsjahren 1864/66 und 1870 durch die zahlreichen „Kommiß"-Lieferungen der Stadt viel Wohlstand brachte. Weniger lohnend war der Betrieb mehrerer Gerbereien, welche die in den benachbarten Wäldern reichlich vorhandene Eichenrinde verarbeiteten. Endlich war die Stadt der Mittelpunkt des Kleinhandels für die ländliche Umgebung und hatte den Vorzug, auf 30 km ohne Konkurrentin zu sein. Diese günstige Lage hat Euskirchen später zum Eisenbahnknotenpunkt gemacht und ihm eine rasche vorzugsweise industrielle Entwicklung gebracht. Der Bau der ersten Eisenbahn, die um das Jahr 1862 eröffnet wurde, war für die ganze Bevölkerung und nicht am wenigsten für die Jugend ein staunenerregendes Ereignis. Dasselbe galt für die Anlage der Gasbeleuchtung, die ungefähr in die gleiche Zeit fiel.

Die Stadt war in früheren Jahrhunderten befestigt und die alten Mauern mit Türmen, Wällen und Wassergräben waren in meiner Jugend noch teilweise erhalten. Sie haben uns viel Kurzweil bei unseren Spielen verschafft. Besonders vertraut war uns der sogen. Judenwall, wo wir uns spielend und auch den Boden durchwühlend öfters herumtrieben und von wo aus wir staunend Einblick in den Betrieb einer tiefer gelegenen Gerberei hatten.

Mein Elternhaus lag einige Minuten vor der alten Stadt, an der nach Cöln führenden Landstraße. Der Gebäudekomplex bestand aus zwei geräumigen Wohnhäusern, wovon das eine von meinem Onkel bewohnt war, einem Geschäftshause, verschiedenen kleinen technischen Betrieben und einigen bescheidenen landwirtschaftlichen Baulichkeiten, die einen großen Hof umschlossen.

Dieses Ganze war umgeben von Gärten, die auf der einen Seite an einen Bach grenzten und von einem kleinen Wassergraben durchzogen waren.

Hier bin ich am 9. Oktober 1852 geboren als letztes und achtes Kind meiner Eltern. Von meinen Geschwistern waren ein Knabe und ein Mädchen vorher gestorben, alle anderen waren Schwestern, von denen die älteste mir um 14 Jahre voraus war.

Man kann sich denken, daß unter diesen Umständen meine Ankunft den Eltern viel Freude gemacht hat, und daß mir auch später ein gewisser Vorzug gewährt wurde.

Die Schwestern haben ihr Interesse an dem einzigen Bruder, den sie nur den „Jungen" nannten, in der mannigfaltigsten Weise bekundet, und ich habe mich ihrer Erziehungskünste des öfteren erwehren müssen, so daß eine gewisse Abneigung gegen junge Damen bei mir über die Knabenjahre hinaus hängen blieb.

Glücklicherweise waren die Verhältnisse in dem Hause meines Onkels, das nicht allein durch den Hof, sondern auch durch den Speicher und einen besonderen Korridor mit unserm Hause in direkter Verbindung stand, gerade umgekehrt; denn es gab dort fünf Söhne und eine Tochter, die merkwürdigerweise wie ich die Jüngste war.

Wenn mir die allzu weibliche Behandlung im eigenen Hause zu viel wurde, so zog ich mit Erlaubnis der Eltern für einige Tage ins Nachbarhaus, bis ich dort durch reichliche Prügel von seiten der stärkeren Vettern belehrt, wieder in die mildere Atmosphäre des eigenen Heims gerne zurückkehrte.

Mein Vater L a u r e n z F i s c h e r betrieb zusammen mit seinem Bruder A u g u s t ein kaufmännisches Geschäft, hauptsächlich in Kolonialwaren, Wein und Spirituosen. Außerdem besaßen sie eine Wollspinnerei, die aber auf einem Dorfe Wißkirchen, etwa eine Stunde von Euskirchen, gelegen war und ursprünglich mit Wasserkraft, später mit Dampf betrieben wurde.

An dem gesamten Geschäft war noch ein anderer Onkel, F r i e d r i c h A r n o l d beteiligt, der in Flamersheim, dem Stammsitz der Familie, wohnte und dort das vom Großvater her stammende Anwesen verwaltete.

In der eben geschilderten Umgebung zu Euskirchen habe ich eine überaus glückliche Jugend verlebt. Der Betrieb des kaufmännischen Geschäfts, das die Krämer der Umgegend bis weit in die Eifel hinein mit

Waren versorgte, brachte reges Leben mit sich. Der Verkehr im Kontor, den Lagerräumen und auf dem Hofe hat mich lebhaft erinnert an die Schilderungen, die Gustav Freytag in dem Roman „Soll und Haben" von dem Kaufmannshause zu Breslau entwarf. Allerdings waren die Verhältnisse bei uns trotz des Wohlstandes der Firma bescheidener. Aber dafür war das Ganze in sehr viel heiterere Farben gekleidet. Die lebhafte Art der rheinischen Bevölkerung und die glückliche humoristische Veranlagung der Geschäftsinhaber gaben sich trotz der sehr geordneten, nach den strengen Grundsätzen kaufmännischer Ehrbarkeit geregelten Geschäftsführung in zahlreichen munter und scherzhaft geführten Gesprächen kund.

Ein kleiner landwirtschaftlicher Betrieb, der im wesentlichen die Bedürfnisse des eigenen Hauswesens befriedigte, trug dazu bei, das Gesamtbild mannigfaltiger und für uns Kinder interessanter zugestalten.

Man denke sich dazu zwölf jugendliche Personen, die auf dem Hof und in den Gärten eine einzige Familie bildeten und die später, als meine Schwestern verheiratet waren, noch durch Enkel vermehrt wurden, und man wird sich eine Vorstellung machen können von der vielfachen Kurzweil, die wir alle in diesem Kreise gefunden haben.

In der früheren Jugend waren es Spiele verschiedener Art, vom Ballspiel bis zum Indianerwigwam, vom Fisch- und Vogelfang bis zum Bivouak, dann Kämpfe der verschiedensten Art unter uns Knaben oder in geschlossener Phalanx gegen feindliche Kräfte. Die Schlachten, in denen man sich nicht allein der Fäuste und Stöcke, sondern auch des Steinwurfs und der Schleuder bediente, arteten zuweilen bis zur Lebensgefährlichkeit aus und mußten dann durch den Eingriff von erwachsenen Personen beendet werden.

Mit dem zahlreichen Dienstpersonal, besonders mit den Knechten, standen wir auf vertrautestem Fuße, und die Unterhaltung wurde hier ausschließlich in dem derben niederrheinischen Dialekt geführt. Selbstverständlich hatten wir alle Spitznamen. Ich wurde „Baron" genannt, ob wegen des üppigen Ernährungszustandes oder wegen der besseren Kleidung, ist mir immer ein Geheimnis geblieben.

Von Verzärtelung war weder in körperlicher noch seelischer Beziehung die Rede. In leichtester Kleidung schlugen wir uns tapfer durch den Winter, und die einzigen Schmerzen, deren ich mich aus der frühen Jugend erinnere, kamen von Panarizien an verletzten Fingern oder von erfrorenen Füßen oder von Stiefeln, die durch Schneewasser hart und eng geworden waren.

Beim Eissport bin ich mehrmals eingebrochen, einmal sogar in eine Jauchengrube bis über den Kopf eingetaucht, und als ich in diesem Zustande nach Hause kam, stark beschmutzt und übel duftend, wurde ich trotz der scharfen Kälte außerhalb des Hauses ausgezogen. Alles das ging ohne Schaden vorüber.

Ein anderes Mal fiel ich von einem mit Wollballen beladenen Wagen kopfüber auf das Pflaster des Hofes und kam mit einer ziemlich tiefen, stark blutenden Kopfwunde ins Haus. Dort begrüßte man mich mit den tröstlichen Worten: „Besser ein Loch im Kopf als in der Hose."

Allerdings gab es auch ernstere Unfälle. Beim Spielen mit Pulver wurde ich durch den Leichtsinn eines Kameraden im Gesicht und am Kopf stark verbrannt. Diesmal gab es einen größeren Schrecken, denn als ich mit geschlossenen Augen, geführt von einer alten Frau vor meiner Mutter erschien, hielt sie mich für erblindet und brach in lautes Schluchzen aus. Glücklicherweise ging aber das Unglück wiederum bei guter ärztlicher Behandlung ohne dauernden Schaden vorüber, und ich hatte noch die Genugtuung, bei hartnäckigem Schweigen keinen Mitschuldigen verraten zu haben. Das Gefühl der Solidarität war bei uns Knaben überhaupt in hohem Maße entwickelt, besonders galt das auch in der Schule, wo jede Lüge in unseren Augen gerechtfertigt war, wenn sie dazu diente, Kameraden vor der Strafe zu schützen.

Es war damals Sitte, die Kinder schon mit fünf Jahren zur Schule zu schicken, und dasselbe Schicksal wurde auch mir zuteil; denn einen Tag nach meinem 5. Geburtstage wurde ich von meinen älteren Schwestern mit zur Schule genommen.

Da die Volksschule meiner Vaterstadt unter dem Einfluß der katholischen Geistlichkeit stand und über den kirchlichen Übungen der eigentliche Unterricht vernachlässigt wurde, so hatte mein Vater eine protestantische Privatschule ins Leben gerufen und dafür glücklicherweise einen ausgezeichneten Lehrer, Herrn Vierkoetter gewonnen. Dieser unterrichtete die Kinder von 5 bis 14 Jahren in einem Raum. Eine strenge Einteilung in Klassen gab es nicht. Trotzdem war der Unterricht in allen Elementarfächern ausgezeichnet, sodaß sowohl meine Schwestern wie ich beim Eintritt in andere Schulen den Altersgenossen im Wissen voraus waren. Der Lehrer ging sogar so weit, die begabteren Schüler und Schülerinnen in euklidischer Mathematik zu unterrichten, und es erregte in späteren Jahren große Heiterkeit, als meine Schwester Fanny ihrem Gatten, einem Holzhändler, in dessen Geschäft die Ausziehung einer Kubikwurzel unerwarteterweise nötig wurde, aus der Verlegenheit half, indem sie diese Aufgabe nach den bei Herrn Vierkoetter erworbenen Kenntnissen löste.

Der gute Unterricht, den wir hier genossen, war allerdings nur möglich bei der geringen Zahl der Schüler, die kaum 20 überstieg; denn die Schule wurde ursprünglich nur von Kindern der wenigen protestantischen Familien und einigen Judenkindern besucht. Erst später baten auch einzelne aufgeklärte Katholiken um die Erlaubnis, Kinder dorthin schicken zu dürfen. Das war zu der damaligen Zeit schon ein kleines Wagnis; denn der Gegensatz zwischen den beiden Konfessionen war recht stark und machte sich namentlich auch für uns Kinder häufig in recht unangenehmer Weise bemerkbar. Als sogen. Blauköpfe, auch kalvinische Kalbsköpfe, wie wir Protestanten genannt wurden, haben wir Fischer-Jungen manche Prügel einstecken müssen, wenn die Überzahl der katholischen Knaben eine erfolgreiche Verteidigung aussichtslos erscheinen ließ.

Bei anderer Gelegenheit, wo wir nicht vereinzelt, sondern als gesammelte Macht auftreten konnten, ist uns aber auch mancher Sieg zuteil geworden. Bei diesen Heldentaten spielte in der Regel die Hauptrolle mein Vetter Lorenz Fischer, der später auch im Kriege 1870 als Soldat und außerdem im Zivilleben als Jäger hervorragende Leistungen aufwies.

Als ich 9 Jahre alt wurde, gab der Lehrer Vierkoetter seine Stellung in Euskirchen auf, weil ihm der viel einträglichere Posten eines Inspektors an der Besserungsanstalt zu Brauweiler übertragen wurde. Ich trat deshalb in die höhere Bürgerschule meiner Vaterstadt über, welche, wie die damaligen Progymnasien, 4 Jahresklassen mit lateinischem, griechischem und französischem Unterricht umfaßte. Sie war in den Nebenräumen der Klosterkirche untergebracht und stand unter Leitung katholischer Priester. Der Rektor, Kaplan Heine, war eine ausgeprägte Persönlichkeit, tyrannisch, jähzornig, aber trotz alledem ein recht guter Lehrer, der auch auf seine Kollegen anregend wirkte und damit der Schule in bezug auf Leistungen eine ebenbürtige Stellung mit den staatlichen Gymnasien verschaffte.

Die Handhabung der Schuldisziplin war allerdings recht willkürlich, und mir passierte eine Ungerechtigkeit, die mir zum erstenmal den großen Wert väterlichen Schutzes zum Bewußtsein brachte.

Ein Mitschüler, namens Flecken, bekannt durch Gewalttätigkeit, hatte mir ohne jeden Grund ein Spielzeug entrissen und in den Schmutz geworfen. Ich beantwortete diese Herausforderung, indem ich ihm die Mütze vom Kopfe riß und in denselben Schmutz hineinwarf. Er ließ sie trotzig liegen und erstattete dann beim Rektor Heine die Anzeige meiner angeblichen Missetat. Meine Entschuldigung, daß ich der Angegriffene gewesen sei, wurde kaum angehört, und ich erhielt den Befehl, die Mütze in ordnungsmäßigem Zustand wieder herbeizuschaffen. Das war aber nicht möglich, da sie inzwischen von fremder Hand entwendet war. Infolgedessen erhielt ich zunächst eine Arreststrafe, so daß ich nicht zum Mittagessen gehen konnte und den ganzen Tag hungernd in der Schule bleiben mußte. Als am nächsten Tage die

Mütze noch immer nicht zur Stelle war, wurde ich von dem Herrn Rektor nach Hause geschickt mit der Bemerkung, ich sei von der Schule entlassen.

Jetzt hielt mein Vater es an der Zeit, einzugreifen, und sandte mich am nächsten Tage mit einem Brief an den Herrn Rektor zur Schule zurück. Ich habe den Inhalt desselben niemals kennen gelernt, aber es mag manches kräftige Wörtlein drin gestanden haben von willkürlicher Behandlung des kindlichen Streites und vom Appell an die Staatsregierung, falls der Rektor seinen ganz ungesetzlichen Ausweisungsbefehl aufrechterhalte.

Die Wirkung dieses Briefes war in der Tat erstaunlich. Ich konnte ruhig meinen Platz in der Schule wieder einnehmen und habe mich seitdem einer guten und gerechten Behandlung erfreut. Ja, ich kann sogar sagen, daß ich mit dem Herrn Rektor später auf recht gutem Fuße stand; denn er liebte wie ich die Mathematik und er freute sich, mir schwierige Probleme klar zu machen. Ich habe ihm ein besseres Gedenken bewahrt als den meisten anderen Lehrern aus der Gymnasialzeit.

Noch nicht 13 Jahre alt wurde ich aus der Tertia der letzten Klasse der Schule mit einem guten Zeugnis entlassen und mußte nun auf ein Gymnasium übertreten.

Bis dahin waren alle Söhne der Familie, die sich höheren Studien zuwenden wollten, in das Gymnasium der früheren Universitätsstadt Duisburg eingetreten, da es unter protestantischer Leitung stand, einen guten Ruf genoß und auch nicht allzu weit von unserer Heimat entfernt war.

Aber unglücklicherweise war es kurz vorher meinem Vetter Ernst Fischer auf der Sekunda dieses Gymnasiums schlecht ergangen; er war bei einem kleinen Verstoß gegen die Schulregeln ertappt und zu einer verhältnismäßig hohen Karzerstrafe verurteilt worden.

Da die Strafe auch von unseren Eltern als zu hart angesehen wurde, so nahm man ihn von der Schule weg und diese kam dann überhaupt für unsere Familie in Verruf. Es wäre nun für uns am einfachsten gewesen, ein Gymnasium in Köln, Bonn oder Aachen zu wählen, aber meine Eltern und ganz besonders meine Mutter legten Wert darauf, daß wir eine protestantische Schule besuchten, und so fiel die Wahl auf Wetzlar, obschon diese Stadt eine volle Tagereise von unserer Heimat entfernt war. Um sicher zu gehen und um unseren starken Drang nach Selbständigkeit zu schonen, rieten uns die Eltern selbst, zuvor Wetzlar anzusehen und ein passendes Quartier auszusuchen.

So reiste ich denn in den Herbstferien 1865 mit meinem Vetter Ernst Fischer, der 4 Jahre älter war, und in Begleitung von Vetter Lorenz über Cöln nach Wetzlar. Die altertümliche Stadt gefiel uns. Wir fanden auch bald ein nach unserem Begriff gutes und vor allen

Dingen freiheitliches Quartier bei einem Bürger, der den größeren Teil seines Hauses an Gymnasiasten und einjährigfreiwillige Soldaten, die im dortigen Jägerbataillon standen, vermietete.

Nach der raschen Beendigung der geschäftlichen Angelegenheit setzten wir vergnügt unsere Reise fort über Gießen nach Frankfurt a. M. Diese alte Stadt mit ihren vielen Sehenswürdigkeiten und historischen Erinnerungen hat auf uns einen großen Eindruck gemacht. Aber unser Aufenthalt erfuhr eine jähe Unterbrechung, veranlaßt durch einen Besuch in einem Mädchenpensionat, wo meine jüngste Schwester Mathilde als 17-jähriger Backfisch zur jungen Dame erzogen wurde. Sie war außer sich vor Freude, den Bruder und die Vettern so unerwartet wiederzusehen und bestürmte uns sofort mit der Bitte, sie in einen Zirkus zu führen, der damals in Frankfurt gastierte.

Die Leiterin der Pension gab nach einigem Widerstreben ihre Einwilligung, stellte aber die Bedingung, daß wir außer der Schwester noch eine zweite junge Dame und eine Lehrerin zur Aufsicht mitnehmen und dann den ersten Platz besuchen müßten.

Mit vornehmer Ritterlichkeit gingen wir auf diese Bedingungen ein. Aber die Auslagen waren für unsere bescheidene Reisekasse zu groß, und am nächsten Morgen beeilten wir uns, die schöne aber teure Stadt zu verlassen. Das übrig gebliebene Geld reichte noch gerade aus, um auf den billigsten Plätzen über Mainz, dann mit dem Schiff rheinabwärts über Bonn nachhause zurückzukehren. Es war meine erste größere Reise, hatte 7 Tage gedauert und hinterließ bei mir einen so günstigen Eindruck, daß die Reiselust mir bis ins Alter treu geblieben ist.

Bald darauf kam der Abschied vom Elternhause, in das ich nun viele Jahre hindurch nur noch in den Schulferien zurückkehrte. Der glücklichste Teil der ersten Jugend war vorüber; denn ich habe es draußen niemals so gut gehabt, wie unter dem Schutze und in der fröhlichen Atmosphäre des Vaterhauses.

Trotz des Wohlstandes, der von dem erfolgreichen Betrieb des Geschäftes herstammte, war unser materielles Leben zwar recht behaglich, aber doch einfach. Verwöhnt waren wir nur durch die gute, kräftige und wohlschmeckende Nahrung, auf deren Zubereitung man im Rheinland damals viel größeren Wert legte als in den meisten übrigen Teilen Deutschlands.

Auch der Wein floß reichlich in unserem Hause, in dem häufig Gäste verkehrten. Aber es war strenge Regel, daß wir Kinder unter 14 Jahren von dem Genuß aller geistigen Getränke ausgeschlossen wurden. Vor allen Dingen stand das Familienleben im Zeichen der Fröhlichkeit, die durch das Wesen meiner Eltern und ihre glückliche Ehegemeinschaft bedingt war.

Die Eltern

Mein Vater Laurenz Fischer wurde am 4. November 1807 in Flamersheim geboren, war also fast 45 Jahre älter wie ich. Seine Geburtsurkunde ist in französischer Sprache ausgefertigt; denn damals stand das linke Rheinufer unter französischem Regiment und die Funktionen der heutigen Standesämter waren schon damals dem „Maire" übertragen.

Flamersheim ist ein kleiner Ort am Fuße der Eifelberge, aber noch in der fruchtbaren Ebene gelegen, etwa 7 km von Euskirchen entfernt. Es besaß eine alte protestantische Gemeinde mit dem Sitz des Pastors, der zu meiner Zeit auch die kleine Gemeinde in Euskirchen mitbesorgte. Die Protestanten waren wie überhaupt in dem katholischen Rheinlande durchgehend gebildeter und betriebsamer als die katholische Volksmasse und erfreuten sich deshalb auch eines verhältnismäßig großen Wohlstandes. Ihre geschäftlichen Unternehmungen hatten auch dem Ort Flamersheim eine über seine Einwohnerzahl weit hinausgehende Bedeutung verschafft. Dazu kam, daß der größte Grundbesitz des Ortes, die sogen. Burg mit dem stattlichen schloßähnlichen Wohnhause, hübschen Park und reichen Ländereien dauernd im Besitz eines protestantischen Herrn blieb.

Der Überlieferung nach war diese Burg aus einer Meierei Karls des Großen hervorgegangen. Sie soll zahlreiche Vorwerke gehabt haben, von deren speziellen Aufgaben Sachverständige die Namen der umliegenden Dörfer, wie Schweinheim (Schweine), Stotzenheim (Stuten), Roitzheim (Rosse), Buellesheim (Bullen), Kuchenheim (Kühe) ableiteten. Ob diese Interpretation historisch berechtigt oder ob sie ein Erzeugnis der lebhaften rheinischen Phantasie ist, vermag ich nicht zu sagen.

Genug, dieses Flamersheim war mehrere Jahrhunderte der Sitz meiner Vorfahren; der älteste in den Kirchenbüchern genannte ist Johann Fischer, der im Jahre 1695 fast gleichzeitig mit seiner Ehefrau geb. Hermany zu Flamersheim starb.

Mit meinem Großvater wäre beinahe der männliche Stamm erloschen. Denn er hat erst mit 49 Jahren geheiratet, und die Familiensage berichtet, daß diese Ehe ein zufälliges Ereignis gewesen sei. Er lebte nämlich mit einer unverheirateten Schwester zusammen, die ihn offenbar tyrannisierte und ihm eines Tages den Schlüssel zum Weinkeller versagte, als er unerwartet einen Gast mitgebracht hatte. Darauf soll er den Entschluß gefaßt haben, sich durch Heirat selbständig zu machen. Seine Wahl fiel auf Fräulein Helene Conrads aus Mülheim am Rhein, ebenfalls einer Kaufmannsfamilie entsprossen. Die Ehe war mit vier Söhnen und einer Tochter gesegnet.

Mein Vater war das zweite Kind. Die beiden letzten, mein Onkel August und Tante Elisabeth kamen 1812 als Zwillinge zur Welt. Ihre Geburt hat der Mutter das Leben gekostet. Auch der Vater ist verhältnismäßig früh mit 61 Jahren an einer Lungenentzündung gestorben.

Die fünf verwaisten Kinder würden wahrscheinlich ein trauriges Lebenslos gehabt haben, wenn sich nicht die unverheirateten Geschwister der Mutter, zwei Brüder und eine Schwester, ihrer in rührender Weise angenommen hätten.

Der älteste, Onkel Hermann Conrads, gab seine behagliche Tätigkeit in Mülheim auf und zog nach Flamersheim, um den Kindern das väterliche Geschäft zu erhalten. Er behielt die beiden jüngsten Kinder (Zwillinge) bei sich. Die drei anderen Brüder Friedrich Arnold, mein Vater und Otto Fischer kamen nach Mülheim am Rhein und wurden hier vorzugsweise von der Tante erzogen. Diese Frau spielte in den Jugenderinnerungen meines Vaters eine große Rolle. Sie war ungewöhnlich begabt, verkehrte nur mit Männern, am liebsten im Disput mit Studierten, besonders Theologen. Sie hatte sich mit der lateinischen Sprache vertraut gemacht und wollte im Alter sogar noch Hebräisch lernen, um die Bibel im Urtext zu lesen, ob schon sie als geistiges Kind der französischen Revolution ganz atheistisch dachte.

Sie führte hauptsächlich das Geschäft in Mülheim und nicht minder das Regiment im Hause, wo der gute Onkel Heinrich, ein ungewöhnlich starker Mann, nichts zu sagen hatte.

Die drei Knaben, die ihrer Obhut anvertraut waren, wurden in keiner Beziehung verwöhnt, sondern frühzeitig abgehärtet und zur Selbständigkeit angehalten.

Als sie bereits erwachsen und verheiratet waren, pflegte die Tante sie noch immer mit „Burschen" anzureden und jeden Widerspruch gegen solche Titulatur oder ihre scharf ausgeprägten Ansichten auf das lebhafteste zu bekämpfen.

Trotzdem waren die Knaben ihr in Dankbarkeit zugetan, genau so, wie sie die beiden Oheims als ihre Beschützer verehrten. Mein Vater und sein älterer Bruder Friedrich Arnold waren von vornherein zum Kaufmann bestimmt, und das entsprach, besonders bei meinem Vater, durchaus der Veranlagung. In der Schule hatte er keinen großen Erfolg, verließ sie deshalb schon mit 14 Jahren und wurde als Lehrling in einem kaufmännischen Geschäft in Luettringhausen bei einem Herrn Moll untergebracht.

Hier erhielt er frühzeitig Gelegenheit, sich als Verkäufer auf Reisen zu betätigen, was ganz seinen Neigungen und auch der folgenden mehr als 70-jährigen Geschäftstätigkeit entsprach.

In späteren Jahren pflegte er oft scherzhaft zu sagen, er habe mehr im Wirtshaus als in der Schule gelernt. Aber er war doch klug genug, die Lücken der Schulbildung zu empfinden und hat sich deshalb später durch Privatunterricht im Französischen und in der Beherrschung der deutschen Sprache, sowie in rechtlichen und gewerblichen Dingen ansehnliche Kenntnisse erworben.

Im Alter machte er noch den Versuch, Englisch zu lernen, da er öfter zum Ankauf von Wolle nach London fahren mußte. Aber dazu war es doch zu spät; sein Englisch hat niemand verstanden.

Bei der Prüfung für den einjährig-freiwilligen Militärdienst, der damals schon in Preußen eingerichtet war, fiel er durch, aber ein vermeintlicher körperlicher Schaden, der sicherlich nichts weiter als eine falsche ärztliche Diagnose war, bewahrte ihn vor der Notwendigkeit, drei Jahre Soldat zu sein. In Wirklichkeit war er ein Mann von ungewöhnlich kräftiger Konstitution und Gesundheit. Es wird davon später noch die Rede sein.

Als er 18 Jahre alt geworden war, übergab der Onkel Conrads ihm und dem Bruder Friedrich Arnold das väterliche Geschäft zu Flamersheim und zog sich nach Mülheim a. Rh. zurück.

Die beiden jungen Leute waren sich ihrer Verantwortung wohl bewußt und haben mit großem Fleiß und kaufmännischem Geschick die für ihre Jahre doch ziemlich schwierige Aufgabe in Angriff genommen. Es ist ein schönes Zeichen geschwisterlicher Eintracht, daß der Geschäftsgewinn nicht den beiden Brüdern allein, sondern allen fünf Geschwistern gleichmäßig zufiel.

Der dritte Bruder Otto war ein ausgezeichneter Schüler, bekundete frühzeitig die Neigung zum Studieren und wurde nach Absolvierung des Gymnasiums zu Duisburg Mediziner. Nach Erledigung des medizinischen Staatsexamens, das damals in Berlin abgelegt werden mußte, ging er auf 1 Jahr nach Paris und entschloß sich dazu, Chirurge zu werden. Nicht lange nachher wurde er der leitende Arzt in der chirurgischen Abteilung des städtischen Hospitals zu Cöln.

Einen später an ihn ergangenen Ruf als Professor der Chirurgie nach Bonn lehnte er ab, weil ihm die dortigen Hilfsmittel und das Krankenmaterial zu dürftig erschienen.

In seiner Stellung zu Cöln hat er sich während einer mehr als 40jährigen Tätigkeit durch glänzende Operationen und überaus sorgfältige Nachbehandlung der Kranken großen Ruf erworben und war ein paar Jahrzehnte hindurch wohl der gesuchteste Arzt am Niederrhein. Selbst aus Holland, Belgien und Frankreich wurde seine Hilfe in Anspruch genommen.

Als Persönlichkeit war er ein Original, ein Volksmann im besten Sinne des Wortes, um den die geschäftige Phantasie der cölnischen armen Leute eine Reihe von Legenden geflochten hat.

Der vierte Bruder August erhielt auch eine gute Schulbildung. Nachdem er die kaufmännische Lehre durchgemacht und bei dem Pionierbataillon zu Cöln als Einjähriger seiner Dienstpflicht genügt hatte, trat er ebenfalls in das Geschäft zu Flamersheim ein.

Ungefähr um dieselbe Zeit heiratete die Zwillingsschwester Elisabeth einen Herrn Dilthey aus Rheydt. Ursprünglich gewillt, Theologe zu werden, war dieser durch den frühzeitigen Tod seines Vaters genötigt worden, mit seinem Bruder Wilhelm das väterliche Geschäft, Seiden- und Sammet-Fabrikation, zu übernehmen, und aus dieser Ehe stammt die zahlreiche Familie Dilthey in Rheydt und Umgegend. Von seinen sechs Söhnen und drei Töchtern sind nur der Jurist Richard Dilthey und eine Schwester unverheiratet geblieben. Alle übrigen erfreuten sich einer zahlreichen Nachkommenschaft, und als meine Tante im hohen Alter von 87 Jahren starb, waren etwa 40 Enkel vorhanden. Die alte Frau bildete bis zum Ende den Mittelpunkt der Familie und erfreute sich infolge ihrer Klugheit, Tatkraft und Güte allgemeiner Verehrung.

Das Geschäft in Flamersheim war durch die gemeinsame Arbeit der Brüder in Blüte gekommen. Mein Vater sorgte durch vielfache Reisen für Vergrößerung des Absatzes und durch die Aufnahme neuer Artikel für Erweiterung des Betriebes. So erzählte er öfters mit Befriedigung, daß er nur wenige Jahre nach Erfindung der Schnellessigfabrikation einen Betrieb dieser Art in Flamersheim angelegt und damit viel Geld verdient habe. Auch eine kleine Mälzerei wußte er schon damals dem Geschäft anzugliedern. Aber gerade durch diese Erweiterungen und durch die Erfahrungen auf seinen Reisen kam er auch zu der Überzeugung, daß Flamersheim nicht der richtige Ort für ein Großgeschäft sei, und er hat wiederholt den Plan erwogen, dieses nach Cöln zu verlegen. Seine Brüder waren aber durchaus dagegen, aus Abneigung gegen die Großstadt und das vergrößerte Risiko. Dagegen setzte er durch, daß das Hauptgeschäft nach Euskirchen, welches gerade zu der Zeit schon gute Landstraßen nach Cöln, Bonn und der Eifel erhalten hatte, verlegt wurde. So entstand, allerdings erst allmählich, der Gebäudekomplex, den ich früher geschildert habe.

Die beiden Haupthäuser wurden, wenn ich nicht irre, im Jahre 1835/36 errichtet. Von der Zeit an ist Euskirchen der Hauptsitz der Firma „Gebrüder Fischer" gewesen. Nur der Onkel Friedrich Arnold blieb in Flamersheim, aber die alte Kompagnie hat noch etwa 30 Jahre fortgedauert und sich auch auf andere Geschäfte, z. B. den Ankauf eines schönen Waldes bei Flamersheim, erstreckt.

Mein Vater und seine Brüder ergänzten sich auf das glücklichste. Er liebte nicht die Kleinarbeit im Kontor und den Lagerräumen, die aber von den beiden anderen Geschäftsinhabern mit allergrößter Sorgfalt und Sachkenntnis besorgt wurde.

Andererseits ging er gerne auf Reisen, wobei er nicht allein verkaufte, sondern auch in früheren Jahren Geld einkassierte und Gelegenheiten nach neuen Geschäften aufspürte. Er nannte sich deshalb scherzhaft den Minister des Äußern.

Zu seinen Aufgaben gehörte natürlich die Vertretung der Firma bei Gerichten, Verwaltungsbehörden, bei öffentlichen Verkäufen usw.

Seiner Initiative war es auch in der Regel zuzuschreiben, wenn neue Unternehmungen begonnen wurden, unter denen an Bedeutung der schon erwähnte Waldankauf hervorragte.

Die Gemeinde Flamersheim besaß in dem benachbarten Vorgebirge der Eifel einen schönen, meist aus Eichen und Buchen bestehenden Hochwald, aus dem sie bei dem damaligen Betriebe nur geringe Einkünfte bezog. Die Mitglieder der Gemeinde hatten sogen. Gerechtsame, die ihnen Anrecht auf den Bezug von Bau- und Brennholz gaben, das damals nur wenig Wert hatte. Infolgedessen ging der allgemeine Wunsch auf eine Teilung des Gemeindewaldes hinaus, konnte aber erst wegen des Widerstandes der königl. Regierung im Jahre 1848/49 durchgesetzt werden. Der Wald wurde dann öffentlich versteigert und der Erlös in bar an die einzelnen Gerechtsamen verteilt.

Von diesem Walde kaufte mein Vater für die Firma ein ansehnliches Stück, etwa 750 Hektar.

Da einige Jahre später infolge des Baues von Eisenbahnen der Preis des Eichenholzes außerordentlich stieg, so wurde der Wald bald abgeholzt, die schönen Eichenstämme meist zu Eisenbahnschwellen verarbeitet und auf dem Boden Eichenschälwald zur Gewinnung von Lohe gezogen. Durch die Anstellung eines Försters sorgte mein Vater rechtzeitig für forstmäßigen Betrieb des schönen Besitzes.

An diesem Walde haben auch wir jüngeren Mitglieder der Familie viel Freude gehabt. Er war reich an Wild, namentlich an Rehen, Sauen und Waldschnepfen. Zur Bequemlichkeit wurde darin ein kleines

Jägerhaus auf dem schön gelegenen Heidberg mit prächtigem Blick in die Eifelberge errichtet. Wir haben dorthin manchen Ausflug, meist zu Jagdzwecken, unternommen, konnten auch über Nacht bleiben. Ein gut ausgestatteter Weinkeller sorgte für fröhlichen Trunk, und unser Förster war ein vortrefflicher Koch, der die meist von Euskirchen mitgebrachten Fleischspeisen zusammen mit Bratkartoffeln, Butterbrot und Kaffee zu einem köstlichen Mahle zu vereinigen wußte. Den Höhepunkt des Jagdbetriebes bildeten die Treibjagden, die im November und Dezember stattfanden, zu denen mein Vater zahlreiche Einladungen ergehen ließ, und die mit einem höchst lustigen Mahl in einem Gasthause zu Flamersheim zu schließen pflegten.

Nach mehr als 50jährigem Besitz hat mein Vater den Wald mit gutem Nutzen wieder verkauft, und zwar an die Familie Haniel, weil niemand von unserer Familie sich um den Besitz mehr kümmern konnte.

Ein anderes Unternehmen, das für die Wahl meines Berufes mitbestimmend geworden ist, war der Erwerb einer kleinen Wollspinnerei, die in Wisskirchen, einem etwa 4 km von Euskirchen entfernten Dorfe, lag. Meinem Vater, der von dieser Industrie nichts verstand, gelang es, in Aachen einen ausgezeichneten Spinnmeister, Herrn Allmacher, zu gewinnen, und da die Firma auch keine Mittel scheute, um die Fabrik mit den besten Maschinen auszustatten, und anstelle der Wasserkraft Dampfbetrieb einzuführen, so gehörten die Erzeugnisse der Firma an Garn bald zu den besten des Bezirks. Ihr Verbrauch blieb nicht auf die Tuchindustrie in Euskirchen beschränkt, sondern die Garne gingen auch nach dem Bezirk München-Gladbach-Rheydt, wo sie zu halbwollenen Waren verarbeitet wurden.

Den Einkauf der Rohmaterialien, vor allen Dingen der Wolle besorgte natürlich mein Vater, ursprünglich bei den Schafzüchtern der Umgegend und besonders der Eifel. Ich habe in jungen Jahren solche Wollfahrten hier und da mitgemacht. In den armen Dörfern der Eifel wurde das Geschäft zuerst mit dem Pastor abgeschlossen. Sobald das geschehen war, kamen die Bauern von selbst mit ihrer Ware. Diese wurde sofort taxiert, abgewogen, in bar bezahlt und auf den mitgenommenen Wagen geladen. Aber bald reichten die heimischen Quellen für die vergrößerte Fabrik nicht mehr aus, auch waren sie für feinere Tuche zu grob. Damals begann der Import der überseeischen Wolle, besonders aus der Kapkolonie und aus Brasilien. Deutsche Händler besorgten dieses Geschäft, natürlich mit einem entsprechenden Preisaufschlag. Der Markt für diese Wolle war London, wo sie auf grossen Auktionen verkauft wurde.

Sobald dies meinem Vater bekannt geworden war, reiste er nach London, wo er mit Hilfe eines Agenten seinen Wollbedarf deckte. Er gehörte mit zu den ersten Fabrikanten in Euskirchen, die den direkten Bezug des Rohmaterials ausnutzten. Für eine weitere Verbilligung der Garne war die Anlage einer Färberei erforderlich. Auch das hat mein Vater in die Hand genommen, ist dabei aber auf recht erhebliche Schwierigkeiten gestoßen.

Die Teerfarben kannte man damals noch wenig und das Färben mit den natürlichen Farbstoffen wie Indigo, Krapp, Gelbholz, Blauholz war rein empirisch. Es gab zwar einige recht dürftige Bücher der praktischen Färberei und man konnte auch Rezepte von den Färbern kaufen, aber in der Praxis pflegte das alles zu versagen, und die Leitung einer Indigoküpe galt geradezu als eine schwierige Kunst, die nur durch langjährige Übung erworben werden konnte.

Es war darum begreiflich, daß in der neu angelegten Färberei zu Wisskirchen, von der unser sonst so trefflicher Spinnmeister nichts verstand, viele Mißerfolge eintraten und nicht allein zu Verlusten, sondern auch zu ärgerlichen Verhandlungen mit den Kunden führten.

Mein Vater, der von jeher die Selbsthilfe hoch eingeschätzt hatte, fing deshalb eine kleine Versuchsfärberei in Euskirchen an, wo er die von verschiedenen Färbern gekauften Rezepte eigenhändig prüfte und es dabei an Variationen besonders im Beizen nicht fehlen ließ. Aber er spürte doch bei diesen Versuchen bald, wie hinderlich ihm der Mangel an chemischen Kenntnissen war, und er pflegte öfters zu sagen, wenn einer von uns Jungen Chemie studierte, so würden alle Schwierigkeiten spielend überwunden werden. Seine Achtung vor der Chemie steigerte sich noch, je mehr er mit der aufblühenden rheinischen Industrie, besonders der Fabrikation von Eisen und Zement in Berührung kam. Ich erwähne das ausdrücklich, da es die Wahl des Berufes sowohl bei mir, wie bei meinem Vetter Otto beeinflußt hat.

Die bisher besprochenen Geschäfte wurden alle mit den Mitteln und zugunsten der Firma „Gebrüder Fischer" geführt. Als aber die Unternehmungslust meines Vaters mit dem wachsenden Wohlstande sich steigerte, während die konservativer gestimmten beiden Onkel mit zunehmendem Alter jedes größere Risiko vermeiden wollten, blieb er zwar Mitglied der Firma, beteiligte sich aber an anderen Geschäften auf eigene Rechnung. So hat er zusammen mit Cölner Kaufleuten größere Landgüter, die wegen nachlässiger Wirtschaft der in Konkurs geratenen Besitzer öffentlich versteigert wurden, angekauft und in kleinen Teilen an die umliegenden Bauern wieder verkauft.

Die Zeit war damals für die Landwirtschaft am Rhein recht günstig, und die meisten Ankäufer konnten aus dem Verdienst in 9jährigen Terminen die Ankaufsumme für den neuerworbenen Acker bezahlen.

Das geschah fast durchweg in barem Silbergeld, welches die Bauern persönlich brachten und das ganze Geschäft vollzog sich nicht in dem Kontor der Firma, sondern in unserm Wohnhause. Da mein Vater vielfach verreist war, so wurde es von meiner Mutter besorgt mit einer Pünktlichkeit in der Buchführung, die jedem Kontoristen Ehre gemacht haben würde.

In der Abschätzung von Grundstücken sowie im An- und Verkauf hatte mein Vater solche Übung erlangt, daß man mit vollem Vertrauen seinem Urteil glauben konnte. Er erkannte auch rechtzeitig die allgemeine wirtschaftliche Gefahr, die im Rheinland in den 70er Jahren der Landwirtschaft durch den Wettbewerb des überseeischen Getreides erwuchs. Er pflegte mit aufrichtigem Bedauern für die Bauern häufig zu sagen: „Mit der Landwirtschaft ist nichts mehr zu verdienen", und er hielt schon damals einen staatlichen Schutz in Form von Getreidezöllen für nötig.

Seine Geschäftsinteressen wandten sich deshalb immer mehr der Industrie zu. An kleinen Unternehmungen in der Eifel, die auf die Gewinnung von Eisenerzen gerichtet waren, erlebte er wenig Freude. Dasselbe galt bezüglich des finanziellen Ergebnisses für eine Beteiligung an einer Zementfabrik in Obercassel bei Bonn. Sie war gegründet von Dr. Bleibtreu, einem ehemaligen Schüler von A. W. Hofmann in London. Er hatte in England die Herstellung von Portlandzement kennen gelernt und die erste derartige Fabrik in Deutschland bei Stettin ins Leben gerufen. Eine zweite Gründung war das Werk in Obercassel. Es würde wahrscheinlich ebenso wie die Stettiner Anlage guten Gewinn gebracht haben, wenn es nicht mit anderen unrentablen Unternehmungen verknüpft worden wäre. Damals war die Rente der Aktiengesellschaft „Bonner Bergwerks- und Hüttenverein" recht bescheiden, aber mein Vater hat selten eine Aufsichtsratssitzung in Bonn versäumt, da er dort immer mit einer Reihe guter Freunde aus Köln und Düsseldorf zusammentraf und man den oft ärgerlichen Geschäftssitzungen ein fröhliches Mahl im Gasthof Stern folgen ließ. Ich durfte daran zu meiner großen Freude öfters teilnehmen, als ich das Gymnasium in Bonn besuchte, und der Chemiker Dr. Bleibtreu, der nicht allein Zement machen konnte, sondern auch ein Meister im Ansetzen von Pfirsichbowlen war, hat mir bei solchen Gelegenheiten immer zugeredet, Chemiker zu werden.

Die markantesten Persönlichkeiten in diesem Aufsichtsrat waren der Freund meines Vaters, Albert Poensgen, Großindustrieller in Düsseldorf, und ein Herr Muehlens aus Cöln, Fabrikant von Eau de Cologne, ein hervorragender Spaßmacher.

An der Tafel, wo der Aufsichtsrat in Gesellschaft von zwei meiner Schwestern eines Tages Platz genommen hatte, speiste auch das studentische Corps Borussia, dem damals die beiden Söhne Bismarcks angehörten. Die Herren schienen keinen großen Hunger zu haben; denn sie begannen das Mahl mit Sekttrinken und zündeten dazu Zigarren an. Das war den alten rheinischen Herren denn doch zu burschikos, und mit weitschallender Stimme gab Herr Muehlens dem Oberkellner den Auftrag, er sollte den jungen Herren sagen, es sei hierzulande nicht Sitte, in Gesellschaft von Damen das Mahl mit Tabakrauchen zu beginnen. Der Kellner entledigte sich seines Auftrages. Das ganze Corps erhob sich sofort und schritt, dicke Tabakwolken verbreitend, aus dem Saal heraus. Herr Muehlens sah ihnen belustigt zu und begleitete den Durchzug mit den Worten „Sehr gut gemacht". Auch mein Vater war geneigt, solche Verstöße gegen gute Sitten vor aller Öffentlichkeit zu rügen.

Ende der 60er Jahre hatte sich mein Vater mit einer erheblichen Summe an einem Geschäft beteiligt, das ihm zwar viel Sorge und Ärger, aber auch später sehr viel Freude bereitet hat. Es war die Gründung einer Brauerei in Dortmund. Die Anregung dazu ging aus von einem Ingenieur Heinrich Herbertz, der in Dortmund eine große Kokerei besaß und aus dem Aufblühen der anderen dortigen Brauereien den Schluß zog, daß hier noch ein gutes Geschäft zu machen sei. Er war verwandt und befreundet mit meinem Schwager Mauritz und dessen Bruder Heinrich.

Nachdem mein Vater die nötigen Erkundigungen eingezogen hatte, ging er auf den Vorschlag ein, und so entstand die Brauerei Herbertz & Co., an der mein Vater ursprünglich der meist beteiligte war. Sie war von Anfang an ein gutes Geschäft, denn das dort erzeugte Bier erfreute sich bald des besten Rufes und der Absatz stieg mit dem steigenden Wohlstand, besonders nach dem deutsch-französischen Krieg.

Leider wurde das Unternehmen in eine Aktiengesellschaft verwandelt und hat dann infolge schlechter Leitung böse Zeiten durchgemacht. Seine Sanierung wurde nötig und mit dem völligen Wechsel der leitenden Persönlichkeiten trat auch bald wieder ein ordnungsmäßiger Betrieb und allmählich eine Rentabilität ein.

Die Aktien sind noch jetzt zum größeren Teil im Besitz der Familien Fischer und Mauritz, und das Unternehmen hat seit etwa 40 Jahren recht gute geschäftliche Erfolge gehabt.

Mein Vater war mehrere Jahrzehnte Vorsitzender des Aufsichtsrats. Er hat auch lange Zeit den Einkauf von Hopfen für die Brauerei besorgt, und zur Zeit meines Aufenthalts in Erlangen bin ich mehrmals mit ihm auf dem Hopfenmarkt in Nürnberg gewesen. Mit großer Sicherheit wußte er aus rein empirischen Proben, wie Farbe, Geruch,

Gefühl die sehr verschiedene Qualität des Hopfens abzuschätzen und jede Übervorteilung von seiten der Händler zu vermeiden.

Auch sonst war er für das Wohl der Brauerei in hohem Maße interessiert und tätig, besonders was den technischen Betrieb anging. Er pflegte immer zu sagen, „Vor allem muß das Erzeugnis gut sein", und schon bei der Legung des Grundsteins hat er seine Wünsche für die Zukunft in einen hübschen Vers gekleidet, „Brau gutes Bier, das rat ich Dir". Durch seine Vermittlung habe auch ich einige Gelegenheit gehabt, der Brauerei kleine Dienste zu leisten.

In den ersten Jahren meines Aufenthaltes zu München hatte ich zufällig von neuen Eismaschinen des Professor Linde und ihrer Verwendung in der Brauerei von Sedlmeier gehört. Ich erzählte dies bei einem Ferienbesuch meinem Vater, und er beauftragte mich sofort, genauere Erkundigungen einzuziehen. Diese fielen sehr günstig aus und wurden bestätigt durch die Firma Meister, Lucius & Brüning zu Höchst a. Main, welche die Lindesche Maschine in dem Betrieb für Azofarbstoffe verwendete. Die Folge davon war, daß die Dortmunder Brauerei sofort mit der von Linde gegründeten Gesellschaft in Verbindung trat und meines Wissens als erste norddeutsche Brauerei in den Besitz einer Lindeschen Maschine und später auch einer damit betriebenen sehr zweckmäßigen Kellerkühlung gelangte.

Wohl nicht minder wichtig war eine Belehrung, die ich dem Braumeister in bezug auf den Gärprozeß geben konnte.

Während des Wintersemesters 1876/77 hielt ich mich wieder in Straßburg auf und lernte dort durch Dr. Albert Fitz das Buch von Pasteur „Études sur la bière" kennen, das kurz vorher erschienen war. Der geniale Forscher hatte darin seine Erfahrungen über die Verunreinigung der Bierhefe durch andere Mikroorganismen und deren schädlichen Einfluß auf die Beschaffenheit des Bieres niedergelegt. Als ich davon meinem Vater berichtete, bat er mich dringend, die Materie gründlich zu studieren, und da diese mich auch wissenschaftlich interessierte, so erklärte ich mich gerne dazu bereit. Ein feines Mikroskop wurde sofort angeschafft und mit Hilfe von Dr. Fitz und dem Botaniker Prof. de Bary habe ich dann in Straßburg Studien über Schimmel-, Sproß- und Spaltpilze angestellt, die mir später bei den Zuckerarbeiten sehr zustatten gekommen sind. Zunächst mußte ich aber die neuen Kenntnisse praktisch verwerten. Darum bin ich mit meinem Mikroskop für einige Wochen nach Dortmund gezogen, um den Beamten der Brauerei die neuen Errungenschaften klar zu machen. Wahrscheinlich war ich der erste Chemiker in Deutschland, der diesen Versuch unternahm, und ich muß gestehen, daß ich bei den Männern der Praxis auf großes Mißtrauen stieß. Man bemühte sich auf alle mögliche Weise, mich irre zu führen, besonders mit falschen Angaben über den Ursprung und die Beschaffenheit der zu prüfenden Hefesorten.

Als ich aber mit Hilfe des Mikroskops ohne Mühe die verdorbenen Hefesorten herausfand, wurde man ernster. Es ist mir zwar nicht gelungen, einen der Männer zum richtigen Gebrauch des Mikroskops heranzubilden und dadurch eine dauernde Kontrolle der Hefe einzurichten. Aber meine Belehrung über die Art, wie gute Hefe und damit auch das Bier verdorben werden kann, fiel doch auf guten Boden. Z. B. befanden sich direkt neben dem Kühlschiff, wo die fertige Würze an offener Luft abgekühlt wurde, der Pferdestall und ein stattlicher Dunghaufen. Auf meine Vorstellung hin ist hier bald Wandel geschaffen worden. Auch die von mir als besonders wichtig gepredigte Sauberhaltung aller Gefäße, mit denen die kalte Würze und das Bier in Berührung kommen, hat dem verständigen Braumeister gefallen, weil sie mit seinen Erfahrungen in der Praxis wohl übereinstimmte.

Etwa 30 Jahre später bin ich bei einer Jubelfeier des Instituts für Gärungsgewerbe in Berlin zu dessen Ehrenmitglied ernannt worden, und der Leiter des Instituts, Professor Max Delbrück, hat bei dieser Gelegenheit die Wahl nicht allein durch meine chemischen Arbeiten motiviert, sondern auch scherzhaft auf meine Bemühungen in der Dortmunder Brauerei und das dadurch bekundete Interesse für das praktische Braugewerbe hingewiesen.

Was ich damals als Pasteur'sche Lehre zu verbreiten suchte, war inzwischen durch die Studien von Professor Ch. Hansen in Kopenhagen, der auch dem Jubelfeste beiwohnte und zum Ehrenmitglied des Instituts ernannt wurde, außerordentlich vervollkommnet worden und hatte für das Braugewerbe eine grundlegende Bedeutung erlangt.

Wie schon erwähnt, gehörte mein Vater zum Aufsichtsrat der Aktienbrauerei und hat viele Jahre das Amt des Vorsitzenden geführt. Auch andere Aktiengesellschaften, z. B. der Bonner Bergwerks- und Hüttenverein, das Röhren- und Eisenwalzwerk Poensgen in Düsseldorf, eine Glashütte in Stolberg, eine Röhrenkesselschmiede zu Kalk a.Rh. und die Versicherungsgesellschaft Concordia in Cöln wählten ihn in den Aufsichtsrat. Und wenn er wegen zunehmender Schwerhörigkeit irgendwo seinen Austritt erklären wollte, so pflegte man den verständigen und stets heiteren, originellen alten Herrn zur Beibehaltung des Amtes aufzufordern.

So kam es, daß er nach Überschreitung des 90ten Lebensjahres sich scherzhaft rühmen konnte, das älteste Aufsichtsratsmitglied in Preußen zu sein.

Wie die vorangehende Schilderung seiner geschäftlichen Unternehmungen zeigt, war mein Vater ein vielseitiger, kluger Kaufmann, der die Möglichkeiten des gewerblichen Lebens richtig abschätzte und selten ein schlechtes Geschäft begonnen hat.

Den Mangel der Schulbildung hat er später durch die Erfahrungen der Praxis auszugleichen gewußt. Er war kein rascher Denker, aber wenn er eine Angelegenheit, die seinem Ideenkreis nicht zu fern lag, durchstudiert hatte, so konnte man sicher sein, daß er sie völlig erfaßt und auch in den Konsequenzen durchschaut hatte.

Ich habe es mit erlebt, daß er Notaren Verträge oder Verkaufsbedingungen für öffentliche Auktionen diktierte, weil ihre eigenen Entwürfe nicht klar genug waren. Das letzte Beispiel dieser Art bot sein autographisches Testament, das durch dieselbe Klarheit der Form und der Gedanken ausgezeichnet war, wie alle seine Schriftstücke. Kurz vor seinem Tode war es auf seinen Wunsch durch einen Berliner Notar in ein amtliches Dokument umgewandelt worden. Als ich dann für die Erbteilung dieses Schriftstück meinen Schwägern vorlegte, erklärten sie einstimmig, „das hat Großpapa nicht verfaßt; denn so unklar hat er sich niemals ausgedrückt". Die Verwirrung war durch die Bearbeitung des Notars und seine juristischen Redewendungen entstanden. Glücklicherweise konnte ich durch Vorlage des autographischen Testaments alle Zweifel über den Sinn des notariellen Schriftstückes beseitigen.

Mit der Klarheit des Geistes war bei ihm eine ungewöhnliche körperliche Rüstigkeit verbunden, die schon dem scharfen Auge seiner Mutter nicht entgangen war; denn wie er selbst gerne erzählte, hatte diese in einem Brief an ihre Schwester Conrads über den kleinen Lor berichtet: „Er scheint etwas dumm zu sein, aber es ist ein wahres Vergnügen, seinen kräftigen Körper anzuschauen". Es ist deshalb auch kein Wunder, daß er allen körperlichen Künsten zugetan war. Reiten, Tanzen, Turnen, Schießen waren ihm wohl vertraut und die Jagd hat er bis ins höchste Alter getrieben. Hand und Auge waren so leistungsfähig geblieben, daß er mir noch mit 93 Jahren einen selbstgeschossenen Hasen schicken konnte. Dazu kam eine große Heiterkeit des Gemütes, die auch durch herbe Verluste nur vorübergehend gestört werden konnte. Alltäglich Bewegung in frischer Luft und abends 1 bis 2 Stunden Geselligkeit im Gasthause oder Casino bei einem Glase Wein und einer Cigarre oder einer Pfeife Tabak waren ihm Bedürfnis. Und wenn er dann nach Hause kam, so war seine Fröhlichkeit geradezu ansteckend für den ganzen Familienkreis. Das Gelächter in unserem Hause war häufig so laut und anhaltend, daß Passanten auf der Straße erstaunt Halt machten.

Selbstverständlich übte er mit besonderer Freude eine freie Gastfreundschaft, und meine Mutter hatte manchmal Mühe, die plötzlich in großer Zahl herbeigeführten Gäste zu versorgen. Besonders hoch und lustig ging es her an Familienfesten, z. B. bei den Hochzeiten meiner Schwestern. Es sind ihrer nicht weniger als 7 in unserem Hause gefeiert worden, sechs bei meinen Schwestern, von denen eine zum zweiten Mal heiratete. Und als das einzige Töchterlein meines Onkels im Nachbarhause in die gleiche Lage kam, wurde nach alter Gewohnheit das Fest auch von meinem Vater in unserem Hause veranstaltet; nur mußte der Onkel, wie billig, die Kosten tragen. Bei solcher Gelegenheit ließ mein Vater alle Quellen seiner Fröhlichkeit springen. Obschon er keine besondere Rednergabe besaß, so erweckten doch seine Tischreden, die manchmal mit kleinen lustigen Versen geschmückt waren, stets großen Jubel.

Sein rheinischer Humor und seine Freude an Späßen führten ihn auch ziemlich regelmäßig zur Karnevalsfeier nach Cöln. In hohem Alter hat er einmal zu solcher Gelegenheit eine große Gesellschaft von Verwandten und Freunden in ein Gasthaus zu Cöln eingeladen. Niemand wußte, zu welchem Zwecke, bis der Gastgeber in seiner Tischrede das Rätsel löste. „Alle Welt feiert jetzt Jubiläen", so begann er, „Drum habe auch ich geglaubt, eine solche Feier veranstalten zu müssen; denn heute ist es das fünfzigste Mal, daß ich an dem Cölner Karneval teilnehme". Man kann sich denken, welche Stimmung bei diesem Feste geherrscht hat.

Noch charakteristischer und für Nichtrheinländer schwer verständlich war ein karnevalistischer Einfall, den er bald nach dem Tode seines von ihm sehr verehrten Bruders Otto, des Arztes in Köln verwirklichte. Obschon er damals fast 80 Jahre alt war, glaubte er doch dem Fasching nicht ganz entsagen zu dürfen. Er ging deshalb auf den Maskenball, aber zum Zeichen der Trauer in der Maske eines Mohren. Diese beschränkte sich allerdings auf die Schwärzung des Gesichts, die mit einem angebrannten Korken hergestellt war.

Unvermeidliche Dinge, wie den Tod seiner vier Geschwister, mit denen er in treuester und stets hilfsbereiter Freundschaft gelebt hatte, und die alle erst im hohen Alter starben, überwand er sehr leicht. Viel schwerer hat ihn der Tod meiner Mutter getroffen, mit der er 46 Jahre in glücklicher Ehe verbunden war und die er 20 Jahre überlebte. Nach ihrem Tode hat er noch 10 Jahre in Euskirchen verbracht, dann wurde er plötzlich auf eine eigentümliche Art veranlaßt, nach 57-jährigem Aufenthalt diesen Wohnsitz aufzugeben.

Im Jahre 1892 war für die preußische Einkommensteuer die obligatorische Selbstangabe des Einkommens eingeführt worden. Im Kreise Euskirchen stellte sich bei dieser Gelegenheit heraus, daß mein Vater das höchste Einkommen im Kreise besaß, was man nach seiner einfachen Lebensweise nicht erwartet hatte.

Die erhöhte Einkommensteuer kam bei der in Preußen üblichen Veranlagung auch der Gemeinde zugute und der Prozentsatz, in welchem die Gemeindesteuer bis dahin erhoben wurde, hätte vernünftigerweise ermäßigt werden müssen. Das hatte auch mein Vater angeregt; aber statt dessen wurde nur die Gewerbesteuer herabgesetzt. Als mein Vater das als ungerecht bezeichnete und die Möglichkeit andeutete, daß vermögende Leute, die kein Gewerbe mehr betrieben, bei der nunmehr recht hohen Gemeindesteuer die Stadt verlassen könnten, entgegnete man ihm spöttisch: Ein Mann wie er, der im Alter von 84 Jahren stehe, würde sich nicht mehr zu einem Wechsel des Wohnsitzes entschließen können. Dieser Appell an seine Altersschwäche ärgerte ihn, und um den Leuten das Gegenteil zu beweisen, entschloß er sich sofort, Euskirchen zu verlassen. Ohne einem seiner Kinder ein Wort davon zu sagen, löste er den Haushalt auf und verließ die Stadt und Preußen und zog im Sommer 1892 nach Straßburg i. Els. Kurz vorher machte er bei mir noch einen Besuch in Würzburg, wovon später die Rede sein wird. Es war aber damals schon zu spät, seinen Entschluß zu ändern.

Der endgültige Umzug ist in den einfachsten Formen erfolgt. Ein Verwandter traf den alten Herrn zu Cöln am Bahnhof, im primitivsten Anzug, die Jagdflinte auf dem Rücken, den Hund an der Strippe und in der anderen Hand einen bescheidenen Handkoffer. Auf die Frage: „Nun, Herr Fischer, wo geht's hin?" erfolgte die knappe Antwort: „Domizilveränderung nach Straßburg". Hier wohnte er zuerst im Gasthaus und später bei meinem Vetter Ernst Fischer, dem außerordentlichen Professor der Chirurgie, der in seinem geräumigen alten Hause in der Küfergasse eine Privatklinik eingerichtet hatte und dessen Frau aus einer elsässischen Bauernfamilie einen einfachen aber guten Haushalt führte.

Mein Vater hat hier rasch einen ihm zusagenden Bekanntenkreis gefunden, besonders unter den Jägern. Nach wenigen Monaten war er an einer Jagd beteiligt und hatte, wie er lachend erzählte, auch schon einen Prozeß, den er richtig gewann. In späteren Zeiten hatte er Fühlung mit den Offizieren der Garnison und wurde zu deren Treibjagden eingeladen. Wegen seines hohen Alters erhielt er dann den Ehrenplatz neben dem Höchstkommandierenden, und es machte ihm großen Spaß, diesem die Hasen fortzuschießen, was die jüngeren Offiziere sich nicht erlauben durften. Den Verkehr mit Offizieren war er übrigens gewohnt, da in der Umgebung von Euskirchen sehr oft im Herbst militärische Übungen stattfanden. Zweimal wurde sogar das Kaisermanöver dort abgehalten und dann waren immer hohe Offiziere in unserem Hause einquartiert. Auch an Hirsch- und Saujagden in den Vogesen hat er sich beteiligt.

Häufige Reisen führten ihn von Straßburg nach dem Niederrhein und auch nach Berlin, um geschäftliche Dinge, besonders die Teilnahme an den Aufsichtsratssitzungen zu erledigen.

Als er 90 Jahre alt wurde, entschloß er sich zur Rückkehr nach Preußen. Aber bezüglich der Steuern hatte er sich inzwischen feste Grundsätze zurechtgelegt, die nicht ganz leicht zu verwirklichen waren. Die Staatssteuer, so hörte man ihn öfters sagen, wolle er gerne bezahlen, weil der Schutz des Staates für jedermann unentbehrlich sei. Dagegen die Gemeinde habe er nicht mehr nötig und er sehe nicht ein, daß er für diese viel bezahlen solle. Er wolle deshalb den Wohnsitz in einer Gemeinde wählen, die durch niedere Steuern ausgezeichnet sei. So kam er zuerst im Sommer 1898 zu mir nach Wannsee bei Berlin, wo damals nur 40% Zuschlag zur Staatssteuer erhoben wurden. Als er aber aus der ersten Steuerrechnung ersah, daß noch 30 % Kreissteuer dazukamen, meldete er sofort seinen Wegzug an.

Die nächste Wahl fiel auf einen kleinen Ort im Harz, wo sein ältester Enkel Heinrich Mauritz als königl. Bergbeamter tätig war.

Der Zuzug des wohlhabenden Mitbürgers erschien nun den anderen Gemeindeangehörigen als eine günstige Gelegenheit, alle möglichen neuen Bedürfnisse der Gemeinde zu befriedigen, und die Steuern gingen im folgenden Jahre erstaunlich in die Höhe. Sofort war der alte Herr wieder verschwunden und jetzt fand er einen Wohnsitz in Griethausen, einem Dorfe in der Nähe von Cleve, wo man auf die Ausbeutung des Zugvogels verzichtete. Er mietete sich ein Zimmer bei dem Gemeindediener und behielt das Domizil bis zu seinem Tode, obschon er kaum 24 Stunden sich dort aufgehalten hatte.

Statt dessen wohnte er abwechselnd bei seinen Kindern oder Schwiegersöhnen oder Enkeln in Rheydt, Uerdingen, Berlin oder Herdt bezw. Dortmund, durfte aber nirgends länger bleiben als 89 Tage, weil er sonst nach dem Gesetz zur Zahlung von Gemeindesteuern verpflichtet gewesen wäre. Den übrigen Teil des Jahres, namentlich die Sommermonate verbrachte er auf Reisen und machte dabei noch alle möglichen Bekanntschaften. So schrieb er mir eines Tages, daß er in einem Gasthaus zu Heidelberg Robert Bunsen zufällig getroffen und sich ihm als Vater eines Chemikers vorgestellt habe. Dieser sei zwar jünger, aber noch schwerhöriger wie er selbst. Trotzdem habe sich ein langes und interessantes Gespräch zwischen ihnen entwickelt.

Im Winter 1900/01 hat er zum letzten Mal an der Jagd teilgenommen, weil das früher so gute Auge versagte. Das kam von einer leichten Trübung der Linse, während der Sehnerv noch ganz intakt geblieben war. Im übrigen erfreute er sich im Frühjahr 1902 noch voller Gesundheit, wenn man von der Schwerhörigkeit absehen will. Er hat damals noch in Frack und weißer Binde einer Abendgesellschaft in meinem Hause beigewohnt. Als aber die üblichen 89 Tage vorüber waren, zog er mit der gewohnten Pünktlichkeit nach Rheydt zu seinem Schwiegersohn Arthur Dilthey. Hier soll er bei häufigem Verweilen in den Gasthäusern, besonders in jugendlicher Gesellschaft dem Bier- und Weingenuß über das Maß der Zuträglichkeit gehuldigt haben. Dem war das alte Herz nicht mehr gewachsen und das erste Zeichen seiner Insuffizienz gab sich kund in der Schwellung der Beine. Die Mahnung des Arztes, seine Lebensweise zu ändern und den Genuß alkoholischer Getränke zu vermeiden, beantwortete er mit der Frage: „Wieviel Leute in meinem Alter haben Sie schon behandelt?" Er blieb also bei seinen Gewohnheiten. Die Wassersucht wurde im Laufe des Sommers schlimmer, ohne ihn aber stark zu belästigen, und Anfangs Oktober kam er in Begleitung seines Enkels Alfred Dilthey als schwer kranker Mann zu mir nach Berlin. Ohne besonders zu leiden ist er hier am 16. Oktober 1902, 18 Tage vor seinem 95. Geburtstage sanft verschieden.

Einen Tag vor seinem Tode hat er nochmals das Bett verlassen, und 5 bis 6 Stunden auf die Vervollständigung seines Hauptbuches, das er stets auf Reisen mit sich führte, verwendet. Die spätere Prüfung des Buches ergab nur einen einzigen Fehler, der an diesem letzten Tage begangen war.

An seinem Todestage hat er in einem Gespräch mit mir die Bilanz seines Lebens gezogen und sich sehr befriedigt darüber geäußert. Eine Stunde vor seinem Tode trank er noch ein großes Glas Bier, offenbar mit Genuß; denn seine letzten Worte waren: „Es ist ein Glück, daß der gemeine Mann für billig Geld einen so guten Trank haben kann". Er starb, wie er gelebt hatte, als vollkommener Atheist, aber in treuer Anhänglichkeit an Frau und Kinder, an die weitere Familie und an zahlreiche Freunde. Politisch war er früher rheinischer Fortschrittsmann, später nationalliberal und immer in scharfer Opposition zur ultramontanen Partei.

Obschon ganz im kapitalistischen Zeitalter groß geworden, hatte er doch Sinn und Verständnis für die soziale Bewegung der Neuzeit. Vom sozialistischen Zukunftsstaat wollte er gar nichts wissen, aber von den Bestrebungen der Arbeiter pflegte er zu sagen: „Die Leutchen haben ganz recht, wenn sie versuchen, ihre Lage zu verbessern". In der großen Öffentlichkeit ist er weder als Redner noch als Schriftsteller hervorgetreten. Dagegen hat er als langjähriger Stadtverordneter und als Mitglied mancher Kommissionen an der Verwaltung von Gemeinde und Kreis teilgenommen, und wenn es sich darum handelte, allgemein wirtschaftliche Zwecke, wie den Bau von Eisenbahnen in der dortigen Gegend zu fördern, so war er gewöhnlich Mitglied der Abordnungen, die nach Cöln oder Berlin zur Verhandlung mit der königl. Regierung geschickt wurden.

Trotz seiner Tatkraft in allen geschäftlichen Dingen war er von Charakter gutmütig und huldigte dem Grundsatz: „Leben und leben lassen". Gegen Frau und Kinder war er nicht allein sehr gütig, sondern auch rücksichtsvoll und überließ die strenge Seite der Erziehung ganz der Mutter. Meine Schwestern konnten mit einigen Schmeichelworten fast alles von ihm erreichen. Nur in ernsten Dingen, z. B. in der Wahl des Gatten, mahnte er dauernd zur Vorsicht und Vernunft, und phantastischen Heiratsplänen wäre er sicherlich mit großer Entschiedenheit entgegengetreten. Er hat dann auch wirklich das Glück gehabt, mit allen Schwiegersöhnen sehr zufrieden sein zu können.

Ich selbst erinnere mich nicht, von ihm gestraft worden zu sein. Ja, ich habe kaum ein böses Wort von ihm gehört. Zwar hat er öfters sein Bedauern darüber ausgesprochen, daß ich kein Interesse an kaufmännischen Dingen und dem Erwerb von materiellen Gütern besitze, aber er ließ mich doch ruhig meinen Weg gehen und sprach nur Anderen gegenüber mit Bedauern darüber, daß der Sohn die Kunst des kaufmännischen Rechnens nicht besitze. Leider hat er es nicht mehr erlebt, daß dieser unpraktische Gelehrte den Nobelpreis für Chemie erhielt, und sich später durch einige kleine Erfindungen Jahreseinkünfte verschaffte, wie er selbst sie niemals gehabt hat.

Wesentlich verschieden von dem Vater war trotz der sehr glücklichen Ehe in Charakter, Anschauungen und Neigungen meine Mutter. Sie stammte aus der zweiten Ehe des Eisenfabrikanten Johann Abraham Poensgen in Schleiden (Eifel) mit Wilhelmine Fomm und war am 19. Februar 1819 geboren. Wie ich einer recht gut geschriebenen und im Druck erschienenen Geschichte der Familie Poensgen entnahm, haben meine Vorfahren mütterlicherseits Jahrhunderte lang als Erzeuger von Eisen und Eisenwaren im Schleidener Tal gewirkt und vielleicht habe ich von ihrer Seite die Freude an chemischen und technischen Prozessen geerbt. Es ist ein merkwürdiger Zufall, daß ich jetzt im Alter durch die Kaiser Wilhelm Gesellschaft zur Förderung der Wissenschaften mit der Kohlen- und Eisenindustrie wieder in nähere Berührung gekommen und in jüngster Zeit sogar Kuratoriumsmitglied des großartig geplanten Kaiser Wilhelm Instituts für Eisenforschung geworden bin.

Da meine Großmutter nach dem Tode des ersten Mannes den Arzt Dr. Fuß zu Gemuend im Schleidener Tal heiratete, so hat meine Mutter dort den größten Teil ihrer Jugend verlebt. Zur Vollendung ihrer Ausbildung kam sie mit etwa 16 Jahren in die Erziehungsanstalt der

Herrnhutergemeinde in Neuwied a. Rhein und daher stammte wohl die tiefe religiöse Überzeugung, der sie während ihres ganzen Lebens treu blieb.

Sie war sehr klug und wissensdurstig und von dem vielen Lesen hatte sie schon in der Jugend eine ziemlich starke Myopie erworben, die ich von allen ihren Kindern allein geerbt habe. Sie wäre heutzutage wahrscheinlich eine gelehrte Frau geworden, aber das Frauenstudium war zu ihrer Zeit noch nicht üblich. Da sie schon mit 18 Jahren heiratete und in den nächsten 15 Jahren 8 Kinder zur Welt brachte, so war ihre Zeit durch andere Pflichten ausgefüllt. Sie wurde eine tüchtige Hausfrau und half meinem Vater auch in geschäftlichen Dingen. Sie wußte sich überall in Respekt zu setzen und niemand von den Kindern oder dem Dienstpersonal hätte es gewagt, ihre Anordnungen zu mißachten. Sie war ernster wie mein Vater, konnte aber doch recht herzlich über seine Späße lachen. Nur wo sie rohen oder gemeinen Äußerungen begegnete, gab sie ihrer Entrüstung so deutlichen Ausdruck, daß jedermann in ihrer Gegenwart gezwungen war, sich anständig zu benehmen. Bei alledem war sie eine liebevolle Mutter, voller Fürsorge für ihre Kinder und später auch für deren Familien.

In ihrer tiefen Religiösität ließ sie sich durch den Unglauben meines Vaters und seinen gelegentlichen Spott über Kirche und Pfaffen nicht irre machen.

Ebenso selbständig waren ihre politischen Anschauungen, und als eifrige Protestantin verehrte sie Preußen als die Vormacht des evangelischen Glaubens in Deutschland.

Als unter der Ministerpräsidentenschaft Otto von Bismarcks der böse Konflikt zwischen der preußischen Regierung und dem Abgeordnetenhause im Jahre 1863/64 tobte, und fast alle Rheinländer, unter ihnen auch mein Vater, zur Oppositionspartei gehörten, war sie ganz auf Seiten Bismarcks: „Jetzt hat Preußen wieder einen Minister, der seiner würdig ist und ihr seid viel zu dumm, diesen Mann zu begreifen“. So sprach sie wohl zu ihren Kindern und auch meinem Vater machte sie in ihren politischen Urteilen so lebhafte Opposition, daß er sie scherzhaft „Frau Bismarck“ nannte.

Dabei war sie eine recht hübsche Frau, besonders ausgezeichnet durch das üppige tiefschwarze Haar und durch die großen klugblickenden Augen. Als einziger Sohn habe ich mich besonders ihrer Liebe und Fürsorge erfreut. Sie hat auch stets meine Neigung zu wissenschaftlichen Studien gefördert, nur von der Wahl der Chemie als Berufsstudium war sie enttäuscht, weil ihr das zu sehr nach Apotheke schmeckte. Sie hätte es viel lieber gesehen, daß ich Jurist oder Mediziner geworden wäre.

Im allgemeinen erfreute sie sich einer guten Gesundheit, und noch mit 59 Jahren reiste sie nach Meran, um meiner dort zur Kur weilenden kranken Schwester Mathilde während des Winters Gesellschaft zu leisten. Von München aus habe ich sie damals während der Weihnachtsfeiertage besucht. Die Fahrt machte ich in Gesellschaft meines Freundes Dr. Tappeiner, dessen Vater in Meran der bekannteste Arzt war.

Als wir von Bozen in einem besonderen Mietwagen abends nach Meran fuhren, war die Kälte so groß, daß wir uns recht unbehaglich fühlten, und bei unserer Ankunft in Meran war die Temperatur unter 12° gesunken. Ich habe damals meine optimistische Ansicht über das warme Klima der südlichen Alpenorte geändert und bei mancher anderen Fahrt nach dem Süden dieses Urteil bestätigt gefunden. In Meran war der Berg hinter der Stadt so mit Eis inkrustiert, daß man mit gewöhnlichen Schuhen kaum auf die Höhe steigen konnte. In dem Gasthause „Erzherzog Johann“ fand ich das Wasser in der Waschschüssel am nächsten Morgen gefroren. Für die Kranken wurde allerdings durch fortwährende Ofenheizung besser gesorgt.

Von Meran kehrte meine Mutter Mitte Februar nach Deutschland zurück, um ihren 60. Geburtstag zuhause zu feiern. Sie besuchte mich noch in München und freute sich, einige meiner Freunde als ihre Gäste im Hotel „Bayrischer Hof“ kennen zu lernen.

Aber diese lange Reise bei damals noch recht ungenügender Heizung der Eisenbahnen und selbstverständlich ohne Schlafwagen trug ihr einen heftigen Bronchialkatarrh ein. Die dadurch erschwerte Zirkulation in Verbindung mit den seelischen Sorgen um das Schicksal meiner Schwester haben wahrscheinlich bei ihr eine Herzkrankheit ausgelöst, die im Frühjahr 1879 begann und nach $3^1/_2$ jährigem, recht schweren Leiden am 14. September 1882 den Tod herbeiführte. Sie starb in Uerdingen im Hause meines Schwagers Mauritz unter der sorgfältigen Pflege meiner Schwester Bertha. Sie ist ebenso wie mein Vater beerdigt auf dem kleinen protestantischen Friedhof zu Uerdingen in der Familiengruft Mauritz-Fischer. Der Friedhof lag früher ganz hübsch auf freiem Felde, ist aber jetzt leider infolge der raschen industriellen Entwicklung des Ortes von hohen und unschönen Gebäuden umgeben.

Von meinen Schwestern ist schon vorher wiederholt die Rede gewesen. Die kleine Abneigung, die der Bruder bei der Abwehr ihrer Erziehungskünste früher manchmal empfunden hatte, war im Laufe der Zeit ins gerade Gegenteil umgeschlagen, und ein wirklich freundschaftliches Verhältnis hat mich dann sowohl mit meinen Schwestern, wie auch mit ihren Männern verbunden.

Die älteste Schwester Laura habe ich als Mädchen wenig gekannt, da sie bereits im Jahre 1858 einen jungen Kaufmann, Friedrich Mauritz aus Uerdingen a. Rhein heiratete. Infolgedessen wurde mir schon mit 7 Jahren die Würde eines Onkels zuteil, was mir von seiten der Altersgenossen manchen Spott eingetragen hat.

Durch diese Heirat wurde zwischen den Familien Fischer und Mauritz eine Verbindung geschaffen, die sich im Laufe der Zeit erweiterte und vertiefte. Mein Schwager war ein prächtiger, frischer Mann, tüchtig in seinem Geschäft, einem Kohlenhandel, und mein Vater ist zu ihm in ein besonders herzliches Verhältnis getreten. Er liebte es, im Alter monatelang im Hause des Schwiegersohnes zu wohnen, hatte mit ihm mancherlei Geschäfte, z. B. die Brauerei in Dortmund begonnen, ging gerne zusammen mit ihm auf Reisen und hat sich oft dahin geäußert, der Fritz sei ihm so lieb wie sein eigener Sohn.

Ich selbst habe bei dem Schwager als Knabe in den Schulferien mich öfters aufgehalten und in der mir neuen Umgebung des mächtigen Stromes, des eigentümlichen Geschäftsbetriebes und der weit verzweigten und kinderreichen Familie Mauritz viel Unterhaltung und Freude erlebt. Leider starb meine Schwester Laura, eine gesunde und kräftige Frau, bald nach der Geburt des dritten Kindes Alfred, der jetzt als Direktor der Aktienbrauerei, als Stadtverordneter und auch sonst in der Öffentlichkeit vielfach tätiger Mann in Dortmund eine sehr angesehene Stellung hat. Sie scheint das Opfer einer Infektion gewesen zu sein, über deren Natur ich aber keine bestimmte Auskunft erhalten konnte. Fünf Jahre später, mitten im Kriege 1870 heiratete der Witwer meine dritte Schwester Bertha, die dem Gatten auch drei Söhne schenkte. Sie starb leider auch früh an einer Lungenentzündung 1888. Mein Schwager Fritz ist ihr etwa 10 Jahre später im Tode gefolgt, was für meinen Vater im hohen Alter ein besonders schmerzlicher Verlust war.

Von seinen 6 Kindern sind nur noch 2 übrig geblieben, der eben erwähnte Brauereidirektor und aus der zweiten Ehe der Sohn Otto, der in Nürnberg als Ingenieur in einer großen Maschinenfabrik tätig ist.

Meine zweite Schwester Emma hat ebenfalls ziemlich früh geheiratet und zwar einen Arzt Dr. Albert Winnertz in Krefeld, den sie durch meinen Schwager Mauritz kennen lernte.

Der Doktor war ein sehr kluger und liebenswürdiger Mensch, aber leider krank und starb schon nach $3^1/_2$ jähriger Ehe an Tuberkulose. Die junge Witwe kehrte dann mit 2 Kindern, Hedwig und Clara, ins Elternhaus nach Euskirchen zurück, und blieb dort fast 10 Jahre. Das ist der Grund, warum ich sie am besten von allen meinen Schwestern kennen gelernt habe und ihr besonders nahe getreten bin. Sie war hübsch, liebenswürdig und sehr gewandt, sodaß mein Vater sie immer Salondame nannte. Sie spielte recht gut Klavier. Ich erinnere mich manchen Abends, wo ich stundenlang ihrem Spiel zuhörte, weil sie meist klassische Sachen auf einem guten Piano vortrug. Ich glaube, daß ihr Spiel mich hauptsächlich bewogen hat, ebenfalls Musik zu treiben. Im Jahre 1872 entschloß sie sich nach langem Zögern, eine zweite Ehe mit meinem Vetter Carl Fischer einzugehen, der in treuer Geduld um sie geworben hatte. So kam sie nach Rheydt, wo schon zwei andere Schwestern von ihr verheiratet waren. Ich habe sie dort öfter besucht und sie kam wiederholt nach Würzburg und Berlin. Auch sind wir in späteren Jahren häufig zusammen gereist. Sie trieb ihre Freundschaft für den Bruder manchmal so weit, daß ihr Gatte eifersüchtig wurde und sich über Zurücksetzung beklagte. Sie starb im Jahre 1901 am Typhus in Nassau, und mein Schwager Carl überlebte sie um 14 Jahre. Sie hatte den Schmerz, drei Söhne im jugendlichen Alter zu verlieren. Die vier Töchter, zwei Winnertz und zwei Fischer, haben sämtlich geheiratet und sind mir liebe Nichten geblieben.

Meine dritte Schwester Berta war eine recht originelle Person, im Grunde ihres Herzens sehr gutmütig, aber als Mädchen voller Launen, gegen die Männer häufig recht unliebenswürdig, aber sehr tüchtig. Nachdem sie im Alter von 29 Jahren meinen Schwager Mauritz geheiratet hatte, wurde sie in Uerdingen wegen ihrer ungewöhnlichen Kunst im Kochen und der Erziehung von Dienstboten, sowie wegen der Originalität ihres Wesens eine bekannte Frau. Besonders jungen Leuten, z. B. den Söhnen aus erster Ehe und deren Freunden war sie eine richtige Kameradin, mit denen sie Karten spielte, Wein trank und Ausflüge machte, fast wie ein Student. Ihr leider so früher Tod ist schon erwähnt.

Die vierte Schwester Fanny war von kräftigem Körperbau, aber wenig schön von Gesicht. Sie zeichnete sich in der Jugend durch Vorliebe für Turnen, Tanzen und sogar Pistolenschießen aus. Sie zog als Gattin des Holzhändlers Max Friedrichs im Jahre 1865 nach Rheydt, und ist hier im Jahre 1912 im Alter von 70 Jahren gestorben. Sie war klug und als Schülerin ebenso tüchtig, wie später als Hausfrau. Wie schon erwähnt, konnte sie mit den bei Herrn Vierkötter erworbenen mathematischen Kenntnissen noch im Alter eine Kubikwurzel ausziehen. Auch als Dichterin besaß sie in der Verwandtschaft einen gewissen Ruf. Hübsche Gelegenheitsgedichte und kleine Festspiele, um die sie von manchen Seiten gebeten wurde, konnte sie in unglaublich kurzer Zeit verfassen. Von ihren Kindern leben noch zwei Söhne, Ernst und Max, die das vom Vater ererbte bedeutende Holzgeschäft weiterführen, und zwei Töchter, Helene und Adele, die auswärts verheiratet sind. Vom Schwager Max Friedrichs wird später noch ausführlich die Rede sein.

Die jüngste Schwester Mathilde war nur 4 Jahre älter wie ich und hat mir deshalb in der frühen Jugend am nächsten gestanden. Sie war ein liebes, sehr gutmütiges, recht hübsches Mädchen, geistig nicht besonders begabt, aber liebenswürdig und deshalb überall gerne gesehen. Ihr Wesen hat ihr manche Huldigung von seiten der Offiziere eingebracht, die während der Manöver oder während des Krieges 70/71 bei uns im Quartier lagen, aber sie zog es doch vor, beim Zivil zu bleiben und heiratete meinen Vetter Arthur Dilthey in Rheydt. Nach der Geburt des dritten Kindes erkrankte sie und starb im Herbst 1879 in Rheydt nach einer vergeblichen Kur in Meran. Es war nicht allein für meinen Schwager, dem sie zwei Söhne und eine Tochter hinterließ, sondern auch für meine Eltern und ganz besonders für meine Mutter ein schwerer Schlag, der sehr ungünstig auf ihren eigenen Gesundheitszustand zurückwirkte.

Ich habe dieser lieben „Tilla" ein dankbares und warmes Gedenken bewahrt. 5 Jahre später ging mein Schwager eine zweite Ehe ein mit Frieda Weuste, und diese hat es verstanden, den drei Kindern eine wirkliche zweite Mutter zu sein. Das Ehepaar Arthur Dilthey lebt jetzt noch in voller Rüstigkeit in Bonn. Von den drei Kindern ist leider der Jüngste Alfred als Opfer des unseligen Krieges 1915 in Rußland gefallen. Von meinem Schwager Arthur, mit dem mich dauernde Freundschaft verbunden hat, wird später die Rede sein.

Fast ebenso nahe wie die Schwestern standen mir in früher Jugend die Vettern und die Kusine im Nebenhause. Der Älteste, Heinrich Fischer, 5 Jahre älter wie ich, der als einziger Vertreter der Familie in Euskirchen zurückgeblieben ist, war als Junge geneigt, sich von dem übrigen Kreise etwas abzusondern. Er ist unverheiratet geblieben und hat mehr und mehr die Gewohnheiten eines Originals angenommen.

Er wohnt jetzt in unserem früheren Hause und betreibt mehr zu seinem Vergnügen das alte Spinnereigeschäft, aber in beschränktem Umfange.

Von den Vettern Ernst und dem Jüngsten Otto Fischer werde ich später noch Manches zu berichten haben.

Vetter Lorenz ist früher schon wegen seiner hervorragenden Eigenschaften als Jäger und Krieger erwähnt worden. In der Schule war er weniger tüchtig, und infolge unvernünftiger Lebensführung ist er im Alter von 35 Jahren an Tuberkulose gestorben. Sein jüngerer Bruder Hermann war ebenfalls für körperliche Leistungen besser veranlagt als für geistige Tätigkeit. Er war ein sehr hübscher Mann, guter Turner und Reiter und übersiedelte bald nach dem Tode seines Vaters nach

Cöln. Er wurde der Stammvater einer kriegerischen Familie; denn seine beiden Söhne Kurt und Walter sind Berufsoffiziere geworden und seine durch Schönheit ausgezeichneten Töchter sind beide glücklich verheiratet und wohnen in Groß-Berlin.

Auch die drei Söhne meines Onkels in Flamersheim kamen bei der geringen Entfernung von Euskirchen durch häufige wechselseitige Besuche vielfach mit uns in Berührung. Die beiden ältesten Karl und August wurden Kaufleute und haben später ein Geschäft in Baumwollgarn in Rheydt geführt. Dort sind sie auch beide gestorben. Wie schon erwähnt, wurde Karl der zweite Gatte meiner Schwester Emma. Er war trotz ziemlich dürftiger Schulbildung ein kluger und geschäftsgewandter Mann, den man gern in praktischen Dingen um Rat frug und mit dem ich häufig einen Teil der Osterferien in Territet am Genfer See verbrachte.

Der zweite Sohn August war körperlich ungewöhnlich stark, ein trefflicher Jäger und ein gutmütiger, zur Heiterkeit hinneigender Gesellschafter. In geschäftlichen Dingen folgte er gern der größeren Autorität seines Bruders Karl.

Ganz anders als die Brüder war der jüngste Flamersheimer Vetter Julius, ein sehr gescheidter und kritischer Kopf, der am meisten die geistige Regsamkeit und die Originalität seines Vaters geerbt hatte. Er wurde Jurist, schlug die Richterlaufbahn ein und starb verhältnismäßig früh an vernachläßigtem Diabetes in Cöln, wo er Richter beim Oberlandsgericht war. In seinem früheren Aufenthaltsort Cleve hat er sich verheiratet. Aus der Ehe stammt ein Sohn, der die Tochter eines Großschafzüchters in Australien heiratete und dort eine neue Heimat gefunden hat. Ich hoffe, daß er durch Naturalisation und durch den Schutz des Schwiegervaters den Unbilden entgangen ist, die unsere Landsleute während des Krieges in den englischen Kolonien erdulden müssen. Die einzige Tochter hat einen Hans von Eicken aus Hamburg geheiratet, den Sohn meiner Kusine Helene geb. Fischer aus Köln.

Der Vetter Julius besaß nicht allein Begabung für geistige Arbeit, sondern hatte auch das besondere Talent, Erkundigungen einzuziehen. Schon als Junge wußte er alles, was sich in der näheren und weiteren Umgebung zutrug. Auch geschäftliche Dinge interessierten ihn. Man konnte kaum etwas Komischeres hören, als wie eine Unterhaltung zwischen ihm und seinem Vater, der von den Gaben des Sohnes sehr hoch dachte und im vertraulichen Kreise nicht selten die Äußerung tat, „der „Jul" wird noch Minister". Im Alter hatte sich der Erwerbs- und Sparsinn des Vaters bis ins Groteske hinein gesteigert, und er würde direkt materiellen Mangel gelitten haben, wenn nicht der Sohn

Julius zusammen mit seinen Brüdern im geheimen die Hauptkosten des Haushaltes getragen hätten. Charakteristisch ist folgende Geschichte. Eines Tages vermißte der Vater auf seinem kleinen Kontor die Hälfte eines Hundertthalerscheines, und er kam auf den Verdacht, daß ein kleiner Hund, der Liebling vom Sohn Julius, diese Hälfte gefressen habe. Er entschloß sich sofort, das Tier aufschneiden zu lassen und hatte schon einen Operateur bestellt, als Julius von der Gefahr für seinen Liebling Kenntnis erhielt. Es gelang ihm auch, den Vater zu beruhigen, indem er ihm den Hund für 100 Taler abkaufte. Das Geschäft konnte glücklicherweise bald nachher rückgängig gemacht werden, da sich die zweite Hälfte des unglücklichen Hunderttalerscheines wieder vorfand.

Ich bin dem Vetter Julius in meiner Bonner Studienzeit näher getreten. Er war damals Assesor beim Landgericht, und wir haben hier und da im Rheinischen Hof eine gute Flasche Wein zusammengetrunken. Ich mußte immer sein besonderes Talent bewundern, eine Sache, die an und für sich ganz klar schien, von einer anderen Seite zu betrachten, und Möglichkeiten zu entwickeln, an die ein Laienverstand gar nicht denken konnte. Diese Art, die Welt in einem besonderen Spiegel anzusehen, war für mich sehr anziehend, so daß ich nach dem Doktorexamen im Herbst 1874 den Vetter zu einer Reise nach der Schweiz aufforderte. Obschon zum Reisen wenig aufgelegt, ging er darauf ein, und ich habe selten so viel gelacht, wie auf dieser Tour. Schon bei der Abreise ließen ihn die landschaftlichen Schönheiten des Oberrheintals, die lieblichen Bergzüge des Odenwaldes und Schwarzwaldes ziemlich gleichgültig. Dagegen kontrollierte er aufs Genaueste den fahrplanmäßigen Gang des Zuges und registrierte jede Minute Verspätung. Zu dem Zweck hatte er gleich nach Besteigung des Zuges Krawatte, Hemdenkragen und Rock ausgezogen und dafür einen Regenmantel angelegt. Das sei die beste Methode, um sich gegen den Staub und Schmutz zu schützen, den er für ungesund und unästhetisch hielt. In Basel machten wir zuerst Station. Die Stadt und die Bewohner interessierten ihn lebhaft, aber aus den Sehenswürdigkeiten und der Landschaft machte er sich nicht viel. Er proklamierte damals für Reisen gleich den Grundsatz: Die öffentlichen Gebäude von außen, die Berge von unten und die Gasthäuser von innen anzuschauen. Ich habe später auch von anderen Mitreisenden, z. B. meinem Freunde Wilhelm Königs ähnliche Worte vernommen, aber sie wurden doch niemals so gewissenhaft durchgeführt, wie von Vetter Julius. Gegen die Abendstunde hatte er immer das Bedürfnis, ein Glas Wein zu trinken, und beim Hinschlendern durch die Stadt war sein Augenmerk auf die Entdeckung eines geeigneten Weinhauses gerichtet. Dem Reisehandbuch traute er gar nicht, ebensowenig den Aussagen der Hotelbedientesten, aber plötzlich entdeckte er auf der Straße einen Mann mit einer ungewöhnlichen großen und feuerroten Nase. „Der muß es wissen" sagte er sofort, und die Auskunft, die wir von dem Herrn erhielten, bewies in der Tat, daß wir uns an die richtige Adresse gewandt hatten.

Von Basel ging's nach Luzern, wo wir in einem Gasthause zweiten Ranges, einem echten Schweizer Hause Quartier nahmen. Den Vetter interessierte es besonders, mit der einheimischen Bevölkerung Fühlung zu nehmen, ihre Sitten, Einrichtungen und Anschauungen kennen zu lernen. Es gelang ihm auch rasch, mit einheimischen Stammgästen ins Gespräch zu kommen. Das alles war begleitet von dem Konsum ziemlich ansehnlicher Weinmengen. Der Reiseweg von dort aus war von ihm auf das genaueste festgestellt und ist auch bis in Kleinigkeiten hinein inne gehalten worden. Er ging zunächst über Flüelen, Andermatt und das Wallis zum Genfer See. Als wir morgens das Schiff bestiegen hatten, um über den Vierwaldstätter See in einer Landschaft, die wirklich des Sehens lohnt, nach Flüelen zu fahren, fiel der Vetter, trotz prächtigen Wetters und des sehr schönen Platzes auf dem Deck gleich nach der Abfahrt in tiefen Schlaf und erwachte erst am Ende der Fahrt. Er schaute sich dann ganz befriedigt um und sagte trocken: „Die Leute müssen denken, ich sei oft hier gewesen". Von Flüelen hatten wir die Wahl, nach Andermatt entweder mit der Post oder auf Schusters Rappen zu kommen. Vetter Julius entschied sich kurz für letzteres. Es war ein schwüler Tag und die Landstraße von der Mittagsonne recht heiß und staubig. Das Gepäck hatten wir der Post übergeben, die später abfuhr und zogen nun los. Der Vetter meinte, der Mantel trage sich am bequemsten, wenn man ihn anziehe. Das geschah, und trug nicht wenig dazu bei, den üppig genährten und an körperliche Anstrengungen durchaus nicht gewöhnten Vetter in Dampf zu versetzen. Unglücklicherweise machte um dieselbe Zeit eine Schwadron schweizerischer Kavallerie Übungen in der Gegend. Kleine Trupps kamen häufig an uns vorbei und entwickelten eine ungeheure Staubwolke, die der Vetter für sehr schädlich hielt und der er zu entgehen suchte, indem er den Straßendamm hinab ins freie Feld flüchtete. Durch all diese Umstände war er nach 2 Stunden so erschöpft, daß ich Sorge bekam und ihn aufforderte, in das nächste Gasthaus einzutreten und dann den staubigen Teil des Weges mit der Post abzumachen. Er ging darauf ein und hat nun die Post nicht mehr verlassen, bis wir an den Genfer See gelangt waren. Ich selbst habe die schöneren Teile des Weges zu Fuß gemacht, besonders die Strecke von Andermatt über die Furka nach dem Rhonegletscher. Ich befand mich dabei in der Gesellschaft eines jungen Amerikaners, dessen praktische Weise, zu reisen, mir besonders gefiel. Er war von Kalifornien herüber gekommen und hatte kein anderes Gepäck, als ein kleines Täschchen für Geld, Zahnbürste und Seife. Sobald er in eine Stadt kam, kaufte er neue Leibwäsche, und verschenkte die schmutzige. So hatte er sich schon 2 Monate in Europa herumgetrieben, viel gesehen, sich trefflich unterhalten und hoffte den Trip noch einige Wochen zu verlängern. Abends traf ich immer wieder mit dem Vetter Julius im Gasthause zusammen und er wußte dann zahlreiche Schnurren über abenteuerliche Postreisen im Gebirge und sonderbare Reisegefährten zu erzählen. An der Teufelsbrücke hatte er sogar ein neues physikalisches Phänomen entdeckt, d. h. einen umgekehrten Regenbogen. Ich wollte es anfangs nicht glauben, habe mich aber später überzeugt, daß bei Wasserfällen durch die Zerstäubung des Wassers in kleine Tropfen bei richtigem Stand der Sonne und des Beobachters diese Erscheinung in der Tat eintreten kann.

Von den landschaftlichen Schönheiten der Schweiz war wenig die Rede. Dagegen interessierte ihn Genf als die Stadt Calvins und als der Hauptsitz des schweizerischen Finanzgeschäftes. Auch hier erlebte ich mit ihm eine komische Szene. Er behauptete, das Wasser, das er zuhause niemals trank, sei in der Schweiz besonders heilsam, nur müßte man es direkt aus der Naturquelle entnehmen. So liebte er es denn, bei den vorgefundenen Brunnen direkt vom Wasserstrahl zu trinken. Das machte er auch bei einem Brunnen in Genf, mußte sich aber, um ans Wasser heran zu kommen, auf die gemauerte Brüstung legen und pendelte hier nun hin und her, um den Strahl mit seinem Munde aufzufangen. Das Bild des kleinen, stark genährten Mannes, auf der Brüstung des Brunnens liegend und trinkend, war so überaus komisch, daß sich ein großer Kreis von Zuschauern bildete. Als er fertig war und die versammelte Menge erblickte, sagte er ruhig, da sehe man wieder, daß man mit wenig Mühe den Leuten viel Spaß machen könne. Von Genf sind wir mit dem Omnibus nach Chamonix gefahren und ich habe dort mehrere kleine Bergtouren gemacht. Seinem Grundsatze getreu blieb Vetter Julius im Hotel; durch fleißige Erkundigungen bei Touristen, Führern und sonstigem Volk war er aber nachher über Bergtouren viel besser unterrichtet als die große Zahl der Reisenden. Selbstverständlich kontrollierte er auch alle Gäste des Hotels und berichtete eines Abends, draußen beim Wetterhäuschen sei ein aufgeregter deutscher Professor, der die Instrumente kritisiere und das Publikum über den bevorstehenden Wechsel des Wetters belehre. Ich erfuhr nachher, daß es Rudolf Fittig aus Tübingen sei, der mir als trefflicher Chemiker aus der Literatur schon bekannt war. Ich habe damals nicht gewagt, mich ihm vorzustellen, bin aber einige Jahre später zu ihm in ein näheres Verhältnis getreten. Unter den Fußtouren, die ich von Chamonix unternahm, war auch die damals sehr beliebte Überquerung des mer de glace und der Abstieg über den mauvais pas. Für berggeübte Wanderer ist das ein Spaziergang, aber Leute ohne Erfahrung und ohne das richtige Schuhwerk konnten doch recht leicht ausgleiten und verunglücken. Ich habe deshalb beim Übergang über das Eis mit einem gewissen Schrecken mich daran erinnert, daß mein Vater 5 Jahre vorher denselben Weg mit 5 Damen, d. h. meiner Mutter, zwei Schwestern und den beiden Kusinen Marie und Helene Fischer aus Cöln unternommen hatte. Er erzählte auch hinterher, daß es die angstvollsten Stunden seines Lebens gewesen seien.

Die Damen schienen sich dieser Gefahr weniger bewußt zu sein, sie haben im Gegenteil auf dieser Schweizer Reise eine ganze Anzahl von merkwürdigen und lustigen Erlebnissen gehabt, von denen noch lange in unserem Hause geredet wurde.

Der Eindruck, den Vetter Julius von der Schweiz mit nach Hause nahm, war viel nüchterner; denn als wir wieder im Rheintal waren, meinte er, es sei nicht der Mühe wert, nach der Schweiz zu fahren, denn hier gäbe es auch Berge und Wasser und im Winter Schnee und Eis im Überfluß. Er hat meines Wissens nie mehr die Grenzen des deutschen Reiches überschritten. Dagegen bin ich in späteren Jahren oft und gern in die Schweiz und besonders an den Genfer See zurückgekehrt, habe aber das Chamonixtal nicht mehr gesehen.

Gymnasialzeit.

Begleitet von meinem Vater fuhren Vetter Ernst und ich anfangs Oktober 1865 nach Wetzlar, wo mein Vater uns bei dem Direktor des Gymnasiums anmeldete. Ernst wurde auf Grund seines Abgangszeugnisses von Duisburg ohne Anstand in die Prima aufgenommen. Ich mußte aber eine Prüfung durchmachen, weil die Schule zu Euskirchen nicht als gleichberechtigt angesehen wurde. Ich habe sie ohne Mühe bestanden und wurde in die Untersekunda, die mit der Obersekunda in einer Klasse vereinigt war, aufgenommen. Noch heute klingt mir das Gelächter in den Ohren, das die Mitschüler erhoben, als ich zum ersten Mal in meiner markanten niederrheinischen Mundart in der Klasse sprechen mußte. Natürlich erschien mir der Wetzlarer Dialekt, der stark an das Frankfurter Deutsch anklingt, ebenso ungewöhnlich und komisch. Ich habe mich aber bald in dem Kreise zurechtgefunden und wurde nach $^1/_2$ Jahr sogar Primus der Untersekunda.

Die originellste Persönlichkeit unter den Lehrern war der Mathematiker Elstermann. Die stark mit Kreide beschriebene Tafel pflegte er im Eifer mit seinem langen schwarzen Rock zu reinigen, und die dabei beschmutzten Finger steckte er in das lange buschige Haar. Er war ein guter Mensch und ausgezeichneter Lehrer, der die schwierige Aufgabe, mathematisches Denken in so viel jugendliche Köpfe hineinzubringen, geradezu mit Leidenschaft betrieb.

Wenn er beim Abfragen der einzelnen Schüler auf Mangel an Verständnis oder Interesse stieß, so pflegte er zur eigenen Beruhigung einem gerade in Reichweite befindlichen anderen Schüler eine kleine Ohrfeige zu versetzen. Wenn diese mit steigender Erregung heftiger wurden und der Empfänger unwillig den Lehrer anblickte, dann pflegte er zu sagen: „Gib sie später dem, der sie verdient hat, wieder". Er hat es aber mit seiner Methode erreicht, daß der mathematische Unterricht an dem Wetzlarer Gymnasium ungewöhnlich gute Resultate brachte.

Das gerade Gegenteil von diesem Mann war der Religionslehrer, den wir fürchteten und haßten, weil er boshaft war und in vielen Dingen den Spion machte. Auch unser Ordinarius, ein klassischer Philologe, genoß wenig Sympathie. Ich habe nur seinen Spitznamen „Specht" behalten. Er war mir anfangs gewogen, entzog mir aber schon im zweiten Semester seine Gunst, weil er mit dem Vetter Ernst in der Prima Differenzen gehabt hatte. Der Zorn meines Vetters auf diesen Lehrer war so groß, daß er sich zu einem dummen Streich hinreißen ließ, der ihm schweren Schaden hätte bringen können. In Verbindung mit einem anderen Primaner warf er nämlich in tiefer Nacht mehrere Fenster in der Wohnung des Lehrers ein. Ich erfuhr von dieser Tat erst, als sie begangen war, und ich bin glücklicherweise der einzige Mitwisser geblieben. Der Streich erregte großen Lärm in der Schule und der Lehrer Specht machte in den Klassen verlockende Versprechungen für einen Verräter, aber das Geheimnis blieb gewahrt.

Wir lebten in Wetzlar, abgesehen von der Schulaufsicht, ähnlich wie Studenten, denn der Hauswirt kümmerte sich um unser Tun und Treiben gar nicht. Der Hausschlüssel lag jede Nacht in einer Rinne im Steinsockel unter der Haustüre, und wenn wir für Schulversäumnis Zeugnisse über Unwohlsein und dergl. nötig hatten, so unterschrieb er alle Schriftstücke, die wir ihm vorlegten. Wir nannten ihn das Kaffeehäubchen, wegen eines runden Käppchens, das er gewöhnlich trug. Sein Bruder trug den Spitznamen „Der Eisbär".

Vetter Ernst und ich hatten zusammen ein Wohn- und Schlafzimmer. Außerdem waren noch zwei andere Gymnasiasten und ein Einjährig-Freiwilliger in dem Hause einquartiert. Der Pensionspreis war recht billig. Er betrug für Wohnung, Heizung und völlige Verpflegung mit Ausnahme alkoholischer Getränke im Jahre nur 110 Thaler, also pro Tag und Person etwa 1 Mk. Man kann sich denken, daß es dafür keine lukullischen Mahlzeiten gab. Im Gegenteil, die Kohlrübe und Backpflaumen mit Schweinefleisch spielten dabei eine erhebliche Rolle. Ich sehe noch immer den Eisbär in dem Bache, der an unserem Wohnhause vorüber floß, in einer Bütte die Rüben waschen, die für unsere Ernährung bestimmt waren. Ich habe mir damals nicht träumen lassen, daß die Kohlrübe 50 Jahre später ein so viel gebrauchtes und ebenso viel verwünschtes Nahrungsmittel für das deutsche Volk werden würde.

Die Kontrolle unserer Lebensweise von seiten der Schule war nicht allzu streng. Man besuchte sich gegenseitig auf den sogen. „Buden" und unser Haus bildete von je her einen beliebten Versammlungsort. Leider war die Mehrzahl der Mitschüler, mit denen ich verkehren mußte, erheblich älter als ich; denn es befanden sich darunter viele Bauernsöhne, die erst im reiferen Alter zum Studieren bestimmt worden waren. Einer davon, der noch in unserer Klasse saß, zählte schon 21 Jahre. Die Gewohnheiten dieser jungen Männer waren natürlich anders, als sie für einen Knaben von 13 Jahren paßten. Ihre Gesellschaft war deshalb nicht gerade vorteilhaft für mich. Man bemühte sich, mir möglichst bald das Tabakrauchen, Biertrinken und Skatspielen beizubringen und die übliche Unterhaltung war auch in der Regel auf einen rohen und unanständigen Ton abgestimmt. Das alles ist gewiß nicht ohne Einfluß auf meine von Hause aus recht gute Gesundheit gewesen und ich fürchte, daß der Grund zu meiner späteren Magenkrankheit schon damals durch Unmäßigkeit im Rauchen und Trinken gelegt worden ist.

Glücklicherweise bot die Musik ein Gegengewicht gegen die so ungünstigen Einflüsse. Vetter Ernst war musikalisch ungewöhnlich begabt und hatte eine Zeitlang sogar die Absicht, Musiker von Beruf zu werden. Auf dem Klavier war er für unsere Begriffe schon ein halber Virtuose, aber außerdem spielte er Geige, Violoncell und Flöte. Ich selbst war gegen ihn ein Stümper, hatte es aber durch fleißige Übung auf dem Klavier soweit gebracht, daß ich Sonaten von Haydn, Beethoven und Mozart usw. leidlich spielen konnte. Infolgedessen wurde ich vom Vetter Ernst auch beim Zusammenspiel zugelassen, und da sich noch einige Andere musikalische Mitschüler fanden, so haben wir häufig Trio und Quartett auf unserem Zimmer gespielt. Unsere Musik war so gut, daß manche Leute in der Nachbarschaft freudigen Anteil daran nahmen. Den Ansprüchen der Schule konnten Vetter Ernst und ich ohne große Mühe Genüge leisten. Eine für uns ganz erwünschte Unterbrechung erfuhr der Schulbetrieb im Sommer 1866 durch den Krieg Preußens mit Oesterreich und den Süddeutschen Staaten. Der preußische Kreis Wetzlar war damals ganz umgeben von feindlichen Staaten d. h. von Hessen-Darmstadt und Nassau.

Nachdem die preußischen Truppen ausgerückt waren, blieben wir infolge der Unterbrechung des Eisenbahn- und Postdienstes für mehrere Wochen ganz abgeschnitten, und dann kam noch ein viel wichtigeres Ereignis, die Invasion feindlicher Truppen. Es war, wenn ich nicht irre, eine Badische Division, die aus der Umgegend von Gießen in den Kreis einrückte. Sobald wir davon Kenntnis erhielten, trieb uns die Neugier diesen feindlichen Kriegern entgegen, und ich selbst mit einigen Mitschülern genoß das Vergnügen, von einer Feldwache gefangengenommen zu werden. Wir wurden regelrecht verhört und dann wieder weggeschickt. Zu unserer ungewöhnlichen Freude ging es aber dem unbeliebten Religionslehrer des Gymnasiums viel schlechter. Er hatte sich ebenfalls die fremden Soldaten ansehen wollen, die aber gleich in ihm die Spionnatur, die wir Schüler so fürchteten, erkannten und ihn mehrere Tage festhielten. Einige Tage später rückten die Feinde auch in die Stadt Wetzlar ein und belegten die öffentlichen Gebäude, zu unserer Freude auch das Gymnasium. Wir hatten also etwa 8 Tage unerwartete Schulferien und dadurch Muße genug, uns mit den feindlichen Truppen zu beschäftigen.

Die Kriegführung war damals gemütlicher als heute. Die badischen Soldaten ließen sich zwar gut auf Kosten der Stadt verpflegen, betrugen sich aber sehr anständig, und das Verhältnis zwischen ihnen und den Bürgern der Stadt und namentlich uns Gymnasiasten war recht freundlich. Sie zogen wieder ab, weil inzwischen die preußischen Truppen durch kräftige Schläge bei Kissingen und am Main das Hauptheer der süddeutschen Staaten besiegt hatten. Die nächsten Soldaten, die in die Stadt einzogen, waren leider nur Verwundete, meist Preußen und Bayern, denen wir aufrichtige Teilnahme schenkten; dann kam auch alte preußische Landwehr, die in den benachbarten hessischen und nassauischen Dörfern als sogen. Fresskompagnie zur Beruhigung der etwa rebellischen Bevölkerung verteilt wurde. Durch die siegreichen Schlachten in Böhmen ging der Krieg bald zu Ende und der Schulunterricht wurde nun mit verdoppelter Strenge bis zu den Ferien wieder aufgenommen.

Die Herbstferien dieses Jahres wurden mir völlig verdorben durch die Konfirmation, die in Flamersheim stattfand; denn ich mußte in etwa 4 Wochen den ganzen Heidelberger Katechismus und eine Reihe von Kirchenliedern auswendig lernen, da ich vorher an dem Konfirmandenunterricht nicht teilnehmen konnte. Das gelang mir infolge meines guten Gedächtnisses, aber es trug nicht dazu bei, meinen kirchlichen Glauben, der schon vorher durch den Einfluß meiner älteren Mitschüler in Wetzlar, durch das Lesen des Lebens Jesu von David Strauß größtenteils verloren gegangen war, zu kräftigen. Tatsächlich bin ich als Freigeist, aber doch mit einem gewissen Gefühl der Beschämung an den Tisch des Herrn getreten, und ich habe den positiven Glauben, den ich in der frühen Jugend besaß, niemals wieder gewonnen. Dagegen bin ich in späteren Jahren wieder zu der Überzeugung gekommen, daß die Religion ein wichtiger Teil der menschlichen Kultur ist und daß der christliche Glaube dem einzelnen Menschen großen sittlichen Wert und inneres Glück geben kann. Meine eigene Mutter und manche der katholischen Mitschüler oder Studiengenossen lieferten den Beweis dafür, und ich habe später in Berlin bei meinen eigenen Söhnen bedauert, daß der Einfluß der Großstadt und leider auch die Atmosphäre der Berliner Schulen der Pflege religiösen Gefühls so wenig günstig sind.

Ein zweites Jahr in Wetzlar verlief ohne politische Störung und brachte schließlich dem Vetter Ernst das Zeugnis der Reife und mir die Versetzung nach Prima. Ich hatte aber inzwischen Wetzlar satt bekommen und verließ deshalb mit dem Vetter die Schule. Als Erinnerungen an den Wetzlarer Aufenthalt sind mir geblieben die landschaftlichen Schönheiten, die altertümliche Bauart der alten, an einem Bergabhang gelegenen Stadt und die zahlreichen, auch von uns mit Ehrfurcht gepflegten Goethe'schen Überlieferungen. Werthers Leiden war selbstverständlich ein viel gelesenes Buch, ohne daß wir aber von der krankhaften Stimmung des Helden beeinflußt worden wären. Mit den Schulkameraden aus dieser Zeit bin ich niemals wieder in Berührung gekommen und auch die Stadt habe ich nicht wieder gesehen.

In den Herbstferien, die ich gewöhnlich zu Hause verbrachte, wurde als Schule für mich zunächst das Friedrich Wilhelm-Gymnasium zu Cöln in Aussicht genommen. Mein Vater machte auch die nötige Meldung bei dem Direktor, einem Professor Jäger, der als Verfasser einer Weltgeschichte bekannt geworden ist, und der als ausgezeichneter Schulmann aus Württemberg nach Cöln berufen worden war. Die Schule war wegen der guten Leitung überfüllt. Trotzdem hatte mich der Direktor für die Prima vorgemerkt, weil er meinem Onkel, dem Chirurgen, für die Errettung seiner Frau aus schwerer Lebensgefahr sich zu besonderem Danke verpflichtet fühlte. Aber in der brieflichen Antwort, die er meinem Vater schickte, war meine Aufnahme nicht klar ausgesprochen und die Überfüllung der Schule so stark betont, daß wir das Gefühl der Ablehnung unseres Gesuches hatten.

30 Jahre später habe ich den Direktor auf der zweiten Schulkonferenz im Kultusministerium zu Berlin kennen gelernt. Dabei wurde das Mißverständnis über die vermeintliche Ablehnung meiner Anmeldung aufgeklärt, und der alte joviale Herr knüpfte daran die scherzende Bemerkung: „Was hätte aus Ihnen werden können, wenn Sie unter meinen Bakul gekommen wären". Durch die Verhandlungen in Cöln war viel Zeit verloren gegangen. Das Schuljahr hatte schon begonnen und ich mußte nun schleunigst sehen, anderswo unterzukommen. Die Wahl fiel auf das Gymnasium zu Bonn, wo ich auch richtig an meinem 15. Geburtstage Aufnahme fand. Mein Einzug in die Schule vollzog sich unter scheinbar wenig erfreulichen Auspizien. Es war gerade die Stunde die Mathematik und als der alte Lehrer, Professor Zirkel, der Vater des bekannten Leipziger Professors der Mineralogie, mich erblickte, frug er mich zornig nach Namen und Herkunft und erklärte dann, meine späte Ankunft wäre eine Rücksichtslosigkeit; denn jetzt sei er genötigt, seine Liste wieder zu ändern. Bald darauf klopfte es an der Klassentür und der Schüler, der zum Nachsehen hinausgeschickt wurde, kam lachend zurück. Draußen sei der Vater des neuen Primaners und lade seinen Sohn ein, heute Mittag um 1 Uhr zum Essen im Gasthof Stern zu erscheinen. Das brachte

den Zorn des Lehrers von neuem in Wallung und er wurde zum Jubel der ganzen Klasse in seinen Vorwürfen so heftig, daß ich im Begriff war, die Schule wieder zu verlassen und jede Antwort auf seine weiteren Fragen verweigerte. Da merkte er, daß er zu weit gegangen war. Er änderte den Ton, und um mich zu versöhnen, frug er nach Euskirchen und dem Flamersheimer Walde, den er wohl kannte. Nun wurde mir erst klar, daß er im Grunde ein gütiger Mann war, und da er bald mein Interesse für Mathematik entdeckte, so sind wir die besten Freunde geworden. Wenn er sich über die anderen Primaner, von denen manche nicht sehr ehrerbietig gegen ihn waren, geärgert hatte, so pflegte er den Rest der Stunde mit mir allein mit Rechnungen an der Tafel zu verbringen.

Das Gymnasium war im allgemeinen, was den Unterricht anbetraf, der Schule in Wetzlar nicht ebenbürtig. Das lag zum Teil an dem Übergewicht, das der Religionslehrer, ein katholischer Geistlicher besaß. Er übte eine wahre Tyrannei aus, worunter allerdings wir Protestanten nicht litten, wodurch aber der wissenschaftliche Unterricht sicherlich geschädigt wurde. Der zweite Grund waren das Alter und die Krankheit des Direktors. Er besaß von früher her als Philologe und besonders als Horaz-Übersetzer einen recht guten Ruf. Aber zu meiner Zeit war er kaum mehr arbeitsfähig, und ist bald nachher gestorben. Auch seinem Nachfolger, einem kleinen Geist, gelang es nicht, die Schule zu neuer Blüte zu bringen. Von den Lehrern war der tüchtigste ein Herr Deiters, der später in das Provinzialschulkollegium zu Coblenz berufen wurde; aber wir haben ihn wegen seines Sarkasmus und seiner Strenge gefürchtet. Nächst dem Mathematiker Zirkel war der Lehrer des Deutschen Professor Remacly die originellste Persönlichkeit. In lebhafter Erinnerung ist mir auch der Ordinarius geblieben, ein klassischer Philologe, der infolge seiner außerordentlichen Kurzsichtigkeit große Mühe hatte, die Schuldisziplin aufrecht zu erhalten. Als Liebhaberei betrieb er das Sammeln von alten Münzen und man konnte ihm durch Lieferung römischer Münzen, die damals im Rheinland vielfach gefunden wurden, die größte Freude bereiten. Seine Neigung, alle Stücke, die man ihm zur Prüfung vorzeigte, als Eigentum zu behalten, gab eines Tages Anlaß zu einem übermütigen Schabernack. Zwei Spaßvögel aus unserer Klasse hatten Hosenknöpfe, die den Namen eines Bonner Schneiders „Hannes" trugen, durch geschickte mechanische und chemische Behandlung in einen Zustand versetzt, der sie stark verwitterten römischen Münzen ähnlich machte. Diese Kunstprodukte wurden dann dem Herrn Ordinarius als merkwürdige Fundstücke überliefert. Er war aber Kenner genug, um nach einigen Studien die Fälschung festzustellen und gab dann seiner berechtigten

Entrüstung einen sehr energischen Ausdruck. Auch für römische Inschriften besaß er großes Interesse, und er brachte uns die Kunst bei, Abdrücke davon in Pappe zu machen. Es hat mir Freude bereitet, diesen Mann, dessen wissenschaftliche Bemühungen mir Achtung einflößten, Abdrücke von einigen römischen Inschriften aus der Gegend von Zülpich, dem angeblichen Tolbiacum der Merovinger Zeit, und Weingarten bei Euskirchen, wo sich die Überreste eines römischen Kastells befanden, zu liefern.

Auf der Bonner Prima habe ich zum erstenmal einen Religionsunterricht genossen, der mich wirklich interessierte, denn der Lehrer, der gleichzeitig protestantischer Pastor in Bonn war, ließ uns das neue Testament in griechisch lesen und wußte die handelnden Personen so lebendig und im Zusammenhang mit den großen historischen Ereignissen jener Zeit zu schildern, daß ich einen Begriff bekam nicht allein von der großen sittlichen Kraft der christlichen Lehre, sondern auch von der gewaltigen geistigen, sozialen und politischen Bewegung, die sie ausgelöst hat. Aus den Schilderungen dieses Mannes waren mir die alten christlichen Gebräuche, die Katakomben zu Rom und ähnliche Dinge längst bekannt, ehe ich sie im Original gesehen habe.

Außerhalb der Schule war ich in Bonn sehr gut aufgehoben; denn ich wohnte bei einer Familie Kemp in der Bonngasse, wenige Schritte vom Gymnasium entfernt, wo es eine vortreffliche Verpflegung gab. Außer mir wohnte dort noch ein Oberprimaner Fischenich, der Sohn eines Landwirten aus der Nähe von Flamersheim, recht musikalisch und ein guter Kamerad. Es hat mich gefreut, daß mein Sohn Hermann mit dem Sohn dieses alten Schulkameraden an der Schlachtfront in Lothringen bekannt wurde und ihn etwa vor einem Jahre als Gast in Berlin mir zuführte. Dann gab es noch zwei viel jüngere Gymnasiasten der unteren Klassen. Wir alle waren bei den Mahlzeiten mit der Familie und den Angestellten des Geschäfts versammelt. Der Wirt betrieb mit seinem Sohn unter der Firma „Paul Kemp & Sohn“ ein einträgliches Geschäft in Galanteriewaren, studentischen Artikeln jeder Art, in Pelzwaren und sogar zur Faschingszeit in Maskenanzügen. Das brachte einen großen Verkehr und viele lustige Ereignisse ins Haus, und ich habe dort zwei wirklich heitere Jahre verlebt. Mein Vetter Ernst Fischer studierte zur selben Zeit in Bonn Medizin, und obschon seine Bude in einem anderen Hause lag, so haben wir doch manchmal zusammen musiziert.

Selbstverständlich wurden wir auch in Bonn mit dem studentischen Leben, seinen berechtigten und unberechtigten Eigentümlichkeiten, vertraut und ich habe schon damals mit Eifer Unterricht im Säbelfechten auf dem Universitätsfechtboden genommen. Die Folge davon war, daß ich später als Angehöriger der Universität an diesen Dingen kein Interesse mehr besaß.

Im August 1869 bestand ich in Bonn das Abiturientenexamen, und das Abgangszeugnis beweist, daß ich kein schlechter Schüler gewesen bin. Trotzdem habe ich zu meinem eigenen Bedauern dem Gymnasium kein freundliches Gedenken bewahren können. Ich halte mich für verpflichtet, das hier auszusprechen, weil ich das Gefühl habe, daß das humanistische Gymnasium die Anforderungen nicht erfüllt, die man an es stellt, und nicht, wie meist behauptet wird, seinen Zöglingen die allgemeine geistige Reife gibt, die zum Besuch der Hochschule nötig ist. Ich spreche hier nicht als Naturforscher, der es immer beklagen mußte, auf der Schule einen ungenügenden mathematischen Unterricht genossen zu haben. Mein Urteil bezieht sich auch auf den sprachlichen Unterricht, der in der jetzigen Form mit der unmäßigen Betonung grammatikalischer Kenntnisse sicherlich verkehrt ist. Wieviel kostbare Zeit haben wir auf das unsinnige Auswendiglernen von Regeln verwenden müssen! Die seltensten Ausnahmen von einer Deklination oder Konjugation, die selbst dem Berufsphilologen in der Praxis kaum vorkommen, mußte man wissen, um ein guter Schüler zu sein. Von den Schönheiten der klassischen Literatur, von ihrem engen Zusammenhang mit der bewundernswerten allgemeinen griechischen Kultur war beim Unterricht fast nie die Rede. Ich bin fest überzeugt, hätten unsere Lehrer auf solche Dinge den Hauptnachdruck gelegt, die meisten von uns wären ihnen mit Begeisterung gefolgt, während wir so den Geschmack am klassischen Altertum geradezu verloren haben und froh waren, mit dem Abiturium diese Studien aufgeben zu können.

Zur Feier des Schlußexamens konnte ich die ganze Klasse in den Garten meines Wirten einladen, denn mein Vater hatte zu diesem Zweck ein stattliches Faß Bier aus der Dortmunder Brauerei gestiftet.

Im Alter von $16^3/_4$ Jahren stand ich nun vor der Wahl des Berufes. Nach meinem Geschmack wäre ich am liebsten Mathematiker und Physiker geworden, aber mein Vater hielt diese Wissenschaften für zu abstrakt und die Möglichkeit für zu gering, sich mit ihnen eine materiell gesicherte Existenz zu schaffen. Er pflegte deshalb seinen Rat in die Worte zu kleiden: „Wenn du durchaus studieren willst, so wähle die Chemie", deren praktisch nützliche Seite ihm von seinen geschäftlichen Unternehmungen her bekannt sei. Im geheimen mag er aber wohl immer noch die Hoffnung gehabt haben, daß ich mich zu einer kaufmännischen Tätigkeit entschließen würde, da ich der einzige Sohn war und er mich begreiflicherweise gern als Nachfolger in seinem Geschäft gehabt hätte. Da ich zudem noch recht jung war und der unmittelbare Besuch einer Universität für mich keinen besonderen Reiz mehr bot, so machte er mir den Vorschlag, auf 1 bis 2 Semester in ein kaufmännisches Geschäft einzutreten. Damit war ich einverstanden und kam deshalb im Oktober 1869 in das Holzgeschäft meines Schwagers Max Friedrichs zu Rheydt und gleichzeitig als Pensionär zu meiner Schwester Fanny, seiner Frau.

Nach den strengen Regeln des Geschäfts wurde ich als unterster Lehrling betrachtet und behandelt. Zu meinen Funktionen gehörte das Abholen der Post, das Zukleben der Briefe und dergl. und zur Übung mußte ich ein kleines Geschäftsjournal, aber nur zu meiner Belehrung ohne Gültigkeit für das Ganze führen. Hier und da wurde ich auch mit einem kleinen Auftrage zu den Kunden, die zum großen Teil kleine Tischler waren, geschickt.

Auf dem Holzplatz hatte ich nichts zu sagen, durfte aber den verschiedenen Arbeiten, Sägen, Hobeln, dem Transport der Stämme, dem Verpassen des Bauholzes usw. zuschauen. Dieses Technische, das damals noch sehr unvollkommen war und meistens mit der Hand gemacht wurde, hat mich mehr interessiert, wie die pedantischen Kontorarbeiten. Aber das Ganze war doch für meinen Geschmack so langweilig, daß ich schon nach wenigen Wochen auf Rat eines Lehrers der höheren Schule zu Rheydt mir Stöckhardts Schule der Chemie mit den dazu gehörigen Apparaten anschaffte und in einem leerstehenden Zimmer des Geschäftsgebäudes ein winziges Laboratorium einrichtete. Die Versuche waren natürlich höchst stümperhaft, endigten mit einigem Gestank oder beschmutzten und verbrannten Fingern, und wurden dem Geschäftsinhaber wegen der Feuersgefahr recht unbequem. Die Abende verbrachte ich mit Klavierspielen oder im Gasthaus mit Biertrinken, Tabakrauchen und Billardspielen.

Der Schwager war mit meinen Leistungen sehr unzufrieden, erklärte mich für den schlechtesten Lehrling, den er je gehabt, und ließ sich einmal im ärgerlichen Unwillen anderen Mitgliedern der Familie gegenüber zu der Bemerkung hinreißen: „Aus dem Jungen wird nie etwas". Er ist später wegen dieses Urteils viel geneckt worden, aber allmählich kam auch mein Vater zu der Überzeugung, daß die kaufmännische Laufbahn für mich nicht das Richtige sei. Er drückte diese Meinung drastisch aus in dem Satze: „Der Junge ist zum Kaufmann zu dumm, er soll studieren".

Inzwischen hatte ich bei dem erwähnten Lehrer der Chemie einige Privatstunden genommen, und mir dabei eine ganz oberflächliche Kenntnis der Atomtheorie angeeignet. Ich kann aber nicht sagen, daß ich davon besonders ergriffen worden wäre. In der Form, wie sie mir dargestellt und wie ich sie auch aus einem kurzen Lehrbuch der Chemie kennen gelernt hatte, erschien sie mir gegenüber den abgerundeten Lehren der Physik zu dürftig und unsicher.

Im Frühjahr 1870 hatte ich das Unglück, mir, wahrscheinlich durch Erkältung, einen Magenkatarrh zuzuziehen, der infolge von Vernachlässigung und ungenügender ärztlicher Behandlung in einen chronischen Zustand überging. Vorbereitet war vielleicht das Leiden durch das unsinnige Rauchen und Biertrinken, das ich schon in jungen Jahren in Wetzlar begonnen und seitdem dauernd beibehalten hatte. Der akute Ausbruch der Krankheit erleichterte mir den Abschied von Rheydt und ich kehrte nach Euskirchen zurück, um mich auszukurieren. Aber der chronische Magenkatarrh war in der Familie etwas ganz Unbekanntes, und auch die Ärzte verstanden damals von der Behandlung desselben recht wenig. Mit den schönen Methoden der Jetztzeit hätte ich in 4 bis 6 Wochen meine volle Gesundheit wieder erlangen können. So aber habe ich trotz der Ratschläge meines sonst als Arzt so ausgezeichneten Onkels in Cöln fast 2 Jahre mich mit der Krankheit herumgeschlagen, und der Verdauungstraktus ist niemals wieder so kräftig geworden, wie er zuvor gewesen. Den Sommer 1870 verbrachte ich in Euskirchen, beschäftigte mich viel im Garten und soweit es möglich war, auf der Jagd. Auf Anraten der Ärzte ging ich Mitte Juli, am Tage der preußischen Mobilmachung mit meiner Mutter nach Bad Ems, um das dortige alkalische Wasser als Heilmittel für den Magen zu trinken. Es herrschte damals eine begreifliche Unruhe im Rheinland, da man glaubte, daß Frankreich sich für den Krieg längst vorbereitet habe und seine Truppen alsbald in die preußischen Rheinlande schicken würde. Mein Vater hatte sich schon darauf vorbereitet, meine unverheirateten Schwestern, sowie Geld, Wertpapiere und andere leicht transportable Gegenstände auf das rechte Rheinufer zu bringen. Um so merkwürdiger berührte es meine Mutter und mich, daß wir auf der Fahrt nach Ems, wobei wir die Festung Coblenz passierten, gar keine militärischen Vorbereitungen beobachteten. Der einzige Soldat, den wir von der Eisenbahn aus in der Festung sahen, war ein Offiziersbursche, der einen Kinderwagen schob. Aber im Stillen war der preußische Mobilisierungsapparat schon im vollen Gange, und 8 Tage später sah man überhaupt fast nur noch Soldaten. Dann begannen auch die großen Truppentransporte, von denen wir in Ems viel zu sehen bekamen, da es an einer Haupteisenbahnlinie von Ost nach West gelegen ist. Anfangs August fanden die siegreichen Schlachten bei Weißenburg, Woerth, Saarbrücken statt, und in Deutschland hatte man sofort das Gefühl, daß die Gefahr einer feindlichen Invasion beseitigt sei. Zwei Tage vor meiner Ankunft in Ems war König Wilhelm von dort abgefahren und von den Erinnerungen an ihn, sowie an die Verhandlungen mit dem französischen Botschafter war in der Kurgesellschaft noch viel die Rede. Obschon die Zahl der Gäste stark abgenommen hatte, nahm doch das Kurleben seinen ungestörten Fortgang.

Nach 4 Wochen konnten wir ohne das geringste Hindernis nach Hause zurückkehren, da der Eisenbahnverkehr sich wieder in normalem Zustande befand. Für die aktive Teilnahme am Krieg war ich noch zu jung und auch damals nicht gesund genug, da die Kur in Ems ohne Erfolg geblieben war.

Im Nebenhause waren die 3 ältesten Söhne schon dienstpflichtig und deshalb eingezogen. Von ihnen hat aber nur der eine, Vetter Lorenz, kriegerische Lorbeeren gesammelt. Als Einjährigfreiwilliger in einem Jägerbataillon nahm er teil an den Schlachten bei Metz. Bei Belagerung der Festung erkrankte er an Ruhr, kehrte nach Deutschland zurück, ging aber bald — noch nicht ganz wieder hergestellt — freiwillig von neuem ins Feld, wo er an den Kämpfen des Generals Werder gegen die Armee Bourbaki bei Dijon beteiligt war. Er wurde im Felde zum Offizier ernannt. Aber das kriegerische Leben, das ganz seinen Neigungen entsprach, ist ihm später zum Unglück geworden; denn es hat mit dazu beigetragen, seine Neigung, dem Bachus zu huldigen, ins Übermaß zu steigern, so daß er schon im Alter von etwa 35 Jahren gestorben ist.

Vorsichtiger war der Vetter Heinrich, ein gedienter Artillerist; er kam nicht ins Feld, sondern blieb bei einem Ersatzbataillon in Coblenz.

Vetter Ernst, der bereits das erste medizinische Examen bestanden hatte, trat freiwillig als Unterarzt in das Heer ein und hatte in dieser Eigenschaft, wie er später gern erzählte, besonders in der Umgegend von Orléans ein recht vergnügtes und wenig anstrengendes Feldleben geführt. Kurzum, der Krieg 70 ist nach allen Schilderungen seiner Teilnehmer ein ganz anderer, als der jetzige gewesen. Die Verluste auf deutscher Seite waren verschwindend gering, sind doch nicht mehr als 28 000 Mann auf unserer Seite gefallen. Auch die Kriegführung war sehr viel humaner. Ein Teil der Truppen hat sogar mit der französischen Bevölkerung auf ganz gutem, fast freundschaftlichem Fuße gestanden. Dazu kam die rasche Beendigung des Krieges und als schönster Lohn für Deutschland seine politische Einigung im deutschen Reiche.

Den Sieg von Sedan verlebte ich in Euskirchen. Ich erfuhr die Nachricht bei der Rückkehr von der Hühnerjagd. Man meinte damals allgemein, der Krieg sei nun zu Ende. Das war aber ein Irrtum, da die Republik unter Gambetta ihn hartnäckig fortzusetzen suchte. Aber die Widerstandskraft Frankreichs war doch gebrochen, nur dauerte das kriegerische Leben auch bei uns auf dem linken Rheinufer mit Truppenbewegung, Einquartierung fort bis zum Waffenstillstand.

Mein Gesundheitszustand hatte sich inzwischen kaum geändert und mein Onkel in Cöln lud mich deshalb ein, im November zu ihm zu kommen und in seinem Hause einige Zeit zu wohnen, damit er mich dauernd beobachten könne. Seine liebe Frau, die durch besondere Güte ausgezeichnete Tante Mathilde, nahm sich des kranken Neffen in der liebenswürdigsten Weise an. Ich bekam besonders zubereitete Speisen und durfte dazu auf Rat des Onkels so viel guten Bordeaux-Wein trinken, wie ich glaubte vertragen zu können.

Wohltuend wirkte auf mich die gleichmäßige Wärme des Hauses, das schon eine Zentralheizung besaß, was man damals als ungewöhnlichen Luxus ansah. So bin ich den ganzen Winter in Cöln geblieben und nur während des Weihnachtsfestes auf 8 Tage nach Hause zurückgekehrt. Ich habe mich in meinem späteren Leben immer mit großer Dankbarkeit gegen den Onkel und die Tante an diese Zeit erinnert; denn damit trat eine Besserung in meinem Zustande ein.

In Cöln war auch in geistiger Beziehung leichter Anregung zu finden, als in Euskirchen. Ich habe hier Theater und Konzerte besucht. Vor allen Dingen wurde mir aber Gelegenheit geboten, bei einem Engländer Unterricht in der englischen Sprache zu nehmen. Da ich nichts anderes zu tun hatte, so war es mir nicht schwer, in $2^1/_2$ Monaten so weit zu kommen, daß ich kleine englische Aufsätze schrieb und mich mündlich leidlich ausdrücken konnte. Leider habe ich diese Übungen später nicht fortgesetzt und auch niemals längeren Aufenthalt in England nehmen können, so daß ich die wichtige Sprache nie vollkommen beherrschen lernte.

Selbstverständlich bin ich in Cöln auch den Kindern meines Onkels näher getreten. Die älteste Marie war schon an Herrn Eugen Coupienne in Mülheim a. d. Ruhr verheiratet, kam aber öfter mit Mann und Kind zum Besuch nach Cöln. Herr Coupienne brachte bei dieser Gelegenheit nicht selten gefangene französische Offiziere, die in Cöln sich frei bewegen durften, als Gäste in das Haus meines Onkels. Es machte dann meinem Onkel besonders Freude, die französische Sprache, die er in Paris recht gut gelernt hatte, wieder zu gebrauchen.

Die zweite Kusine Helene, ein schönes wohlgenährtes Mädchen, die den Beinamen „Dick" trug, und mir von verschiedenen Besuchen in Euskirchen her wohl bekannt war, ist bald nachher ihrer Schwester gefolgt und hat einen Tabakfabrikanten, Carl von Eicken, ebenfalls aus Mülheim, geheiratet. Sie lebt noch jetzt zusammen mit ihrem Gemahl in Hamburg, in sehr glücklichen Verhältnissen, und ist die Stammmutter von 4 blühenden Kindern und einer ganzen Anzahl von Enkeln geworden. Ihr Sohn Karl, jetzt ordentlicher Professor der Ohren- und Nasenheilkunde in Gießen, war als Student öfters Gast in meinem Hause zu Berlin, und ich hatte gleich den Eindruck, daß er die Eigenschaften eines guten Arztes vom Großvater Fischer geerbt habe.

Die Tochter Helene lebt mit ihrem Mann Dr. Kroehnke und 4 hoffnungsvollen Söhnen in Zehlendorf und ist eine durch Liebreiz und Güte ausgezeichnete Frau geworden.

Eine zweite Tochter des Ehepaares von Eicken ist in Hamburg verheiratet und der zweite Sohn, ein großer Sportsmann, führt zusammen mit dem Vater das bedeutende Tabakgeschäft. Er ist verheiratet mit der einzigen Tochter meines Vetters Julius Fischer.

Die dritte Cölner Kusine, „Tönn" genannt, war klug und wissbegierig und erhielt in der Schule die besten Zeugnisse. Sie hat später einen Fabrikanten Herrn Vorster aus Mülheim a. d. Ruhr geheiratet und ist Schwiegermutter des bekannten Malers Petersen in Düsseldorf geworden.

Der einzige Sohn Fritz, von den Schwestern nur „der Fischer" genannt, wurde von ihnen viel geneckt, war 4 Jahre jünger wie ich und deshalb noch Gymnasiast. Er war der Liebling des Vaters und pflegte sich sehr viel in dessen Studierstube aufzuhalten. Wie leicht begreiflich, ist er in die Fußstapfen des Vaters getreten, hat Medizin studiert und zwar in Straßburg, wo ich wieder viel mit ihm zusammentraf. Er wurde auch an der dortigen Universität außerordentlicher Professor der Chirurgie, starb aber verhältnismäßig früh im Alter von etwa 50 Jahren. Er war verheiratet mit Anni geb. Stinnes aus Mülheim a. d. Ruhr und hat einen Sohn Otto hinterlassen, der wiederum Medizin studiert und z. Zt. als Unterarzt im Felde steht. Er ist in den letzten Jahren meinem lieben Sohn Alfred näher getreten, der von der Tante in Straßburg viele Freundlichkeiten erfahren hat.

Während meines Aufenthaltes in Cöln hatte der Onkel seinen Kindern den strengen Auftrag gegeben, den kranken Vetter in keiner Weise zu ärgern und ihm womöglich nicht zu widersprechen, selbst wenn er in seiner etwas aufgeblasenen Gelehrsamkeit gewagte Behauptungen aufstelle. So schwer das auch den mundfertigen Kusinen wurde, so haben sie sich doch alle Mühe gegeben, mir den Aufenthalt in Cöln angenehm zu machen. Mit dem Vetter stand ich selbstverständlich als früherer Gymnasiast auch auf gutem Fuße. So steht der Winteraufenthalt in Cöln bei mir in bester Erinnerung und dem trefflichen Onkel, der mir nicht allein in seiner ärztlichen Kunst, sondern auch als vornehmer Charakter und als origineller Mensch stets ein Vorbild geblieben ist, sowie der herzensguten Tante habe ich mich Zeit meines Lebens zu warmem Danke verpflichtet gefühlt.

Im Frühjahr 1871 war meine Gesundheit soweit wieder hergestellt, daß ich die Universität beziehen konnte, um Chemie zu studieren. Das Leben im Gasthause war mir allerdings noch nicht zuträglich, aber ich wurde in freundlicher Weise wieder aufgenommen in meinem alten Gymnasiastenquartier. Dort waren inzwischen einige persönliche Veränderungen vorsichgegangen. Das alte Ehepaar Kemp hatte sich zurückgezogen und das Geschäft dem Sohn übergeben. Der hatte inzwischen die zu meiner Zeit als Verkäuferin im Geschäft tätige Fräulein Marie geheiratet und die junge, liebenswürdige Frau erklärte sich sofort bereit, die Verpflegung für mich ganz nach meinen Wünschen einzurichten. Das ist auch in gewissenhafter Weise geschehen. Ich lebte also so weiter, wie ich es in der Familie meines Onkels gewohnt war, und der Hausherr, der sich bei seinem früheren Aufenthalt in den Vereinigten Staaten von Nordamerika infolge von Malaria und übermäßigem Chiningenuß den Magen verdorben hatte, lernte von mir die für solche Zustände angemessene Diät. Er wurde in der Tat auf diese Weise auch wieder gesund und hat mir später wiederholt für die guten Ratschläge gedankt.

Im Sommersemester 71 habe ich mich in Bonn nicht überanstrengt, sondern in aller Behaglichkeit Vorlesungen über Physik, Botanik und nur wenig Chemie gehört. Das botanische Kolleg von Professor Hanstein fand schon morgens um 7 Uhr im Schloß zu Poppelsdorf statt, war sehr populär und nur auf die Bedürfnisse eines Mediziners zugeschnitten. Systematik und Anatomie standen im Vordergrund und die Physiologie, die mich viel mehr interessiert hätte, spielte nur eine geringe Rolle.

Die Physik wurde von dem berühmten Thermodynamiker Clausius gelesen, war recht langweilig und in experimenteller Beziehung geradezu dürftig. Meine alte Vorliebe für die Physik ist dadurch etwas gedämpft worden und erst als ich später in Straßburg den glänzenden Vortrag und das gute Praktikum von Kundt besuchte, wurde mir erst wieder klar, welch herrliche Wissenschaft sie ist.

Das gerade Gegenteil von Clausius war der Chemiker August Kékulé, ein ausgezeichneter Redner, guter Experimentator und eindrucksvolle Persönlichkeit. In dem Sommersemester las er allerdings nur ein zweistündiges Kolleg über organische Chemie, aber in dem darauf folgenden Winter habe ich auch seine große Vorlesung über anorganische Experimentalchemie besucht.

Nachdem ich in den Herbstferien 71 zusammen mit dem Vetter Lorenz, der inzwischen aus Frankreich zurückgekehrt und noch ganz von den Erlebnissen des Feldzuges erfüllt war, das Seebad Blankenberghe bei Ostende besucht hatte, war mein Gesundheitszustand wieder so gut, daß ich im Winter 71/72 mit voller Kraft an das Studium herantreten konnte. Zu den Vorlesungen kamen jetzt auch die praktischen Übungen im Laboratorium. Das Institut lag in Poppelsdorf, war 1864/65 von Hofmann erbaut und machte äußerlich einen pompösen Eindruck. Viel weniger schön waren die inneren Einrichtungen, z. B. Raumverteilung und Beleuchtung. Besonders die Ventilation ließ sehr zu wünschen übrig. Der Direktor hielt die großen Vorlesungen und leitete die praktischen Arbeiten in der organischen Abteilung. Der analytische Unterricht war dem außerordentlichen Professor Engelbach anvertraut. Ohne irgendwelche Vorbereitung in der Anstellung von Experimenten wurde man sofort vor die Aufgabe gestellt, nach den Tafeln von Will eine qualitative Analyse auszuführen. Dazu war ich nicht imstande und hatte gegenüber den anderen Praktikanten, die meistens Apotheker waren, einen schweren Stand, denn diese pflegten meist auf illegitimen Wege die Aufgaben zu lösen, während ich ohne jede Anleitung viele Fehler beging und dann unbarmherzig zur Wiederholung angehalten wurde. Dazu kam, daß die Vorlesung über analytische Chemie, die ich pflichteifrig besuchte, äußerst trocken und langweilig gehalten wurde. Mein Eintritt in die Wissenschaft vollzog sich also unter ziemlich ungünstigen Verhältnissen, die mich sehr niederdrückten. Und als ich im darauffolgenden Sommer an die quantitative Analyse herankam, wurde die Sache noch schlimmer. Engelbach war inzwischen gestorben und Th. Zinke, der bisherige erste Assistent in der organischen Abteilung und gleichzeitig Privatdozent, wurde sein Nachfolger. Er war gewiß ein geschickter und auch schon verdienter Chemiker, aber der Unterricht in der analytischen Chemie lag ihm nicht besonders nahe, und wenn er sich auch große Mühe gab, und mir manchmal einen guten Rat erteilte, so war es doch auch für ihn schwer, die Hindernisse zu beseitigen, die der Ausführung einer guten quantitativen Analyse im Bonner Institut entgegen standen. Schon allein die Ungenauigkeit der Wagen war groß genug, um das Innehalten der üblichen Fehlergrenzen bei den Analysen unmöglich zu machen. Der Gebrauch der Wasserluftpumpe, die schon längst von Bunsen erfunden war, fand in Bonn eine ungünstige Beurteilung und man begnügte sich noch immer mit der alten primitiven Wäsche der Niederschläge auf einem gewöhnlichen Filter. Das waren harte Geduldsproben. Als ich schließlich einen Niederschlag von Aluminium- und Eisenhydroxyd, der wohl nicht unter den richtigen Vorsichtsmaßregeln gefällt war, 8 Tage lang gewaschen hatte, ohne die Mutterlauge ganz verdrängen zu können, war ich so verzweifelt, daß ich die Chemie aufgeben und mich wieder der Physik zuwenden wollte. Daß ich es nicht getan habe, ist wohl hauptsächlich dem Einfluß meines

Vetters Ernst zuzuschreiben, der mir riet, es in einem anderen Institut nochmals zu versuchen.

Inzwischen war auch Vetter Otto mit der Schule fertig geworden, hatte ebenfalls das Studium der Chemie in Berlin begonnen und war dann auch nach Bonn gekommen. Er nahm die Schwierigkeiten keineswegs so tragisch wie ich, trotzdem gefiel auch ihm der Gedanke, die Hochschule zu wechseln, nicht allein der Studien halber, sondern um in der Welt sich umsehen zu können. So kam es, daß wir am Ende des Sommersemesters 1872 Bonn verließen und dann halb durch Zufall nach Straßburg i. E. kamen. Das war für uns beide ein rechtes Glück, denn wir sind so mit Adolf Baeyer in Berührung gekommen, und ich wüßte nicht, was mir Besseres hätte passieren können, um die chemische Experimentierkunst zu lernen. Den Aufenthalt im Straßburger Laboratorium habe ich versucht in einem kleinen Aufsatz zu schildern, den Baeyer seiner Selbstbiographie eingefügt hat. Auf diese Art ist er in der Einleitung zu Baeyers gesammelten Abhandlungen zum Druck gelangt. Ich habe ihm nur wenig zuzufügen, um noch ein Bild von unserer Lebensweise und dem Freundeskreise außerhalb des Laboratoriums zu geben.

Im ersten Semester wohnte ich mit Vetter Otto zusammen bei zwei sehr alten Frauen in einem ebenso alten Hause, wie denn überhaupt die ganze Stadt damals einen recht alten und verkommenen Eindruck machte. Verschärft wurde dieser durch die starken Verwüstungen, die von der Belagerung herstammten. So lag das sogen. Steinviertel noch größtenteils in Trümmern.

Mein Magenkatarrh war inzwischen geheilt, aber er hatte doch eine Empfindlichkeit des Organs hinterlassen, die mir eine gewisse Vorsicht in der Diät auferlegte. Die gute Straßburger Küche und der noch bessere, in großen Vorräten vorhandene französische Wein kamen mir zugute, und ich habe im Wintersemester im badischen Hof nach Verabredung mit dem sehr verständigen Wirt ein stattliches Faß ausgezeichneten Burgunder konsumiert, war dadurch in einen recht guten Ernährungszustand gekommen, so daß ich von den Professoren unter dem Titel „der dicke Fischer" von dem Vetter unterschieden wurde.

Bei dem nächsten Besuch während der Osterferien erregte mein Aussehen bei der Mutter große Freude, während der erfahrene Vater mich mit den Worten begrüßte: „Du hast einen ziemlich vertrunkenen Kopf gekriegt." Infolgedessen habe ich das Burgundertrinken aufgegeben und mich dem guten aber viel unschuldigeren Elsasser Landwein zugewandt. Auch die Mehrzahl der Kollegen, die vielfach aus dem Rheinland stammten, besuchten die Weinhäuser. Zwar gab es auch Bier in mannigfaltiger Qualität, sowohl einheimisches, in dem Orte Schiltigheim gebraut, wie auch bayerisches, aber diese Häuser pflegten wir nur im Sommer zu besuchen, der in Straßburg ungewöhnlich heiß ist und deshalb zum Bierverbrauch anregt.

Das studentische Leben, wie es auf anderen deutschen Hochschulen üblich ist, war zu unserer Zeit wenig entwickelt. Es gab zwar einige Corps, aber sie spielten keine große Rolle, weil ihnen der Resonanzboden in der Bevölkerung und der ganzen Tradition der Stadt fehlte. Dagegen war der zwanglose Verkehr der Studenten in den Gasthäusern und auch in den Vorlesungen sehr angenehm. Die vielen fremdländischen Elemente, Russen, Polen und andere Ausländer störten denselben nicht. Die Elsässer hielten sich anfangs zurück, kamen aber doch in den späteren Semestern in die Institute. Es waren darunter manche fein gebildete junge Männer mit sehr angenehmen Umgangsformen. Mit den gebildeten Familien in Straßburg sind wir nicht in Berührung gekommen, wohl aber habe ich das Volk und speziell auf mancherlei Ausflügen in die Vogesen die Landbevölkerung kennen gelernt und einen recht guten Eindruck erhalten. Mit Ausnahme vom Südelsaß, wo die Tuchindustrie sehr entwickelt und in ihren geschäftlichen Interessen auf das übrige Frankreich angewiesen war, verriet sich in der Bevölkerung durchweg die alte deutsche Abstammung. Die Leute sprachen ihr Elsasser Deutsch, vermischt mit französischen Brocken, waren aber zum erheblichen Teil garnicht imstande, französisch zu sprechen. Sie waren zwar nichts weniger als deutsch-freundlich und schimpften gerne auf die „Schwoben". Aber wenn man die Politik beiseite ließ, so konnte man mit ihnen recht gut auskommen. Es ist mir ein Rätsel, daß die deutsche Verwaltung es nicht fertiggebracht hat, in 40 Jahren bei dieser im Grunde gutmütigen, allerdings demokratisch gesinnten Bevölkerung mehr Sympathie zu erwerben.

In Straßburg selbst waren noch manche Überreste französischer Einrichtungen und Sitten geblieben. In öffentlichen Tanzlokalen belustigte man sich noch am Cancan, und es war drollig anzusehen, wie selbst auf dem Lande beim sonntäglichen Tanz einzelne Bauernburschen die beim Cancan üblichen Sprünge versuchten, aber selten einen durchschlagenden Erfolg erzielten. Auch im sittlichen Verkehr zwischen den beiden Geschlechtern herrschte für deutsche Begriffe eine zu freie Form.

Der Verkehr zwischen Studenten und Professoren, die fast alle noch im jugendlichen Alter standen, war behaglicher und anziehender als auf irgendeiner anderen Hochschule. Besonders profitierten davon die Naturforscher und die Mediziner. Da auch die Institute von der deutschen Regierung reichlich mit Geld ausgestattet wurden und die Zahl der Studierenden in den ersten Jahren gering blieb, so war der praktische Unterricht ausgezeichnet und gab sich auch bald in den wissenschaftlichen Leistungen kund. Dieser Vorzug von Straßburg übte allmählich seine Anziehungskraft aus, und wenn nicht die Stadt als so teuer verschrieen gewesen wäre, so würde auch bald die Studentenzahl derjenigen der altdeutschen Universitäten gleichgekommen sein.

Unsere Schilderungen von dem akademischen Leben an der jungen Universität in der alten Reichsstadt haben auch in der eigenen Familie ihre Werbekraft ausgeübt. Schon nach einigen Semestern kam der Vetter Ernst Fischer dorthin, nachdem er das medizinische Staatsexamen bestanden und einige Zeit als Assistent in der chirurgischen Abteilung des Cölner Bürgerhospitals tätig gewesen war. Er besuchte in Straßburg nicht allein die chirurgische Klinik, die unter Leitung von Professor Luecke stand, sondern interessierte sich auch für Studien auf der Anatomie bei Waldeyer und auf der pathologischen Anatomie bei von Recklinghausen. Bei dieser Gelegenheit hat er eine kleine Erfindung gemacht, mit der sein Name wohl dauernd verbunden bleiben wird, und die ich erwähnen will, weil ich dazu die Anregung geben durfte. Damals war das Färben von anatomischen Präparaten, das von meinem Schwiegervater Josef von Gerlach mit der Anwendung des Carminrots in die Wissenschaft eingeführt wurde, schon allgemein üblich. Aber die Zahl der benutzten Farbstoffe war gering. Als ich davon durch den Vetter Ernst hörte, riet ich ihm, das Eosin zu versuchen, das kurz zuvor von Baeyer und Caro entdeckt worden war und das ich bei meiner Doktorarbeit über Fluorescein nicht allein als chemisches Präparat, sondern auch an meinen Händen als prächtiges Färbemittel für tierische Gewebe kennen gelernt hatte. Die Versuche vom Vetter fielen dann auch so gut aus, daß er den Farbstoff in die anatomische Praxis einführen konnte, worin er sich bis jetzt erhalten hat.

Nebenher interessierte er sich besonders für die Antisepsis in der chirurgischen Praxis, die durch den Lister'schen Verband schon eine hohe praktische Bedeutung erlangt hatte. Er bemühte sich, diesen zu verbessern und war schon damals der Ansicht, daß der Antisepsis in der Chirurgie die Asepsis folgen würde. Zunächst bemühte er sich aber, anstelle der Carbolsäure unschuldigere Antiseptica zu finden und so kam er auf das Naphtalin, dessen Giftigkeit für Ungeziefer man schon kannte. Er hat für den medizinischen Gebrauch dieses Kohlenwasserstoffes Propaganda gemacht und es auch bei der damals in Frankreich so stark wütenden Phylloxera (Reblaus) empfohlen, ohne aber in der Praxis durchzudringen.

An diese Bemühungen knüpfte sich eine andere Erfindung, deren Zustandekommen wie eine Anekdote anmutet. Infolge der Empfehlung des Naphtalins als Antiseptikum von Ernst Fischer kamen 2 Assistenten der Klinik für innere Medizin zu Straßburg auf den Gedanken, den Kohlenwasserstoff zur Desinfektion des Darms im krankhaften Zustande zu verwenden. Sie bestellten das Mittel bei einer Straßburger Drogenhandlung mit dem Bemerken, daß es ganz rein sein müsse, und sie erhielten dann auch ein Präparat, von dem das Geschäft behauptete, es sei so rein, daß es keinen Geruch mehr habe. Als sie nun dieses Mittel bei fiebernden Darmkranken anwendeten, beobachteten sie das rasche Sinken der Temperatur. Glücklicherweise war der eine Mediziner, ein Bruder des Chemikers Eduard Hepp, auch chemisch gut gebildet, und er kam bald zu der Überzeugung, daß das von ihnen angewendete Mittel kein Naphtalin sei. Die chemische Untersuchung durch Eduard Hepp ergab dann auch, daß es sich um Acetanilid handelte, welches der Drogist offenbar durch Verwechselung der Flaschen als Naphtalin geliefert hatte. So ist das Acetanilid unter dem Namen Antifebrin ein bekanntes und noch jetzt namentlich in Ostasien viel gebrauchtes Fiebermittel geworden. In Europa wurde es verdrängt durch das Phenazetin, das im wesentlichen ein modifiziertes Antifebrin ist, und die gleiche fieberstillende Atomgruppe besitzt.

Es scheint mir nützlich, solche Zusammenhänge zu schildern, nicht allein aus historischen Gründen, sondern auch als neues Beispiel dafür, daß häufig der Zufall bei Erfindungen mitspielt.

Vetter Ernst hat sich später wieder ausschließlich der Chirurgie zugewandt, ist Privatdozent und außerordentlicher Professor an der Universität geworden und besaß eine chirurgische Privatklinik in der Küfergasse, von der früher schon die Rede war. Er starb während des Krieges an Typhus und Lungenentzündung. Bald nachher ist sein einziger Sohn an der Westfront gefallen. Zwei von seinen Töchtern sind in Frankreich verheiratet. Seiner Frau und seinen Kindern zuliebe ist Ernst zum Katholizismus übergetreten. Daß dabei religiöse Überzeugung mitgewirkt hat, glaube ich nicht; denn ich habe ihn nur als vollkommenen Freigeist gekannt.

Etwas später kam noch ein zweiter Vetter Fritz Fischer, Sohn des Chirurgen in Cöln als Student der Medizin nach Straßburg. Gleich nach Beginn seiner Studien erkrankte er nicht unbedenklich an Scharlachfieber, wahrscheinlich infolge einer Infektion auf der Anatomie. Wie schon erwähnt, ist er ebenfalls dort geblieben und auch als außerordentlicher Professor der Chirurgie gestorben.

Von den sonstigen studentischen Bekannten muß ich noch erwähnen Josef von Mering, mit dem zusammen ich später mehrere chemisch-medizinische Arbeiten ausgeführt habe. Er war schon damals ein Original, als echter Sohn Cölns ein Freund des Humors, mit ungewöhnlicher Körperkraft ausgestattet und als gefährlicher Säbelfechter gefürchtet.

Die Sommermonate waren in Straßburg ungewöhnlich heiß und brachten infolge der durch Ill und Rhein stark versumpften Umgebung eine recht lästige Mückenplage. In diesem Sumpf gediehen auch die Frösche ausgezeichnet und diese waren zweifellos wieder die Vorbedingung für die Existens der zahlreichen Störche, die auf den hohen Giebeln der alten Häuser ihre Nester angelegt hatten. Es war ein drolliger Anblick, wenn man von der Plattform des Straßburger Münsters auf die Stadt herabschaute und in das Getriebe von Hunderten von Storchnestern Einblick erhielt. Die warmen Sommer und die Lage der Stadt im Rheintal zwischen Schwarzwald und Vogesen brachten auch ungewöhnlich schwere Gewitter mit sich, und ich habe oft von meiner Wohnung in der Kalbsgasse, die ganz nahe beim Münster lag, das elektrische Wechselspiel zwischen dem Blitzableiter am Münsterturm und den Gewitterwolken bewundert. An solchen heißen Tagen fuhren wir gerne nach Kehl, um dort im Rhein zu baden und später in einem guten Gasthaus badischen Landwein zu trinken. Hier haben wir Chemiker auch im Juli 1875 ein kleines Abschiedsfest für Professor Baeyer veranstaltet, bei dem ich meine erste Tischrede hielt und glücklich stecken blieb. Das erregte großen Jubel, war aber für mich eine Mahnung, bei späteren Gelegenheiten mich besser vorzubereiten.

Das Baden im Rhein war übrigens keine ungefährliche Sache; denn die Strömung betrug in der Mitte des Stromes 3 m in der Sekunde und das Geschiebe der schweren Kieselsteine am Boden war so stark, daß man es im Wasser leicht belauschen konnte.

Im Sommer verlockte die schöne Umgebung Straßburgs verführerisch zu Ausflügen. Wir haben sie meistens nach dem Badischen ausgeführt. Schwarzwald und Odenwald sind mir auf diese Weise früh bekannt geworden. Zweimal führte mich der Weg auch nach der Schweiz. Die eine Reise, die ich mit Vetter Julius unternahm, ist früher geschildert. Ein Jahr zuvor hatte ich aber schon zusammen mit Vetter Otto Fischer und einem Studierenden der Pharmazie aus Bayern eine lustige Tour durch das Berner Oberland gemacht. Den Schluß dieser Rundfahrten in der Süd-Westecke Deutschlands machte ein Besuch der südlichen Vogesen in Gesellschaft von Vetter Ernst Fischer. Wir sind dabei auf dem Kamm des Gebirges, der die Grenze zwischen Deutschland und Frankreich bildet, marschiert und kamen schließlich in die Schlucht beim großen Belchen, wo sich ein sehr gutes Gasthaus gleichen Namens befand. Hier war eine große, rein französische Gesellschaft versammelt, die einen Ausflug über die Grenze gemacht hatte. Bei Tisch blieben wir ziemlich isoliert. Die Unterhaltung mit der Nachbarschaft beschränkte sich auf einige höfliche Phrasen. Aber plötzlich erkrankte eine Dame der Gesellschaft an einer Unterleibs-

MÜNCHENER ZEIT

blutung und man verlangte dringend nach einem Arzt. Da kein anderer in der Gesellschaft zugegen war, so bot Vetter Ernst seine Hilfe an und hatte rasch die Gefahr beseitigt. Von nun an war die Stimmung der Gesellschaft gegen uns wie umgewandelt. Man lud uns ein, an den Spielen teilzunehmen und überbot sich in Liebenswürdigkeiten gegen uns. Aber bald mußten wir uns verabschieden, um unser Ziel, die Stadt Mülhausen, noch am gleichen Tage zu erreichen.

Am nächsten Tage trennte ich mich von Vetter Ernst, der die Krankenhäuser in Mülhausen kennen lernen wollte, und fuhr allein nach Straßburg zurück. Bei der Gelegenheit habe ich auf komische Art die Bekanntschaft von Otto N. Witt gemacht. Er saß in der Eisenbahn mir gegenüber und richtete nach kurzer Zeit an mich die Frage: „Mein Herr, sie sind gewiß Chemiker?" Ich bejahte das und meinte lachend, er habe das wohl an meinen stark gefärbten Händen gesehen, er selbst scheine aber auch nicht weit vom Handwerk entfernt zu sein. Die Diazokörper, mit denen wir beide vorher gearbeitet hatten, waren die Verräter und brachten uns in kollegiale Verbindung. Wir stellten uns vor und erzählten uns gegenseitig von unseren Arbeiten. Ich war mit dem Phenylhydrazin beschäftigt gewesen und Witt hatte damals gerade das Chrysoidin entdeckt. Er kam von Zürich und fuhr nach England, um dort seine Entdeckung praktisch zu verwerten. Ich lud ihn ein, in Straßburg Station zu machen, worauf er auch einging und wir haben dann einen sehr vergnügten Tag zusammen verlebt. Als ich viele Jahre später nach Berlin kam, war es mir ein Vergnügen, diese alte Bekanntschaft zu erneuern. Am nächsten bin ich ihm getreten, als wir beide Kurgäste in Kissingen im Sanatorium von Dapper waren und genug Muße fanden, uns in zwanglosen Gesprächen nicht allein über wissenschaftliche Dinge gegenseitig zu belehren und zu belustigen. Witt war eine Künstlernatur, sehr gewandt in Schrift und Sprache, humoristisch veranlagt und eine impulsive Natur. Nur mußte man sich sehr in acht nehmen, seine Eigenliebe zu verletzen. Begleitet war er damals von seiner ebenso hübschen wie liebenswürdigen Gattin, einer Engländerin.

Einige Wochen nach der Vogesenfahrt habe ich Straßburg verlassen und mit Vetter Ernst eine Fahrt durch Süddeutschland gemacht, wobei wir zuerst Tübingen, Stuttgart und dann München kennen lernten, und viele lustige Erlebnisse hatten. In Tübingen herrschten damals noch patriarchalische Zustände. Als wir in ein Weinhaus eingekehrt waren, setzte sich sofort das Wirtstöchterlein zu uns, um uns über die Stadt und das städtische Treiben Auskunft zu geben. Ebenso ging es uns im chemischen Institut, dessen Leiter Professor R. Fittig schon als Nachfolger von Baeyer nach Straßburg berufen war. Als wir dort nach einem Assistenten frugen, der uns das Institut zeigen könne, meldete sich eine junge Dame, die im Hause von Fittig tätig war. Sie wußte so gut Bescheid im Institut wie ein Fachchemiker und machte in liebenswürdigster Weise den Führer.

In Stuttgart gab es natürlich sehr viel mehr an Merkwürdigkeiten zu sehen, und in der bayerischen Hauptstadt trafen wir Professor Baeyer, der schon einige Wochen vorher übergesiedelt war. Es machte ihm besonderen Spaß, uns abends in einige originelle Wirtshäuser zu führen, um namentlich mir die Stadt sympathisch zu machen. Ich erinnere mich noch, daß eine der Kneipen den klassischen Namen „Orlando di Lasso" führte und in der Nähe des Hofbräuhauses gelegen war.

Die Kunstschätze Münchens haben wir natürlich gewissenhaft angeschaut, aber ich muß offen gestehen, daß im Gegensatz zu dem großen Eindruck, den ich von den Bildern der Pinakothek hatte, die bemerkenswerten Skulpturen der Glyptothek mich ziemlich kalt ließen. Auch zum Kunstgenuß müssen die meisten Menschen erzogen werden. Die Bedeutung der Antike für Skulptur und Architektur ist mir erst bei späteren Besuchen Italiens klar geworden. Was mich damals am meisten in München interessierte, waren doch das wissenschaftliche und das materielle Leben; denn es handelte sich um die Frage, ob ich meinem Lehrer folgen und mich für längere Zeit dort binden sollte. Meine Eltern und besonders meine Mutter waren entschieden dagegen, weil München damals in gesundheitlicher Beziehung einen recht schlechten Ruf genoß. Der Typhus war so verbreitet, daß junge Leute aus dem Rheinlande, falls sie nicht durch vorherige Erkrankung immunisiert waren, ziemlich sicher darauf rechnen konnten, in München infiziert zu werden. Und zwei Jahre vorher, im Jahre 1873 hatte dort auch die Cholera sehr stark gehaust. Zudem war das Klima der hochgelegenen Stadt mit rasch wechselnden Temperaturen im Rheinland gefürchtet.

Über alle diese Dinge beruhigte mich aber Professor Baeyer und versprach mir den hygienischen Schutz der Pettenkofer'schen Schule. Das ist auch später in ausgiebigem Maße der Fall gewesen. Wir lernten durch Baeyer die Assistenten Pettenkofers kennen, unter ihnen besonders Dr. Forster, einen sehr liebenswürdigen und kenntnisreichen Mann, der später Professor der Hygiene in Amsterdam und zuletzt in Straßburg i. Els. war. Er besaß eine Typhuskarte der Stadt, in der nicht allein die verdächtigen Straßen, sondern die einzelnen Häuser mit Typhuserkrankung bezeichnet waren. Wir haben nach dieser Karte unser Quartier in der gesunden Umgebung des chemischen Instituts gewählt und sind auch sonst immer dem hygienischen Rat von Forster gefolgt.

Als ich damals von München aus weiter reiste, war ich entschlossen, im Spätherbst dorthin zurückzukehren, um im Baeyer'schen Laboratorium meine Studien über Hydrazinverbindungen fortzusetzen.

Zunächst ging ich allein über Salzburg nach Wien, von dessen Vorzügen die österreichischen Studiengenossen in Straßburg so viel erzählt hatten. Meine Erwartungen sind in der Tat nicht enttäuscht worden. Die prächtige Lage in der Nähe des mächtigen Stroms und am Fuße lieblicher, bewaldeter Berge, die an historischen Erinnerungen reiche Altstadt und die prächtigen Neubauten am Ring, die großen Kunstschätze und die ausgezeichneten Theater, dazu das liebenswürdige naive Wesen des Volkes, die vielen schönen Frauen und die gute Natural-Verpflegung waren wohl geeignet, einen jungen Menschen wie mich zu bestechen, und ich würde vielleicht München gegen Wien ausgetauscht haben, wenn nicht der Besuch des chemischen Instituts mich ernüchtert hätte. Es war zwar ein Neubau kostspieliger Art, machte aber in keiner Beziehung den Eindruck der Zweckmäßigkeit. Viel besser gefielen mir die Menschen, die darin auch in den Ferien beschäftigt waren. Zunächst wandte ich mich an einen alten Bekannten, den Straßburger Studiengenossen Dr. O. Zeidler, der mich dann den anderen Assistenten vorstellte. Unter ihnen ragte hervor Dr. Zdenko Skraup, mit dem ich mich sofort befreundete.

Beim Mittagstisch lernte ich noch Professor Weidel kennen, der aber 10 bis 15 Jahre älter war als wir, und im Verkehr mit jungen Leuten sich ziemlich zurückhaltend zeigte. Professor A. Lieben, der die eine Abteilung des Instituts leitete, habe ich nur aus der Ferne gesehen. Erst später bin ich mit diesem fein gebildeten und liebenswürdigen Fachgenossen in nähere Berührung gekommen.

Skraup und Zeidler waren die Führer in denjenigen Teilen des Wiener Lebens, das in die Nachtzeit fällt und nicht in den Reisebüchern beschrieben wird. Wir haben aber auch zwei kleine Ausflüge in die Umgebung Wiens gemacht und mit Skraup zusammen besuchte ich die große Oper. Er war durch seine materielle Lage zu bescheidener Lebensweise gezwungen, und wir gingen deshalb in das billige Stehparkett, was mir auch sehr sympathisch war.

Vernünftigerweise durften auch die Offiziere dorthin gehen, und sie hatten sogar den Vorzug, nur die Hälfte des Eintrittspreises zu bezahlen. Da Skraup Reserveoffizier war, so erklärte er sofort aus Sparsamkeitsgründen die Uniform für diesen Theaterbesuch anlegen zu wollen. Dazu mußte er sich aber rasieren. Da die Zeit sehr knapp war, so erklärte ich mich bereit, ihm diesen Liebesdienst zu erweisen. Zeidler lieh dazu sein Rasiermesser, das, wie er später gestand, von ihm für viel niederere Zwecke benutzt wurde. Als ich nun meine Barbieroperation noch nicht zur Hälfte ausgeführt hatte, erklärte Skraup schimpfend, er wolle sich lieber einen Zahn ausziehen lassen, als solche Tortur länger zu dulden, sprang auf und lief halb rasiert und noch mit Seife beschmiert zum ziemlich weit entfernten nächsten Rasierladen. Seitdem habe ich nie mehr Leibesoperationen an anderen Menschen vollzogen. Nach etwa achttägigem höchst befriedigendem Aufenthalt kam der Abschied von Wien, wohin ich später noch zweimal zum Besuch von Naturforscherversammlungen zurückgekehrt bin. Die Rückreise ging über Nürnberg, Würzburg, Frankfurt a. M. nach Euskirchen, wo ich noch einige Wochen meinen Vater in der Hühner- und Hasenjagd unterstützen konnte. Gleichzeitig suchte ich meinen Eltern klar zu machen, daß ich mit dem Phenylhydrazin eine hübsche Entdeckung gemacht hätte und daß es mein Wunsch wäre, diese weiter zu verfolgen in dem Münchener Laboratorium als freier Privatgelehrter. Obschon mein Vater behauptete, die Sache nicht zu verstehen, und überhaupt die Nützlichkeit dieser Erfindung, die man nicht praktisch verwerten könne, anzweifelte, so ließ er mir doch vollkommen freie Wahl. Zudem hatte er mich schon selbständig gemacht; denn sobald ich mit Vollendung des 21. Lebensjahres mündig geworden war, übergab er mir dieselbe Summe, die meine Schwestern als Heiratsgut erhalten hatten. Die Zinsen reichten bei meinen nicht übertriebenen Ansprüchen zum Lebensunterhalt vollkommen aus. Das Kapital überließ ich gerne meinem Vater zur Verwaltung, und er hat diese auch noch später weiter geführt, bis ich 40 Jahre alt wurde und nach Berlin übersiedelte. Dann meinte er, ich sei nun alt genug, um solche Dinge selber zu besorgen. Die Bedenken meiner Mutter bezüglich der gesundheitlichen Gefahren ließen sich beschwichtigen und so zog ich Mitte Oktober nach München. Das Laboratorium, in dem Baeyer sich provisorisch eingerichtet hatte, war nach den Plänen von Liebig 20 Jahre früher erbaut worden.

Bei seiner Berufung nach München hatte der große Forscher ausdrücklich die Verpflichtung abgelehnt, praktischen Unterricht an Studierende zu erteilen. Er hielt deshalb nur die großen Vorlesungen über Chemie und im Laboratorium widmete er sich ausschließlich seinen eigenen Forschungen. Die Zeit der großen Experimental-Untersuchungen war übrigens bei ihm schon vorüber und die Münchener Zeit hat er mehr literarischen Arbeiten gewidmet. Infolgedessen war das Laboratorium, in dem sich auch die Wohnung Liebigs befand, recht klein. Mit dem Einzug von Baeyer wurde die Dienstwohnung geopfert und im alten Gebäude ein provisorisches Laboratorium eingerichtet. Gleichzeitig begannen die Vorarbeiten für den Neubau des Instituts.

Von Straßburg waren die Assistenten Dr. E. Hepp und Dr. C. Schraube, ferner der Inspektor Kamps und der Diener Gimmig mit übergesiedelt. Als einzigen früheren Insassen des Instituts trafen wir den außerordentlichen Professor an der Universität Jacob Volhard. Er hatte seit vielen Jahren seinem Onkel Liebig sowohl als Vertreter bei den Vorlesungen, als auch bei der Redaktion der Annalen der Chemie Hilfe geleistet. Er war ein kluger, feingebildeter Mann und eine anziehende Persönlichkeit. Durch ungewöhnliche Größe und körperliche Schönheit ausgezeichnet, formgewandt und humoristisch veranlagt, hatte er sich namentlich im Verkehr mit Künstlern einen flotten, von jeder Pedanterie freien Umgangston angeeignet. Seine offene Art, sich auszusprechen, die im allgemeinen Vertrauen erweckte und namentlich uns jungen Leuten sympathisch war, konnte sich aber auch zu heftiger Grobheit steigern. Das geschah, wenn er in Jähzorn geriet, wovon ich selbst eine Probe erfahren sollte. Eines Tages traf er mich nämlich bei der Ausführung einer Elementaranalyse und begann eine gemütliche Plauderei, bis sein Blick auf eine Pipette fiel, die mir von einem Diener als wertloses Stück übergeben und zu niederen Zwecken benutzt worden war. In Wirklichkeit gehörte sie zu einem Satz von Pipetten, die Volhard sich eigens für analytische Zwecke kalibriert hatte. Und nun erfolgte ein so plötzlicher und furchtbarer Zornesausbruch, daß ich fürchten mußte, es würde zu Tätlichkeiten kommen, und mich deshalb in Verteidigungszustand setzte. Glücklicherweise ging der Anfall bald vorüber. Die Heftigkeit tat ihm dann leid, wir haben uns leicht versöhnt und sind von da an immer gute Freunde geblieben. Ich bin sogar zweimal, in München und auch in Erlangen, sein Nachfolger geworden.

Die Vorlesungen, die Professor Baeyer rechtzeitig in München begann, erfreuten sich von Anfang an eines recht guten Besuches. Auch in der kleinen analytischen Abteilung des Instituts, deren Leitung Professor Volhard übernommen hatte, herrschte ziemlich reges Leben. Bescheiden blieb dagegen im ersten Semester der Besuch der organischen Abteilung. Außer mir gab es vielleicht 6 Praktikanten. An die einzelnen Personen kann ich mich nicht mehr erinnern, aber wenn ich nicht irre, waren Paul Friedländer und Wilhelm Königs darunter. Dieser präsentierte sich gleich als eine originelle Persönlichkeit; denn er erschien im Institut mit einem Schuh und einem Pantoffel bekleidet und blieb auch bei dieser Ausrüstung wochenlang, obschon scheinbar kein zwingender Grund dazu vorhanden war. Leute, die sich so wenig um die Verwunderung anderer Menschen kümmern, haben mir von jeher gefallen und so habe ich denn auch die Bekanntschaft von Königs gerne gesucht, wobei sich noch das erfreuliche

Resultat ergab, daß er als Cölner ein Landsmann von mir sei. Die äußere Hülle war bei ihm nicht schön, sie umschloß aber eine um so schönere Seele. Mir sind nicht viele Leute im Leben begegnet, die bei hervorragendem Verstand und ausgesprochener Begabung für Witz eine so vornehme Denkweise und ein so mildes Urteil über die Fehler ihrer Mitmenschen hatten. Dieser Eigenschaft verdankte er auch die allgemeine Beliebtheit und ich selbst habe ihn immer gerne zu meinen Freunden gezählt.

Theodor Curtius hat ihm eine liebenswürdige Biographie gewidmet, aber das kann mich nicht abhalten, auch hier seiner zu gedenken und einige gemeinsame Erlebnisse zu schildern. Königs besaß einen offenen Kopf, hatte für wissenschaftliche Dinge volles Verständnis und war keineswegs arm an guten experimentellen Ideen. Dagegen fehlte ihm die praktische Gewandtheit. Wie er selbst öfter beklagte, war es ihm nicht möglich, verschiedene Versuche zu gleicher Zeit anzustellen oder auch nur zu beaufsichtigen. Dazu kam eine gewisse Langsamkeit in der praktischen Arbeit, an der zum Teil wohl seine frühere chemische Erziehung schuld war. Er kam nämlich von Bonn, wo er auch promoviert hatte und wo das rasche Arbeiten offenbar nicht sehr gepflegt wurde. So erzählte er bei seiner Ankunft, daß er zu einer Elementaranalyse in Bonn gewöhnlich 2 bis 3 Tage gebraucht habe. Als ich ihm darauf lachend erwiderte, daß ich 5 an einem Tage machen könne, wollte er es nicht glauben, bis ich ihm den tatsächlichen Beweis dafür lieferte. Allerdings ist dafür doppelte Apparatur erforderlich. Ohne die Langsamkeit des Experiments hätte Königs bei seiner Begabung und seinem Ideenreichtum der Wissenschaft noch viel größere Dienste leisten können.

In der Geselligkeit junger Leute wurde Königs durch seinen schlagfertigen Witz rasch der Mittelpunkt. Daneben besaß er die Gabe, recht hübsche Gelegenheitsgedichte und kleine Festspiele zu verfassen. Manche davon hätten wohl verdient, im Druck zu erscheinen. Sie würden zweifellos, ähnlich den Werken des Berliner Chemikers Jacobsen eine wertvolle Bereicherung der humoristischen chemischen Literatur bilden.

Dem Inhalt, leider nicht dem Wortlaut nach ist mir ein Gedicht in Erinnerung geblieben, das er zu einem kleinen Feste der chemischen Gesellschaft in München bei meinem Abschied von dort beisteuerte. Es war dem Guanolied von Scheffel nachgebildet und bezog sich auf die von mir aufgefundene Bereitung des Kaffeins aus dem Xanthin und Guanin. Die Kunde davon war durch ein Heft der Berichte zu den Vögeln an der Guanoküste gedrungen. Um nun der deutschen Konkurrenz die Spitze zu bieten, macht ein alter Vogel zur Beruhigung der jungen Kollegen den Vorschlag, „von nun an in homologen Reihen zu scheißen", und „wenn dann schließlich gelungen, das homologe Produkt, so wird eine besondere Probe Herrn Fischer zu Ehren gedruckt". Diesem Liedchen fügte der Dichter noch einem Spottvers zu: „So sangen 2 muntere Vögel auf Kosten des Herrn Präsident und tranken dazu von dem Weine, den nie vor Anderen er nennt". Damit hatte es folgende Bewandnis: Königs und ich wohnten damals bei derselben Wirtin im gleichen Stock und besuchten uns spät abends häufig, um gemeinsam ein Glas Wein zu trinken. Kurz vor dem Feste traf ich Königs bei meiner Heimkehr auf seinem Zimmer in Gesellschaft eines jungen Chemikers, beide offenbar etwas betroffen durch meinen Eintritt. Königs hatte nämlich eben das Guanolied verfaßt und sich dazu aus meinem Weinkeller eine der besten Sorten kommen lassen, die bei meiner Ankunft aber rasch durch Kutscherwein ersetzt worden war. Der junge Fachgenosse war der Conkneipant; denn Königs hatte beim Dichten Wein und Gesellschaft nötig und ließ sich von solchen Assistenten den Versfuß vortreten, um nicht zu entgleisen. Als ich mich dann entfernt hatte, fügte er dem Gedicht noch den oben erwähnten Vers hinzu.

Meine Weinvorräte waren damals nicht ganz gering und es befanden sich darunter einige recht gute Sorten. Beim Umzug nach Erlangen Ostern 1882 hatte ich der Wirtin den Auftrag gegeben, denselben nachzuschicken. Der Wein kam aber nicht und eine Anfrage ergab, daß Königs ihn mit seinen Freunden getrunken hatte. Statt dessen erschien ein großes Faß Münchener Hofbräu als Ersatz für den verschwundenen Wein. Er wußte genau, daß ich über den Spaß ebenso lachen würde, wie er es im gleichen Fall getan haben würde.

Mit Königs bin ich auch wiederholt gereist, das letzte Mal 1902 nach Nervi an der Riviera di Levante. Der dritte in unserem Bunde war S. Gabriel, ebenfalls ein richtiger Witzbold. Wenn wir abends zusammen bei einem Glase Bier saßen, so bemühten die beiden Kollegen sich in Witzen zu überbieten. Es ging wie ein Raketenfeuer und ich hatte stundenlang nichts weiter zu tun, als zu lachen. Fast noch komischer als die Witze war das außerordentliche Vergnügen, das die beiden Herren an ihren gegenseitigen Produkten hatten. Es waren heitere Tage, die wir zu prächtigen Fußwanderungen in dieser herrlichen Landschaft benützten. Auch hierbei passierten komische Dinge. Eines Tages kehrten wir auf einem sehr steilen, gepflasterten Fußweg nach der Stadt zurück und begegneten dabei drei italienischen Knaben im Alter von ungefähr 10 bis 12 Jahren, die sofort ihre Bemerkungen über die Forestiere machten, ohne zu ahnen, daß wir etwas von der Sprache verstanden. „Che banda senile" meinte der erste, „gleich wird er hinfallen" meinte der zweite, und dann schlossen sie sofort eine Wette, wen von den dreien dieses Schicksal treffen würde. Es dauerte auch nur noch wenige Sekunden, da saß Gabriel wirklich auf dem Boden, was der jungen Bande natürlich einen unbändigen Spaß machte. An der Heiterkeit hat sich aber auch die alte Bande gründlich beteiligt.

Eine andere Italienfahrt, die ich mit Königs von München aus machte, führte bis Neapel. Infolge seines roten Haares und seiner doppelten Brille wurde er von Droschkenkutschern, Bettlern und ähnlichem Volk schon aus weiter Ferne als Fremder erkannt und dementsprechend bestürmt. Das führte auch zuweilen zu komischen Auftritten. Die Droschkenkutscher hatten damals in Neapel die üble Angewohnheit, den Fremden nicht allein zur Benutzung ihres Fahrzeuges einzuladen, sondern auf offener Straße zu folgen und überall in den Weg zu fahren, wo man eine Straße kreuzen wollte. Unserem Freund Königs schien das eine Gelegenheit, die Zudringlichkeit mit einer Neckerei zu erwidern. Er stieg also ruhig in das Fahrzeug hinein und auf der anderen Seite ebenso rasch wieder hinaus. Wenn der Kutscher dann los fuhr, war der Wagen leer. Kaum aber hatte er dieses Experiment einige Male ausgeführt, als die Kutscher auch schon eine Gegenlist erfanden. Sobald er nämlich seinen Fuß in den Wagen hineinsetzte, war dieser auch schon in Bewegung, dann gab es ein großes Gelächter und Königs mußte bezahlen.

Über solchen Dummheiten wurde aber doch der Hauptzweck der Reise niemals versäumt, denn Königs war gebildet genug, um für die ungeheuren Kunstschätze Italiens und für die Fundstätten der antiken Kultur Verständnis zu besitzen. Für gute Bilder konnte er sich sogar begeistern. Dazu mag auch wohl das Beispiel seines einen Bruders beigetragen haben, der in Berlin Bankier war und seine stattlichen Einkünfte für den Ankauf von guten Bildern verwendete. Der größte Teil der wertvollen Sammlung ist nach dem frühzeitigen Tode des Bankiers von den Geschwistern der Nationalgalerie geschenkt worden.

In späteren Jahren hat Königs mich in Berlin regelmäßig wenigstens einmal im Jahre besucht und dann manchmal mit seinen Geschwistern, besonders mit seiner klugen und fein gebildeten Schwester Fräulein Elise Königs in Berührung gebracht, die den Berliner Gelehrten als weiblicher Mäcen wohl bekannt ist und von der Akademie der Wissenschaften durch Verleihung der goldenen Leibniz-Medaille geehrt wurde.

Eine ganz andere Natur als Königs war Paul Friedländer, der Sohn eines Universitätsprofessors in Königsberg, mit dessen Persönlichkeit meine Münchener Erinnerungen auch eng verknüpft sind.

Er kam als Student in das Baeyer'sche Laboratorium, wurde aber wegen seiner Begabung auch von uns älteren Chemikern als gleichberechtigt angesehen. Nebenher war er sehr musikalisch und konnte recht schwierige klassische Sachen auf dem Klavier aus dem Gedächtnis vortragen. Durch seine späteren Arbeiten über Thio-Indigo und über den antiken Purpur, den er als Dibromindigo erkannte, hat er sich in der Wissenschaft einen geachteten Namen geschaffen. Auch mit ihm habe ich mehrere Reisen nach Italien gemacht, sogar meine erste, die über Verona, Venedig, Padua, Florenz und Mailand ging und bei der er mir wegen seiner besseren Sprachkenntnisse ein wertvoller Führer war. Er ist 6 Jahre jünger wie ich und noch in voller Rüstigkeit an der technischen Hochschule zu Darmstadt tätig. Ich bin durch die Kriegswirtschaft, woran er sich auf Veranlassung von Professor Haber beteiligte, wieder öfter mit ihm in Berührung gekommen.

Im Frühjahr 1876 kehrte auch der Vetter Otto Fischer in Baeyers Laboratorium zurück, nachdem er das Wintersemester bei Liebermann über Methylanthracen gearbeitet hatte, und wir haben bald nachher die gemeinsame Untersuchung über Rosanilin begonnen, wovon später ausführlich die Rede sein wird.

Ungefähr um dieselbe Zeit gesellte sich zu unserem Kreise als älterer Student der Chemie Hans Andreae aus Dresden, dessen Mutter eine geborene Dilthey aus Rheydt und deshalb meine Kusine war. Der Vater, ein Kunstmaler, stammte aus der rheinischen Familie Andreae, die in Mülheim a. Rh. eine große, sehr bekannte Sammetfabrik hatte. Dieser Hans war ein frischer, lustiger Geselle, aber stark verbummelt und dem Gambrinus ergeben. Gearbeitet hat er in München kaum, aber sehr viel Bier getrunken, Skat gespielt und Späße gemacht. Bei seiner unvernünftigen Lebensweise waren Unfälle nicht selten. Schon in Leipzig hatte er bei einer Rauferei einen Messerstich in die Lunge bekommen. In München zog er sich eine langwierige Verrenkung des Fußknöchels zu und ein Semester später erkrankte er an einem schweren Typhus. Da er auch nach dieser Krankheit die alte Lebensweise beibehielt, so riet ich seinem Vater, der sich an mich gewandt hatte, ihn wieder ins Elternhaus zurückzunehmen und auf der technischen Hochschule zu Dresden seine Studien fortsetzen zu lassen. Das war seine Rettung; er wurde nun vernünftig, machte seine Examinas und ist dann wohlbestallter Fabrikant in Burgbrohl in der vulkanischen Eifel geworden, wo er eine natürliche Kohlensäurequelle ausnutzt und hauptsächlich zur Bereitung von Alkalibicarbonat verwendet. Sogar zum kirchlichen Würdenträger hat er es gebracht, denn bei seinem letzten Besuche in Berlin erzählte er mir, daß er als Vertreter der niederrheinischen Synode an einer kirchlichen Versammlung teilnehme.

In dem provisorischen Laboratorium zu München haben wir es trotz mancherlei Mängel der Einrichtungen recht behaglich gehabt; denn die Arbeiten waren von Erfolg begleitet, wie ich in einem besonderen Kapitel noch ausführen werde, und Professor Baeyer tat alles, uns zu fördern und den Aufenthalt angenehm zu machen. Ebenso wie in Straßburg habe ich mich auch hier seiner besonderen Gunst erfreut. Obschon ich nur einfacher Praktikant ohne jede Verpflichtung gegen das Institut war, so räumte er mir doch mancherlei Rechte ein, die sonst nur den Assistenten zustanden. Auch im persönlichen Verkehr sind wir uns näher getreten. Er lud mich öfters zu Gesellschaften in seiner Familie und im Sommer, wo diese frühzeitig auf das Land ging, sind wir auch manchmal zusammen ins Gasthaus gegangen. Aus den hier geführten Gesprächen habe ich manches gelernt, das außerhalb des wissenschaftlichen Ideenkreises lag. Einer Unterhaltung aus dem Sommer 1876 erinnere ich mich noch deutlich, weil ich Baeyer darin meine Absicht kund gab, bei der wissenschaftlichen Laufbahn zu bleiben. Zu dem Zweck wollte ich im nächsten Winter wieder nach Straßburg gehen, um bei Rose den analytischen Unterricht näher kennen zu lernen. Er hielt das für ganz vernünftig, weil Robert Bunsen doch wohl schon zu alt sei, knüpfte daran aber lachend die Bemerkung, daß ich mit großer Überlegung auf mein Ziel lossteuere.

Zunächst habe ich aber die Herbstferien 76 zu einer Reise nach Berlin, Kopenhagen und Hamburg benutzt, die sehr genußreich und zum Schluß belehrend für mich ausfiel. Der erste Teil der Reise ging über Dresden und Leipzig. Mein Begleiter war Vetter Otto. Die sächsische Hauptstadt hat uns durch die schöne Lage, die prächtigen Bauten und die wunderbaren Kunstsammlungen sehr imponiert. Wir hörten auch manches über die Bewohner durch die dort ansäßige und früher erwähnte Familie Andreae, ein Ehepaar mit 10 Kindern, das uns Vettern sehr freundlich aufnahm.

Leipzig hat uns nicht allein als Universität, sondern auch als Handelsstadt gut gefallen. In Berlin, wo wir uns 8 Tage aufhielten und mit Wilhelm Königs zusammentrafen, der sich dann als Reisegefährte anschloß, konnten wir noch viel mehr Interessantes sehen und erleben, da der Bekanntenkreis von Königs dort groß war und wir durch ihn auch in die nächtlichen Geheimnisse der Großstadt rasch eingeweiht wurden. Trotzdem war der Gesamteindruck, den die Stadt mir hinterließ, nicht besonders freundlich. Infolge der Gründerperiode war sie in einem Umwandlungsprozeß begriffen, der sich durch gewaltiges Wachstum und viele häßliche Bauten auszeichnete. Aber auch von den unschönen Einrichtungen des alten Berlins war manches übrig geblieben, und ich erinnere mich noch jetzt mit Schaudern, nächtlicherweile in einen Rinnstein gefallen zu sein, der wenigstens $^1/_3$ m tief und mit keiner angenehmen Flüssigkeit gefüllt war. Auch der Grundzug der Bevölkerung, ihre schnarrende und schnoddrige Redeweise, mit dem befehlshaberischen, an das Militär erinnernden Ton, waren mir recht ungewohnt.

Die chemischen Institute, die wir ansahen, konnten ebenso wenig unseren Beifall erwecken, und wenn mir jemand damals prophezeiht hätte, daß ich einmal nach Berlin berufen würde, so hätte ich sicherlich in Gedanken ein solches Angebot energisch abgelehnt. Von dort ging die Reise weiter über Stettin, wo wir sofort das Schiff nach Kopenhagen bestiegen. Vetter Otto trennte sich für diese Zeit von uns und fuhr in der zweiten Kajüte, weil er annahm, daß hier die Gesellschaft besser sei. Es war meine erste Meerfahrt, und sie ist mir dauernd im Gedächtnis geblieben, weil sie recht stürmisch verlief. Schon auf dem Haff wurden viele Personen seekrank, und als wir auf das offene Meer kamen, geriet das kleine Schiff in bedenkliche Schwankungen. Königs und ich haben uns tapfer gehalten, bis wir am späten Abend, durch die Kälte gezwungen wurden, in die Kajüte zu gehen. Hier haben wir dann ebenfalls Neptun unser Opfer bringen müssen, und als wir am nächsten Morgen in Kopenhagen ankamen, war die ganze Schiffsgesellschaft in ziemlich trostlosem Zustand.

Der Aufenthalt in der prächtig gelegenen und durch die große Kunst Thorwaldsens geweihten dänischen Hauptstadt hat uns für die erlittene Mühsal reichlich entschädigt.

Die Dänen sind ein sehr höfliches Volk und trotz der politischen Abneigung, die sie damals gegen Deutschland hegten, haben sie uns mit großer Zuvorkommenheit behandelt. Nur einmal stießen wir mit ihrem Vorurteil zusammen. In einem Cafehaus, wo wir die einzigen Gäste waren, zog Königs ein Kartenspiel hervor, das er immer auf Reisen mitführte, und wollte ein kleines unschuldiges Skatspiel beginnen. Dagegen erhob aber der Wirt entschiedenen Widerspruch, weil in den besseren Gasthäusern der Stadt das Kartenspiel Anstoß errege.

Natürlich haben wir auch die schöne Umgebung von Kopenhagen, das Seebad Klampenborg, die kleine Stadt Helsingoer, mit dem alten aus Hamlet bekannten Schlosse, und endlich das im Innern gelegene prächtige Schloß Frederiksborg besucht. Ich bin später nicht mehr nach Dänemark gekommen, habe aber dem schönen Seeland eine dauernde freundliche Erinnerung bewahrt.

Die Überfahrt über Korsoer nach Kiel ging ohne Zwischenfall von statten, da die See ruhig war. Kiel machte damals den Eindruck einer kleinen schmutzigen, ziemlich unbedeutenden Seestadt. Von der großen deutschen Marine mit den gewaltigen Werftanlagen der Neuzeit war noch sehr wenig vorhanden, dagegen hatten wir bei der Einfahrt, die früh morgens erfolgte, das Vergnügen, die prächtige Kieler Bucht zu bewundern.

Ebenso gab uns die Eisenbahnfahrt von Kiel nach Hamburg willkommene Gelegenheit, das hübsche und freundliche Holstein'sche Land mit manchen kleinen Seen und prächtigen Waldungen zu sehen. Hamburg war damals noch nicht die gewaltige Handelsstadt wie heute. Aber als Seeplatz nahm es doch schon den ersten Rang in Deutschland ein, und das Leben im Hafen bot für uns Landratten viele Überraschungen. Auch die Stadt selbst mit den schönen Bauten an der Alster, mit der prächtigen Kunsthalle und den guten Theatern war wohl geeignet, die Aufmerksamkeit des Reisenden zu erwecken. Als junge Leute, die den ganzen Tag auf der Schau sein konnten, haben wir das alles in ein paar Tagen gesehen.

Inzwischen hatte die Tagung der deutschen Naturforscher und Ärzte begonnen, an der wir als Mitglieder teilnahmen. Die chemische Sektion war nicht sehr besucht, aber ich habe doch einige für mich interessante Bekanntschaften gemacht. Zunächst A. Ladenburg, damals Professor der Chemie an der Universität Kiel, dann der außerordentliche Professor Michaelis von Karlsruhe, der gerade aus England kam und sich ebenso durch einen ungeheuren Bart wie durch einen altersschwachen Cylinderhut auszeichnete. Endlich Dr. E. Noelting, ein Freund und späterer Schwager von Witt. Er interessierte sich besonders für Farbstoffe und interpellierte mich sofort über das Rosanilin, über das ich mit dem Vetter Otto eine kleine Publikation gemacht hatte. In einer Sitzung der chemischen Abteilung habe ich meinen ersten wissenschaftlichen Vortrag gehalten. Er war ziemlich kurz und handelte von dem asymmetrischen Diphenylhydrazin, das ich kurz vorher gefunden hatte, und dessen Isomerie mit dem Hydrazobenzol interessante Vergleiche gestattete. Der Präsident der Sitzung, Professor Ladenburg, hat mir dazu einige freundliche Worte der Anerkennung gewidmet.

Von den allgemeinen Sitzungen, die wir ebenfalls besuchten, ist mir nur ein Vortrag des Zoologen Moebius aus Kiel in Erinnerung geblieben, weil er ein chemisches Kuriosum brachte. Es war die Identität des mysteriösen Urschleims (Bathybius Haeckelii) mit amorphem Gyps, der beim Vermischen von Seewasser mit Alkohol ausfällt.

Die Stimmung der Naturforscher, die im Hamburger Volksmund „die forschen Naturen" hießen, war die denkbar beste, gehoben durch das prächtige Herbstwetter, die vielen Vergnügungsgelegenheiten der Seestadt und die flotte Gastfreundschaft der Hamburger Bürgerschaft. Von den Festlichkeiten ist mir in Erinnerung geblieben eine Dampferfahrt nach Blankenese. Bei der Rückkehr hatten wir Gelegenheit, die Ausgelassenheit und Rohheit der Hafenbevölkerung kennen zu lernen. Es war schon halb dunkel und Scharen von halbwüchsigen Burschen vergnügten sich nun damit, unserer Gesellschaft und besonders den Damen kleine, aber doch scharf explodierende Feuerwerkskörper unter die Füße zu werfen. Anderwärts hätte man sich über diesen Unfug entrüstet, die Hamburger aber waren daran gewöhnt und blieben ganz gelassen.

Von Hamburg sind wir drei ohne Unterbrechung an den Rhein gefahren, und ich habe mich dann im Oktober von den beiden anderen für ein Semester getrennt, da ich nach Straßburg ging, um bei meinem früheren Lehrer Professor F. Rose in der analytischen Chemie und ganz besonders im Unterricht der Anfänger Erfahrungen zu sammeln. Angemeldet hatte ich mich natürlich schon während des Sommers und dabei den Wunsch ausgesprochen, als Volontärassistent von Rose in der analytischen Abteilung des Laboratoriums zu Straßburg mitwirken zu dürfen. Der Direktor des Instituts Professor Fittig ließ mir aber sagen, er könne meinen Wunsch nur erfüllen, wenn ich die Stelle eines bezahlten Assistenten annehme. Das habe ich auch getan, mußte nun aber nach einigen Monaten die Überraschung erleben, daß mein schon an und für sich recht bescheidenes Gehalt herabgesetzt wurde. Es war eine Maßregel, die aus Verwaltungsgründen vielleicht gerechtfertigt war, aber es hat doch auf die anderen Angestellten des Instituts, besonders die Diener einen merkwürdigen Eindruck gemacht, weil sie es als eine Degradation für mich ansahen. Ich habe natürlich darüber gelacht, aber es passte zu der Persönlichkeit von Fittig, der durch seine Anordnungen, ohne es zu wollen, andere Leute leicht vor den Kopf stieß. Ich hatte vor Fittig wegen seiner ausgezeichneten Arbeiten die größte Hochachtung, aber bei näherem Verkehr mit ihm merkte ich doch den großen Unterschied im Vergleich zu Baeyer. Trotz seiner scharfen Beobachtungsgabe und seiner Geschicklichkeit im Experimentieren sowie seiner gesunden Kritik, entbehrte er der Genialität und war neuen Ideen, z. B. der Stereochemie und den physikalischen Lehren schwer zugänglich. Obschon ein guter Lehrer für Anfänger, besaß er auch nicht die Fähigkeit, ältere Chemiker an sich zu fesseln und dadurch eine größere wissenschaftliche Schule zu bilden. Von Tübingen her war er gewöhnt, auch den Unterricht in der chemischen Analyse zu erteilen. Das hätte er am liebsten auch in Straßburg getan, wenn nicht die von Baeyer schon getroffene Organisation, die den Extraordinarius Rose zum selbständigen Leiter in der Abteilung machte, vorhanden gewesen wäre. Ich war als Assistent nur in dieser Abteilung beschäftigt. Leider blieb mir dabei eine gewisse Enttäuschung auch nicht erspart. Der Unterricht von Rose, dem ich als Student so viel verdankte, erschien mir jetzt mehr wie eine einseitige, allerdings mit den vielen praktischen Erfahrungen des Bunsen'schen Laboratoriums geschmückte Dressur. Neue Methoden, die in der Literatur erschienen, wurden garnicht versucht und die geringste Abweichung von der Schablone galt als Fehler. Trotzdem ist mir das, was ich in bezug auf Unterricht bei Rose lernte, später sehr zustatten gekommen, als ich selbst die analytische Abteilung des Münchener Laboratoriums übernehmen mußte.

Die Zahl der Studierenden im Straßburger Laboratorium war nicht so groß, daß meine Zeit durch den Unterricht ganz in Anspruch genommen worden wäre. Es blieb mir deshalb die Möglichkeit, nebenher andere Dinge zu treiben. So habe ich mich ziemlich regelmäßig an dem physikalischen Kolloquium bei Kundt beteiligt, und vor allen Dingen konnte ich einige Studien über die Morphologie und Physiologie der niederen Pilze machen. Wie früher schon erwähnt, erhielt ich die Anregung dazu von Dr. A. Fitz, der mir von früher her flüchtig bekannt war, und dem ich jetzt auch persönlich näher treten durfte. Er war ein wohlhabender Weingutsbesitzer aus der Pfalz, schon in reiferen Jahren, unverheiratet und führte damals interessante Versuche über Spaltpilzgärung aus, durch die er sich in der Geschichte der Gärungschemie einen geachteten Namen gemacht hat. Durch ihn lernte ich einige Schriften von Pasteur, vor allem das Buch „Etudes sur la bière" kennen. (Welche praktische Folge das für eine Brauerei in Dortmund gehabt hat, ist früher geschildert). Die Pilzchemie hat mich damals so interessiert, daß ich bei längerem Aufenthalt in Straßburg sicherlich eigene Forschungen auf diesem Gebiete angestellt hätte. Zunächst war es aber für mich nötig, morphologische Kenntnisse zu erwerben und Übung in der Handhabung des Mikroskops und der Pilzkulturen zu bekommen. Die Gelegenheit dazu bot das botanische Institut, das unter Leitung von de Bary stand, der einer der besten Kenner der niederen Pilze in Deutschland war. Mit geringer Mühe habe ich dort eine Reihe von Schimmelpilzen und Hefearten kennen gelernt, die meist auf trockenen Nährböden wie Kartoffeln, Rüben, Möhren gezogen wurden. Es ist mir deshalb immer ein Rätsel geblieben, daß erst Jahre nachher für die Kultur der Spaltpilze Robert Koch die festen Nährböden in die Mykologie einführen mußte, und seitdem als Erfinder dieser Methode angesehen wird. Tatsächlich hat Pasteur die Kultur von Mikroben immer nur in Flüssigkeiten, d.h. in seinem bekannten Kolben angestellt und auch seine Nachfolger wie Fitz usw. waren nicht auf den Gedanken gekommen, den festen Nährboden zu benutzen.

Außerhalb der Laboratorien hat es mir an Unterhaltung und Gesellschaft in Straßburg nicht gefehlt, und als junge Leute durften wir uns auch hier und da kleine Streiche erlauben. So erinnere ich mich eines Spaßes, wobei wir Dr. C. Wurster, der damals in Straßburg sein einjährigfreiwilliges Jahr abdiente, und bei einer nächtlichen Kneiperei den Urlaub überschritten hatte, vor der Festnahme durch eine Patrouille schützen mußten, indem wir ihn auf der Straße rasch in einen Zivilisten verwandelten.

Von neuen Bekanntschaften, die ich in ziemlich großer Zahl machte, will ich nur zwei erwähnen. Einmal den Vorstand der Apotheke am Bürgerhospital Dr. Musculus, ein Freund von Mering. Er war erheblich älter wie wir, ein geborener Elsässer, ein liebenswürdiger Gesellschafter und guter Chemiker. Ihm verdankt man die erste Synthese künstlicher Dextrine aus Traubenzucker, mit denen ich mich 15 Jahre später eingehend beschäftigt habe. Noch viel origineller als Mensch war der Physiker Dr. Fuchs, ein Kneipgenie und nebenher ein Mann von umfassender Bildung. Er hatte ursprünglich Philologie studiert und war dann Mediziner geworden. Nach Abschluß aller dazu nötigen Examinas hatte er kurze Zeit als Nervenarzt praktiziert, war dann aber zur Physiologie übergegangen. Als auch diese Wissenschaft ihm noch nicht exakt genug erschien, wurde er Physiker, und infolge der Unterhaltungen, die er öfters mit mir führte, bekam er Lust, auch noch Chemiker zu werden. Dazu ist es aber nicht mehr gekommen, weil inzwischen sein bescheidenes väterliches Erbe zur Neige ging, und er genötigt war, Geld zu verdienen. Er hat sich später in Bonn habilitiert und ist, wie ich zu meinem großen Vergnügen hörte, zum Schluß als Pseudoleibarzt zu Friedrich Krupp in Essen gekommen. Ich kann mir denken, daß es diesem Schöpfer des größten deutschen Industrieunternehmens am Schluß eines arbeitsreichen Lebens Vergnügen bereitet hat, mit dem vielseitig gebildeten, witzigen und kindlich-naiven Gelehrten zu verkehren.

Im Frühjahr 1877 hatte ich meinen Zweck in Straßburg erreicht und kehrte schleunigst nach München zurück, um die liegengebliebenen Arbeiten über die Hydrazinverbindungen und das Rosanilin fortzusetzen. Dort war inzwischen der Neubau des Instituts mächtig gefördert worden und man bereitete sich schon darauf vor, im Herbst desselben Jahres den Umzug aus dem alten Hause zu bewerkstelligen.

Der Sommer verlief wie sein Vorgänger für uns als eine zweckmäßige Kombination von fleißiger Arbeit und heiterem Leben. Die Zahl der Studierenden und der älteren Chemiker war gewachsen und auch unser gesellschaftlicher Kreis hatte sich dementsprechend vergrößert. Wir waren alle inzwischen gute Münchener geworden und hatten uns dem Bierleben ganz angepaßt. Besonders unterhaltend waren die Abende auf den sogen. Bierkellern, d. h. den Ausschanken der großen Brauereien in hübschen Gärten, die zum Teil auf der Theresienwiese, zum Teil auf dem rechten Isarufer lagen. Hier herrschte ein gemütliches Treiben, eine aus allen Teilen der Bevölkerung zusammengesetzte Gesellschaft erfreute sich an dem köstlichen Bier. Das Abendessen dazu brachte man sich mit, indem man unterwegs in einem Fleischer- und Bäckerladen die nötigen Einkäufe machte. Ganze Familien, von kleinen Kindern bis zum Greise konnte man hier versammelt sehen. Das Bierleben bot auch sonst manche Merkwürdigkeiten. So gab es in gewissen Stadtlokalen eine Bierbettelei, d. h. einzelne Leute, die behaupteten, kein Geld zum Ankauf von Bier zu besitzen zogen mit einem leeren Krug umher, um ihn von mildtätigen Menschen in kleinen Beiträgen füllen zu lassen. Der sonderbarste Auswuchs von Biergemütlichkeit entwickelte sich aber beim Ausschank des Salvatorbiers, das von einer einzigen Münchener Brauerei hergestellt wurde und gewöhnlich in 5 bis 6 Wochen auf dem Salvatorkeller ausgetrunken wurde. Meinen ersten Besuch machte ich dort gemeinschaftlich mit Professor Baeyer. Es war ein fürchterlicher Betrieb. Trotz der kalten Jahreszeit lagen die Menschen dutzendweise betrunken im Garten und an den Abhängen des Berges. Das Hereinkommen in den ungeheuren geschlossenen Raum bot erhebliche Schwierigkeiten, weil fortwährend Ruhestörer oder Betrunkene hinausgeworfen wurden. Als wir schließlich drin waren, wurden wir durch eine Persönlichkeit, die Baeyer kannte, aufgefordert, in einem reservierten Zimmer Platz zu nehmen. Dort war eine sonderbare Gesellschaft versammelt, die man für Leute niederen Standes hätte halten können. Auch die Unterhaltung belehrte darüber zunächst nicht. Ich erinnere mich noch, daß einer der Männer einen alten stark riechenden Käse aus der Tasche zog, mit dem dolchartigen Messer, das die Bayern immer tragen, zerlegte und dann seinen Nachbarn zum Konsum anbot. Auf meine Erkundigung nach seiner Person erfuhr ich, daß es ein hoher Offizier sei. Mein Nachbar, den ich nach dem Äußeren auch sehr unterschätzt hatte, war der bekannte, feinfühlige Dichter Lingg. Und so entpuppte sich allmählich die ganze Gesellschaft als hochgeachtete Männer, Künstler, Gelehrte, Offiziere hohe Staatsbeamte, Großkaufleute usw. Ich bin damals zu der Überzeugung gekommen, daß kein Mittel in der Welt existiert, welches ähnlich dem Bier alle Standes- und sozialen Unterschiede verwischt und die Menschen gleich macht.

Ein großer Vorzug von München ist die Nähe des Gebirges. Man konnte schon damals mit der Eisenbahn in einigen Stunden nach Tegernsee oder in die Nähe des Walchensees gelangen, und wir haben die in

Bayern recht häufigen katholischen Feiertage des Sommers vielfach zu kleinen Touren in die Berge benutzt. In den Pfingstferien, die länger dauerten, führte uns der Weg in der Regel nach Tirol, weil dort Küche und Wein erheblicher besser waren. Großer Bergsteiger bin weder ich, noch die anderen Mitglieder unseres Kreises gewesen, aber auf mittlere Höhen, wie auf Herzogenstand, Wendelstein haben wir uns öfters hinaufgetraut. Daß auch hier durch ungünstige Umstände Gefahr entstehen kann, hat mich die Besteigung des Kitzbüchlerhorns während eines Pfingstausfluges gelehrt. Die Spitze des Berges war noch mit ziemlich viel Schnee bedeckt. Unsere Gesellschaft bestand aus etwa 5 Personen. Glücklicherweise hatten wir einen Träger, der gleichzeitig als Führer diente. Als wir oben angekommen waren, erkrankte Dr. Boesler, von dem später noch die Rede sein wird, an Bergkrankheit so stark, daß er sich nicht mehr auf den Beinen halten konnte. Wir haben ihn in unsere Mäntel gewickelt und in den Schnee legen müssen und uns dann selbst in dem eisigkalten Wind durch starke Bewegung, Freiübungen usw. warm halten müssen. Da die Sorge entstand, daß der Patient nicht mehr imstande sein würde, den Heimweg anzutreten, so richteten wir an den Führer die Frage, was nun zu tun sei. Er erwiderte kaltblütig, daß er den Herrn in den Rucksack stecken und bergabwärts tragen werde. Vor dieser sicherlich wenig bequemen Beförderungsweise blieb aber glücklicherweise Dr. Boesler verschont, weil er sich bald wieder so weit erholte, daß er mit unserer Unterstützung den Rückweg machen konnte.

Derartige Ausflüge haben mich im Laufe der 7-jährigen Münchener Periode fast nach allen schönen Punkten von Oberbayern und Südtirol geführt. Manche davon sind mir noch in Erinnerung. Besonders eine Bergfahrt, die ich zusammen mit dem Physiologen Dr. Tappeiner, dem Zoologen Dr. Graf und einem Professor der Anatomie an der tierärztlichen Hochschule zu München Dr. Frank unternahm. Sie ging zunächst per Eisenbahn über den Starnberger See bis Pensberg, wo sich das einzige Kohlenbergwerk in Bayern befindet, und von dort zu Fuß nach Kochel. Hier gerieten wir in die Herbstmanöver bayerischer Truppen, welche die Gegend gänzlich ausgegessen hatten. Es gab schon kein Brot mehr, wohl aber noch reichlich Bier. Quartier mußten wir auf dem Heuboden in Gesellschaft von Soldaten nehmen, wobei es mir passierte, daß ich nachts durch das Heuloch sanft in den darunter gelegenen Kuhstall hinabfiel. Am Abend vorher hatte sich im Gasthause vor unseren Augen eine charakteristische Szene aus dem bayerischen Volksleben abgespielt. In unserer Nähe saß ein Mann, wahrscheinlich Holzfäller, der, wie später bekannt wurde, den Versuch gemacht hatte, die Kellnerin zu betrügen. Diesem Mann flog plötzlich, ohne daß ein Wortwechsel stattgefunden hatte, vom Ausschank her ein leerer Maßkrug an den Kopf. Gleichzeitig trat ein starker Mann aus derselben Ecke hervor — es war der sogen. Zapfbube — prügelte den Getroffenen mit gewaltigen Schlägen und verschwand dann wieder ganz ruhig in die dunkle Ecke des Ausschankes. Der Geprügelte verließ ebenso ruhig die Gaststube und die Sache war erledigt.

Am nächsten Morgen sind wir, nachdem am Brunnen flüchtig Toilette gemacht war, den Kochelberg hinauf, zum Walchensee und von dort immer per pedes bis nach Partenkirchen. Den Führer der kleinen Gesellschaft machte Professor Frank; denn er hatte vor seiner akademischen Laufbahn lange hier als Tierarzt gewirkt und war deshalb mit Land und Leuten auf das Genaueste vertraut. So kamen wir mit der Bevölkerung in enge Fühlung und ich habe damals einen recht guten Eindruck von dieser frischen, tatkräftigen und sich sehr natürlich und frei gebenden Rasse gehabt. Leider ließ die Küche, namentlich in den Gasthäusern zweiten und dritten Ranges, wohin Frank uns führte, viel zu wünschen übrig und die Folge davon war, daß ich mir schon am vierten Tage in Partenkirchen den Magen und Darm gründlich verdarb und allein nach München zurückkehren mußte.

Eine andere Fahrt, die für mich besser verlief, habe ich zusammen mit Dr. Fluegge gemacht, der jetzt Professor der Hygiene an der Universität Berlin ist. Er war vorher praktischer Badearzt gewesen, hatte sich aber dann der frisch aufblühenden Hygiene zugewandt und war nach München gekommen, um bei Pettenkofer zu arbeiten. Wir haben auf der langen Fußtour, die von Tegernsee über den Achensee nach dem Inntal und von dort über Mittenwald, Partenkirchen, Walchensee und Herzogenstand nach Benedictbeuren ging, uns trefflich unterhalten. Er war ein lustiger Gesellschafter, flotter Fußgänger mit offenen Augen für landschaftliche Schönheiten und Eigentümlichkeiten der Bevölkerung. Auch an wissenschaftlichen Gesprächen hat es nicht gefehlt, und ich habe mich damals bemüht, Fluegge von der Wichtigkeit der Pasteur'schen Untersuchungen und von der Bedeutung der Stoffwechselprodukte der pathogenen Mikroben bei den Infektionskrankheiten zu überzeugen.

Im September desselben Jahres fand die Versammlung der deutschen Naturforscher und Ärzte in München statt. Baeyer hatte die Organisation der chemischen Abteilung übernommen und die Folge davon war, daß ich zusammen mit Dr. Wilhelm von Miller von der technischen Hochschule zum Schriftführer ernannt wurde. Die Geschäftsführung geriet aber in einige Schwierigkeiten, weil Frau Baeyer während eines Ferienaufenthaltes im bayerischen Gebirge gerade zu jener Zeit den Herrn Gemahl mit einem Söhnchen, dem jetzigen Professor der

Physik zu Berlin Otto von Baeyer beschenkte und infolgedessen der Gatte mit einer kleinen Verspätung bei der Versammlung erschien. Die Tagung war für uns Chemiker recht interessant, weil ungewöhnlich viele Fachgenossen zusammengeströmt waren. Ich wurde hier zuerst bekannt mit: Victor Meyer, C. Liebermann, J. Wislicenus, F. Tiemann, C. Scheibler, C. A. Martius und Peter Gries. Dieser war unstreitig die originellste Persönlichkeit in unserem Kreise, wie man nach dem allerliebsten Nekrolog, den A. W. Hofmann ihm später gewidmet hat, gerne glauben wird. Wegen Überfüllung stieß er in München auf allerlei Unbequemlichkeiten, besonders bei den festlichen Veranstaltungen. Er pflegte dann stumm eine Weile in den Wirrwarr hineinzuschauen und hinterher seinem Gefühl im richtigen kurhessischen Dialekt mit den Worten Luft zu machen: „Die Leute verstehen keine Massen zu bewältigen".

In der chemischen Sektion wurde er mit großer Aufmerksamkeit behandelt und auch zum Vorsitzenden gewählt. Hinterher hat ihm die Münchener Universität noch den Titel eines Dr. phil. h. c. verliehen, denn es war gerade die Blütezeit der Benzolchemie und die von Gries aufgefundenen Reaktionen wurden in ausgiebigster Weise zur Lösung von Stellungsfragen benutzt. Zudem hatte eben die Industrie der Azofarben einen großen Aufschwung genommen. Endlich waren Leute, die wie Gries in einem praktischen Berufe stehen und trotzdem in den Mußestunden vortreffliche wissenschaftliche Untersuchungen anstellen, in Deutschland eine große Seltenheit.

Von den Vorträgen der chemischen Sektion ist mir keiner in Erinnerung geblieben, wie denn überhaupt solche Versammlungen für die Publikation von fachwissenschaftlichen Dingen keine große Rolle spielen. Ihre Bedeutung liegt vielmehr in dem persönlichen Verkehr der Teilnehmer, und in dem Austausch von Erfahrungen, die man der öffentlichen Rede oder Abhandlung nicht anvertraut. Dann geben sie auch den jüngeren Gelehrten eine willkommene Gelegenheit, sich den Alten zu präsentieren und ihre Kunst im Vortrag zu zeigen. Ich bin in späteren Jahren häufig auf die Naturforscherversammlungen gegangen hauptsächlich zu dem Zwecke, jüngere Fachgenossen kennen zu lernen. Bei dem zwanglosen Verkehr wird auch der Grund zu mancher Freundschaft gelegt. Von der Münchener Versammlung her datieren z. B. meine freundschaftlichen Beziehungen zu Victor Meyer und F. Tiemann.

In der ersten allgemeinen Sitzung der Tagung erregte ein Vortrag von Haeckel großes Aufsehen, weil er aus den Fortschritten der Biologie weitgehende Schlüsse für das ganze geistige und sittliche Leben der Welt zog, und mit scharfen Angriffen auf Kirche und staatliche Unterrichtsanstalten verband. Wenige Tage darauf, in der zweiten allgemeinen Sitzung antwortete ihm Virchow und führte in interessanter und geschickter Weise die Theorien und Forderungen Haeckels auf das legitime Maß zurück. Man hatte allgemein den Eindruck, daß nur wenig deutsche Gelehrte in der kurzen Frist von einigen Tagen eine so musterhafte wissenschaftliche Kritik in Form einer populären Rede abfassen könnten.

Natürlich fehlte es bei einer so großen Versammlung in München nicht an mannigfachen Festlichkeiten. Unter anderem hatte der Magistrat der Stadt im alten Rathause eine zwanglose Begrüßung veranstaltet. Dabei wurde auch das politische Gebiet gestreift, was sonst bei diesen Tagungen nicht gerade üblich ist. Ein alter bayerischer Professor hieß nämlich die in ziemlich großer Zahl erschienenen Deutschschweizer herzlich willkommen, knüpfte aber daran die politische Aufforderung, sich von dem Welschtum abzuwenden und dem im neuen Reiche geeinten Deutschtum anzuschließen. Das wurde von den Schweizern höflich, doch recht entschieden abgelehnt mit der zutreffenden Bemerkung, daß die Schweiz nach ihrer politischen und kulturellen Struktur, sowie nach der geschichtlichen Entwicklung es für ihre Pflicht ansehen müsse, mit allen benachbarten Völkern und Staaten in einem freundschaftlichen Verhältnis zu bleiben.

In der chemischen Sektion wurden mein Vetter und ich privatim öfters nach dem Stande unserer Arbeit über das Rosanilin gefragt, mußten aber mit einer gewissen Beschämung gestehen, daß wir darin seit einem Jahre nicht vorwärts gekommen seien. Das mag wohl unsere weiteren Bemühungen in dieser Frage beschleunigt haben; denn im darauffolgenden Wintersemester ist uns tatsächlich ihre Lösung geglückt. Nebenher hatte ich die Untersuchung über die Hydrazinverbindungen zu einem gewissen Ende gebracht, so daß sie als zusammenfassende Abhandlung in Liebigs Annalen erscheinen konnte. Die direkte Veranlassung dazu war meine bevorstehende Habilitation als Privatdozent. Ich wurde dazu von Professor Baeyer geradezu gedrängt, weil inzwischen das neue Institut fertig geworden und bezogen war und deshalb das Bedürfnis nach Privatdozenten deutlicher zutage trat. Zudem hatte ein anderer Kollege, Dr. Aronstein seine Absicht kund getan, Privatdozent in München zu werden. Da er aber ein Mann von nur mittleren Fähigkeiten war, so lag Professor Baeyer viel daran, ihn nicht als ersten Privatdozenten mit dem neuen Institut verbunden zu sehen. Um das zu vermeiden, wurde ich vorgeschoben, obschon ich viel lieber noch eine Weile in meiner unabhängigen Stellung als Privatgelehrter geblieben wäre. Meine Habilitation fand statt am Ende des Wintersemesters im Frühjahr 1878. Da die Universität München den anderwärts erworbenen Doktortitel nicht als vollberechtigt anerkannte, so mußte ich zuvor noch eine Prüfung in Form eines Kolloquiums zur sogen. Nostrifikation des Doktortitels ablegen. Als Examinatoren waren bestellt Baeyer und ein zweiter aus der Liebig'schen Zeit herstammender Professor der Chemie Dr. Vogel, ein sehr unbedeutender Mann.

Da für Baeyer, der mich seit Jahren kannte, die Prüfung nur eine Formfrage war, so stellte er scheinbar in vollem Ernste an mich die Frage: „Herr Doktor, können Sie mir einiges über Hydrazinverbindungen mitteilen?" Für den Entdecker der Hydrazine war die Antwort nicht schwer. Und dann stellte Baeyer mit demselben Ernste an den Kollegen Vogel, der offenbar von den Hydrazinen nie etwas gehört und auch meine Habilitationsschrift nicht gelesen hatte, die Frage: „Sind Sie zufrieden?" Das war der Fall. Vogel ergänzte dieses Geständnis noch durch einige ziemlich törichte Fragen, womit die Prüfung schloß.

Sehr viel schwerer waren die Anforderungen für den Habilitationsakt selbst, denn die Fakultät stellte ein Thema, das nach einer Pause von 3 Tagen in freier etwa $^{3}/_{4}$ stündiger Rede behandelt werden mußte. Mein Thema lautete: „Die heutigen Aufgaben der Chemie". Da ich bis dahin niemals einen größeren öffentlichen Vortrag gehalten hatte, so kann man sich denken, daß ich in der 3-tägigen Frist angestrengt arbeiten mußte, um eine brauchbare Rede zustande zu bringen. Dabei ist mir mein gutes Gedächtnis zustatten gekommen, denn ich konnte den Schriftsatz später in der Aula der Universität fast wörtlich ohne Manuskript vortragen. Außer der Fakultät waren natürlich alle Bekannte aus dem Institut dort versammelt und viele davon haben mich hinterher beglückwünscht, daß ich ohne Stocken den Vortrag zu Ende führen konnte. Diesem folgte noch eine Diskussion über von mir aufgestellte Thesen, an der sich aber nur die Mitglieder der Fakultät beteiligten. Dabei hatte ich einen kleinen Zusammenstoß mit dem Senior der Fakultät, dem Mineralogen Kobell, der aber nach einer kleinen Konzession meinerseits zu einer Einigung führte.

Bei dieser Gelegenheit will ich nicht unterlassen darauf hinzuweisen, daß in München die philosophische Fakultät vernünftigerweise in zwei Sektionen geteilt war, und daß für alle Geschäfte, die uns Naturforscher betrafen, nur die mathematisch-naturwissenschaftliche Sektion in Betracht kam. Ich halte das für einen großen Vorzug gegenüber den preußischen Universitäten, wo die philosophischen Fakultäten nicht geteilt sind und deshalb in toto alle Geschäfte erledigen müssen. Das bringt z. B. in Berlin, wo die Zahl der Ordinarien das halbe Hundert längst überschritten hat, eine recht mühsame Geschäftsführung mit sich, die zahllose, langdauernde Sitzungen nötig macht, und führt von

Zeit zu Zeit zu heftigen und sehr überflüssigen Auseinandersetzungen über prinzipielle Fragen zwischen den Vertretern der Natur- und Geisteswissenschaften.

So war ich denn glücklich Privatdozent der Chemie und zwar als Erster der neuen Ära an der Universität München geworden, und im Sommersemester trat die Verpflichtung an mich heran, eine Vorlesung zu halten. Ich wählte als Thema die Teerfarbstoffe und hatte das Glück, eine ziemlich große Anzahl von Zuhörern zu bekommen, weil solche speziellen Vorlesungen damals in München etwas Seltenes waren. Der Vortrag selbst hat mir anfangs rechte Mühe gemacht. Obschon ich an die freie Rede nicht gewöhnt war, habe ich prinzipiell von Anfang an auf die Benutzung eines Manuskripts verzichtet. Das war aber nur möglich, wenn ich vorher den Vortrag vollständig ausarbeitete und ihn dann gut memorierte. Auf diese Weise ist es mir gelungen, schon nach einem Semester eine so große Übung in der freien Rede zu erhalten, daß ich mir später nur noch den Inhalt und die Gedankenverbindung zurecht zu legen brauchte, dagegen die Form in der Vorlesung selbst erfinden konnte. Da man mir häufig gesagt hatte, daß mein Vortrag klar und den Bedürfnissen des Zuhörers angepaßt sei, so glaube ich, das von mir eingeschlagene Verfahren jungen Dozenten empfehlen zu können. Daß man die Materie beherrschen muß, ist ja selbstverständlich, aber das gedankliche Gerippe muß auch dem Redner klar vor Augen stehen. In reiferen Jahren hat es mir auch Freude gemacht, das Mienenspiel der Zuhörer zu beobachten, um herauszufinden, ob der Gegenstand interessiere und ob der Ton richtig gewählt sei. Ferner schien es mir im Eifer der Rede manchmal erlaubt, freie Exkursionen im Nachbargebiete zu unternehmen oder Ideen zu entwickeln, die ich mir selbst von ungelösten Fragen gemacht hatte, und die ich nur als Hypothese darbieten konnte. Wenn der Zuhörer merkt, daß der Redner nicht nur von allgemein anerkannten Dingen, sondern auch von seinen eigenen geistigen Produkten Einiges hergibt, so wird die persönliche Fühlung mit dem Redner enger, die Aufmerksamkeit gespannter und der geistige Gewinn größer.

In der ersten Vorlesung zu München ist mir allerdings der Erfolg dieser Fühlungnahme mit der Zuhörerschaft zunächst nicht beschieden gewesen; denn nach einigen Stunden erschien eine Abordnung bei mir, die in wohlwollendem Ton mir erklärte, daß sie weder den Sinn noch die Worte verstanden hätten wegen allzu starkem Dialekt des Redners. Ich habe über diese Kritik herzlich gelacht und versucht, mich zu bessern, und die Zuhörer haben es mir gelohnt dadurch, daß sie in verhältnismäßig großer Zahl bis zum Ende des Semesters erschienen.

Als zweite Vorlesung habe ich ein viel schwereres Thema gewählt, nämlich ausgewählte Kapitel der theoretischen Chemie. Damals war gerade das bekannte Buch „Moderne Theorien der Chemie" von Lothar Meyer in neuer Auflage erschienen und als glückliche Ergänzung dazu hatte sein Bruder, der Physiker eine ziemlich populäre Darstellung der kinetischen Gastheorie herausgegeben. Diese Bücher habe ich fleißig studiert und versucht, den Inhalt meinen Zuhörern in gedrängterer und noch etwas populärer Form wieder zu geben. Das schien auch zu gelingen, aber in der Folge habe ich einen großen Fehler begangen, indem ich die vorgetragenen Lehren einer Kritik unterzog, um das Bleibende als gesetzmäßig Erkannte zu trennen von allem hypothetischen Beiwerk. Das Ganze lief auf eine Kritik der Atomtheorie nach den damaligen Kenntnissen hinaus. Für mich selbst war dieser Versuch sicherlich sehr belehrend, aber bei meinen Zuhörern habe ich Unheil angerichtet und einige davon konnten bittere Klagen darüber nicht verschweigen: „Wo soll das hin", sagten sie zu mir, „wenn man nach den Darlegungen eines Professors sich solche theoretischen Kenntnisse mühsam erwirbt, und hinterher erfahren muß, daß vieles doch noch zweifelhaft ist". Ich habe daraus die Lehre gezogen, daß man in Vorträgen für Studierende mit der Kritik sehr vorsichtig sein muß und am besten nur Dinge bringt, die als sicherer oder vermeintlich sicherer Besitz der Wissenschaft gelten.

Im Frühjahr 1879 wurde Professor Volhard, der Leiter der analytischen Abteilung, als Ordinarius nach Erlangen berufen und auf Vorschlag von Baeyer bot das Kultusministerium in München mir eine außerordentliche Professur an der Universität an, wenn ich gewillt sei, die Funktionen von Volhard im chemischen Institut zu übernehmen. Ich habe mich gerne dazu bereit erklärt und bin am 1. April desselben Jahres zum außerordentlichen Professor in der philosophischen Fakultät mit einem Gehalt von Mk. 3160.— angestellt worden.

Die Kunde von dieser festen Position und dem ersten Geldverdienst des teuren Sohnes hat in Euskirchen großen Jubel erweckt und mein Vater schrieb sofort einen Glückwunschbrief mit der Bemerkung, daß er mit der Mutter zusammen das Ergebnis mit einer feinen Flasche Wein gefeiert habe. Zugleich erkundigte er sich aber angelegentlich, ob bei der neuen Professur auch auf ein ansehnliches Honorar für Vorlesungen zu rechnen sei, da ihm das Gehalt nicht übermäßig hoch vorkam. Im bayerischen Kultusministerium war man in diesem Punkte allerdings anderer Ansicht; denn der Referent Ministerialdirektor Dr. Voelk hatte mir gesagt, daß man bei diesem glänzenden Gehalt von mir außerordentliche Anstrengung in bezug auf den Unterricht erwarte. Ich habe das alles lachend versprochen, konnte aber die

Bemerkung nicht unterdrücken, daß es für den Chemiker Gelegenheit gäbe, sehr viel mehr Geld zu verdienen, als in der akademischen Laufbahn.

Durch die Übernahme der neuen Stellung, die ich natürlich in erster Linie dem Wohlwollen von Baeyer verdankte, erfuhr meine Tätigkeit im Institut eine radikale Änderung. Bis dahin war ich freier Forscher gewesen, ohne jede Verpflichtung für den Unterricht. Dadurch war es mir möglich gewesen, alle für meine wissenschaftlichen Arbeiten nötigen Experimente allein auszuführen. Nur bei der Rosanilinarbeit war ich mit dem Vetter Otto verbunden, und dieses Zusammenarbeiten hatte sich sehr glatt abgespielt. Von nun an konnte ich in dieser Weise nicht mehr wirtschaften, da der größte Teil meiner Zeit durch den praktischen Unterricht in der chemischen Analyse beansprucht war. Die Hilfe von Assistenten auch für die Privatuntersuchungen wurde unentbehrlich. Einen schwachen Versuch dieser Art hatte ich allerdings schon ein Semester vorher gemacht, als ich Dr. Erhardt einlud, mit mir zusammen die gemischten Azoverbindungen, speziell das Phenyl-Azoäthyl zu bearbeiten. Aber von dieser Hilfe habe ich wenig Freude gehabt, da sie in allen entscheidenden Dingen versagte und ich schließlich das Ganze fast allein machen mußte. Ich bin deshalb mit einem gewissen Zagen an die Wahl neuer Mitarbeiter herangegangen und habe auch das Unglück gehabt, bei der Wahl des ersten Privatassistenten einen Mißgriff zu tun; denn ich ließ mich damals durch Empfehlung bewegen, einen Herrn von auswärts, den ich nicht kannte, Dr. Troschke aus Berlin als Assistenten anzunehmen, war aber froh, ihn nach etwa einem Jahre wieder los zu werden, denn er hat mich eher gehindert, als gefördert. Von da an ist mir das Glück hold gewesen. Fast ausnahmslos sind meine Privatassistenten tüchtige, fleißige und gewissenhafte Männer gewesen, deren Hilfe ich mit wärmsten Dank anerkennen muß. Ihre Reihe beginnt mit Magnus Boesler aus Königsberg, einem vortrefflichen Menschen, der mir zunächst bei der Kaffein-Arbeit geholfen und gleichzeitig seine Doktorarbeit über das Anisoin und Cuminoin unter meiner Leitung ausführte. Bald kam dazu Emil Besthorn aus Frankfurt a. M., ein humoristisch angelegter Herr, den sich deshalb Königs bald für seinen engeren Kneipkreis ausersah. Erheblich größer war die Zahl der Unterrichtsassistenten, die ihren Platz in den großen Arbeitssälen hatten. An ihrer Spitze stand der leider so früh verstorbene Clemens Zimmermann, ein talentvoller und außerordentlich strebsamer Chemiker, auch für den Unterricht in hohem Maße begabt. Er hatte seine Studien unter Volhard absolviert und fing eben mit eigenen Untersuchungen an, als ich in die analytische Abteilung eintrat.

Das periodische System der Elemente kam damals in der anorganischen Chemie immer mehr zur Anerkennung, und so war es natürlich, daß Zimmermann sich Aufgaben zuwandte, die damit im Zusammenhang standen. Dahin gehört die Bestimmung der Dichte des Urantetrachlorids und der spezifischen Wärme des metallischen Urans. Bei besserer Gesundheit hätte er sicherlich die Mineralchemie um viele hübsche Entdeckungen bereichert.

Zwei andere Unterrichtsassistenten, die unter mir ihre Doktorarbeiten anfertigten, waren die Herren Lehnert und Renouf. Der erste hat sich später als Angehöriger des Patentamts in Berlin eine einflußreiche Stellung verschafft und der zweite ist Professor an der John Hopkins Universität in Baltimore geworden. Er war in mancher Beziehung ein Original und durch so große Vergesslichkeit ausgezeichnet, daß ich ihm häufig sagen mußte, welche Präparate in seinen verschiedenen Flaschen und Schalen enthalten seien. Er war einige Jahre älter als ich und mit einer amerikanischen Landsmännin verheiratet, der er alle Sorge um die Behütung der beiden Kinder abnahm, wenn sie Lust bekam, allein einen Ausflug ins Gebirge zu machen. Aber auch er liebte die Berge und hatte einmal das Unglück, bei einer allein unternommenen Besteigung zu stürzen und ein Bein zu brechen. Er blieb 2 Tage dort ohne Hilfe liegen und die Folge war, daß bei der nachträglichen Heilung das Bein eine erhebliche Verkürzung erlitt. Das hat seine Freude am Bergsport aber keineswegs abgekühlt; denn er wurde nun Mitglied des Alpenvereins, schaffte sich Kniehosen an und humpelte eifriger denn je in den Bergen umher.

Zuletzt war auch Krüss, ein Sprosse der bekannten Hamburger Optikerfamilie, in der Abteilung tätig. Er ist ebenso wie Zimmermann frühzeitig zugrunde gegangen und zwar an einer pernitiösen Anämie, die bei jungen Männern in Deutschland außerordentlich selten vorkommt. Ich vermute deshalb, daß es sich mehr um eine chronische Vergiftung handelte, wahrscheinlich durch Schwefelwasserstoff; denn die Ventilation im Institut ließ zu wünschen übrig und bei der großen Zahl von Praktikanten herrschte in der analytischen Abteilung recht häufig eine schlimme Atmosphäre.

Da an der Universität München nur ein einziges chemisches Institut bestand, so wurde dasselbe nicht allein von Chemikern, sondern noch viel mehr von Apothekern und Medizinern in Anspruch genommen. Das galt namentlich für den anorganischen Teil. Zu meiner Zeit betrug die Zahl der Praktikanten, die allerdings meistens nur halbtägig arbeiteten, in dieser Abteilung etwa 150. Da mir die Aufsicht über das Ganze anvertraut war, so konnte ich dem einzelnen Studierenden immer nur einige Minuten widmen und selbst mit dieser Einschränkung dauerte der Rundgang durch beide Säle etwa zwei Tage. Mein Hauptaugenmerk mußte darauf gerichtet sein, die Assistenten zu einer verständigen

Tätigkeit anzuhalten. Außerdem habe ich nicht selten einen größeren Kreis von Studierenden um mich versammelt und ihnen einen kleinen Vortrag gehalten, oder eine Prüfung improvisiert. Besonders die Mediziner waren dafür sehr dankbar. Zimmermann hat dieses System angenommen und mit großer Geschicklichkeit weiter ausgeführt.

Nur eine kleine Zahl der damaligen Schüler ist mir dauernd im Gedächtnis geblieben. Dazu gehören in erster Linie Ludwig Knorr und Reisenegger, von denen noch die Rede sein wird, dann ein Herr Ehrensberger, den ich einige Versuche über die Bestimmung von Arsen in Nahrungsmitteln nach der von mir ausgearbeiteten Methode und über die Bestimmung der Salpetersäure als Stickoxyd anstellen ließ. Der junge Mann hatte durch Verstand, Frische und lebhaftes Interesse für wissenschaftliche Dinge auf mich einen besonders guten Eindruck gemacht, und als er die Absicht kund gab, Gymnasiallehrer zu werden, entgegnete ich ihm lachend: Dafür sei er zu schade, er solle bei der Chemie bleiben, und wenn er zu frühzeitigem Verdienst gezwungen sei, so möge er in die Industrie gehen.

Er ist diesem Rate gefolgt und kam durch Vermittlung eines Onkels in den Dienst der Firma Friedrich Krupp. Hier hat er eine rühmliche Laufbahn gemacht; denn er war später ein sehr einflußreiches Mitglied des Direktoriums und hat wesentlich mit dazu beigetragen, die Stahlfabrikation bei Krupp zur höchsten Leistungsfähigkeit zu bringen. Als ich ihn 30 Jahre nach der Münchener Studienzeit in Essen besuchte, konnte er mir mit aufrichtigem Stolze einen Teil der Fabrikation und vor allen Dingen das imposante wissenschaftliche Laboratorium für chemische und physikalische Untersuchungen zeigen. Ich war ihm schon vorher bei den Vorbereitungen zur Errichtung einer chemischen Reichsanstalt begegnet, wo er nicht allein sein persönliches, sondern auch das Interesse seiner Firma an der Förderung der wissenschaftlichen Chemie stets betonte. Er selbst hat sich jetzt von der industriellen Arbeit zurückgezogen und will auf seinem Landsitz in Traunstein (Oberbayern) die Muße des Alters für astronomische Beobachtungen benutzen. Als Student kam er eines Tages von einer kleinen Reise nach seinem Heimatsorte Berchtesgaden stark zerschunden in München wieder an, und als ich ihn scherzhaft frug, ob er sich an einer Rauferei beteiligt habe, erwiderte er, daß er beim Blumensuchen mit einer jungen Dame einen nicht ganz ungefährlichen Absturz erlitten habe. Als ich bei der Wiedererneuerung unserer Bekanntschaft mich nach dem Schicksal dieser jungen Dame erkundigte, erwiderte er lachend, sie sei seine Frau geworden und habe ihn mit einer großen Anzahl prächtiger Kinder beschenkt.

Ein anderer Lehramtskandidat, G. Brandl, ersuchte mich um ein Thema für eine kleine Arbeit, die er beim Lehrerexamen einreichen wollte. Da ich damals gerade von dem Professor der Mineralogie zu Bonn G. vom Rath ersucht worden war, einige von ihm gesammelte Fluormineralien zu analysieren, so ließ ich durch Brandl das alte Wöhler'sche Verfahren für die Bestimmung des Fluors durch Umwandlung in Siliciumfluorid und dessen Absorption in gewogenen Gefäßen vervollkommnen, und wir konnten dann mit gutem Erfolge die gewünschten Analysen durchführen. Das Resultat ist in einer Abhandlung der bayerischen Akademie der Wissenschaften publiziert. Dieser Herr Brandl hat mir später, ohne daß ich es wußte, einen wertvollen Gegendienst geleistet, indem er mich dem Führer der ultramontanen Landtagsmehrheit in München Dr. Daller bestens empfahl und dadurch die Bewilligung eines Neubaues für das chemische Institut zu Würzburg stark beeinflußte.

Auch unter den Medizinern, die ich damals unterrichtete, war einer, der durch Begabung und Interesse für chemische Arbeiten meine besondere Aufmerksamkeit erregte. Er hat sein Verständnis für chemische Probleme später durch eine beachtenswerte Untersuchung über die Mucine bewiesen und ist jetzt der berühmte Professor der inneren Medizin an der Universität München, Friedrich Müller. Für die eigenen Untersuchungen stand mir ein sehr schönes Privatlaboratorium mit Nebenräumen zur Verfügung, in der Größe genau entsprechend den Räumen, die Baeyer benutzte. Hier habe ich alle freie Zeit, die mir der Unterricht ließ, zugebracht, nicht selten auch die Sonntage, wo man im Winter wegen Stillstand der Zentralheizung tüchtig frieren mußte. Dagegen waren die Abende meistens der Erholung gewidmet; denn wer den ganzen Tag über experimentell tätig gewesen ist, pflegt abends zu müde zu sein, um literarische Studien zu treiben.

Die Stadt und die mit ihr eng verknüpfte Behaglichkeit des Lebens lud auch gebieterisch zu Geselligkeit und lustiger Unterhaltung ein. Eine Zeitlang hatten wir sogar im kleinen Kreise die englische Tischzeit angenommen und speisten als Quartett, d. h. Dr. Königs, Dr. Tappeiner, mein Vetter und ich, in dem als vornehm verschrieenen Restaurant Schleich. Die Gesellschaft außer uns bestand aus zwei serbischen Prinzen und dem Präsidenten der bayerischen Reichsratskammer. Wir wurden wegen dieser Gewohnheit als Protzen angesehen. In Wirklichkeit aber war es eine Maßnahme der Zeitersparung; denn wir blieben den ganzen Tag meistens von morgens 8 Uhr bis $5^1/_2$ Uhr im Institut, und ich half mir über den Hunger mit einem Butterbrot oder durch vermehrten Tabakgenuß hinweg. Auf die Dauer war aber diese Zeiteinteilung doch zu anstrengend, und wir sind zu der in Deutschland üblichen Mittagsstunde zurückgekehrt. Ja, ich habe diese Gewohnheit sogar in Berlin beibehalten; denn wer mit anstrengender

Experimentalarbeit, die meist im Stehen verrichtet wird, Morgens zwischen 8 und 9 Uhr beginnt, hat um 1 oder 2 Uhr das Bedürfnis, eine Ruhepause zu machen, die am besten mit der Hauptmahlzeit verbunden wird. Dann kann man von neuem 4 bis 5 Stunden arbeiten, und das ist nach meiner Erfahrung für den Chemiker die beste Ausnutzung des Tages, da man in der Mittagszeit eine Reihe von Operationen gehen lassen kann, die lange dauern und keine besonders sorgfältige Beaufsichtigung verlangen.

Der Wechsel der Tischzeit brachte für mich auch einen Wechsel der Tischgesellschaft mit sich. In dem Verein der Privatdozenten, der sich von Zeit zu Zeit in behaglichen Zusammenkünften betätigte, hatte ich zahlreiche Vertreter anderer Wissenschaften kennen gelernt und so bin ich zu einer neuen Gesellschaft im Künstlerhause gekommen. Zu ihr gehörten der Philosoph Jodl, der später Professor in Wien war und sich einen guten Ruf in seiner Wissenschaft verschafft hat, dann die beiden Historiker Stieve und von Druffel, beide Westfalen. Auch diese waren wohl unterrichtete und wissenschaftlich verdiente Männer, deren Anschauungen und Ziele mich in mancher Beziehung interessiert haben. Zudem besaß Stieve ein wunderbares Talent, humoristische Reden (sogen. Bierreden) in beliebiger Länge zu halten. Sein sarkastischer Witz und seine Neigung, den alten Professoren etwas am Zeuge zu flicken, trugen viel zur Belustigung unseres jungen Kreises bei. Wir waren alle unverheiratet, bis Stieve sich eines Tages mit einer Bonnerin verlobte und uns dann zuweilen einlud. Durch die beiden Historiker bin ich auch mit der Malerfamilie Kaulbach in Berührung gekommen, wo es höchst lustig zuging. Als ich mich später in Würzburg verheiratet hatte, war der erste, der auf dem Plan erschien, Freund Stieve, um einmal nach dem Rechten zu sehen, wie er sich ausdrückte. Sie sind jetzt alle drei tot.

Zeitweise habe ich auch, um ganz einfache zuträgliche Nahrung zu bekommen, wieder in einem Bierhaus gegessen, dem sogen. Abentum, wo es gut und billig war, und wo Hunderte von jungen Leuten, besonders viele Künstler, verkehrten. An unserem Tisch war eine kunterbunte Gesellschaft versammelt, Künstler, Gelehrte, Techniker, junge Kaufleute, ja es befand sich sogar ein echter deutscher Prinz darunter. Hier bin ich in nähere Berührung mit zwei Kunstgelehrten gekommen. Der eine war der Privatdozent der Archäologie Dr. Julius. Sein Eifer, die Chemiker in kunstverständige Männer zu verwandeln, steigerte sich soweit, daß er für einen kleinen Kreis von uns in der Glyptothek eine Privatvorlesung hielt. Das hatte sogar praktische Folgen. Ein Künstler, der dazu gehörte, entschloß sich, eine berühmte, aber stark beschädigte Nike zu restaurieren, und ich selbst bin indirekt dadurch mit dem Gipsformator der Museen in Berührung gekommen. Er bat mich um ein Mittel, schwarze Flecke, die spontan beim Aufbewahren seiner Gipsfiguren entstanden, zu beseitigen. Ein Mittel dafür fand ich in der Wirkung des Chlorgases, welches die Flecken, die von Pilzen herrührten, zerstörte, ohne die Figuren sonst zu beschädigen. Der Formator hat das Verfahren im großen Maßstabe angewandt, wozu er sich eine kleine Bude für die Operationen mit dem Chlorgas errichten mußte. Wie lange das Verfahren im Gebrauch geblieben ist oder ob es später von einem besseren ersetzt wurde, vermag ich nicht zu sagen.

Der zweite Kunstgelehrte an dem Biertisch war Dr. Dehio, ein Balte aus Reval. Er ist später durch seine großen und prächtig ausgestatteten Werke über die Geschichte der Architektur ein berühmter Mann geworden, war zuletzt Professor in Straßburg, lebt aber seit einigen Jahren im Ruhestand. Es hat mich sehr gefreut, ihn im letzten Herbst 1917 in Baden-Baden nach langer Pause wiederzusehen und zu erfahren, daß seine Gemahlin eine Schwester von Paul Friedländer sei. Die kriegerischen Ereignisse im Osten hatten ihn so stark bewegt, daß er sich gerade damit beschäftigte, eine öffentliche Propaganda für die Loslösung des Baltenlandes von Rußland und seinen Anschluß an Deutschland ins Werk zu setzen.

Durch den vielfachen Verkehr mit Künstlern und Kunstgelehrten habe ich mehr und mehr Verständnis für die bildende Kunst gewonnen, und mehrfache Reisen nach Italien waren wohl geeignet, diese Einflüsse zu verstärken. Äußerlich gab sich das kund sowohl bei mir, wie auch bei anderen Mitgliedern unseres chemischen Kreises durch fleißigen Besuch der Kunstausstellungen, die im Glaspalast, der direkt neben unserem Institut lag, veranstaltet wurden, und in den letzten Jahren des Münchener Aufenthaltes bin ich auch ziemlich regelmäßig Sonntags nachmittags in den Ausstellungsraum des Künstlervereins unter den Arkaden des Hofgartens gegangen. Man wurde dort recht gewahr, welch' günstige Atmosphäre in München für die Künstler herrschte. Der größte Teil der gebildeten Einwohnerschaft interessierte sich für ihre Werke, nahm teil an ihrer Entwicklung und ihren persönlichen Schicksalen, vermied es aber, sie durch übertriebene Lobhudelei oder durch protzenhafte Heranziehung zu Diners und dergl. zu verderben. Auch das Leben, das die Künstler untereinander führten, bot mannigfache Vorteile. Im allgemeinen hörte man die Lehrer loben, wenn auch die Ansichten der Jungen in bezug auf die Ziele und Mittel der Kunst nicht selten ganz andere waren. Die realistische und in der Freilichtmalerei häufig bis zur Karrikatur übertriebene Malweise, die, wenn ich nicht irre, von Frankreich herkam, begann gerade in München Fuß zu fassen, und ich erinnere mich noch des Aufsehens, das ein Bild des jugendlichen Max Liebermann, „Christus im Tempel", auf einer Ausstellung in München erregte. Bei denjenigen Leuten, die am alten System festhielten, galt es für eine Verirrung, und von kirchlich gesinnten Organen wurde es als eine Blasphemie bezeichnet. Aber eine große Zahl der jungen Künstler und von ihnen wohl nicht unbeeinflußt auch wir jungen Chemiker sahen darin das ernste Bestreben, die Malerei mehr der Wirklichkeit anzupassen und die geistige Durchdringung des Gegenstandes nicht in einer übertriebenen Idealisierung der Figuren zu suchen.

Mit Recht berühmt waren in München die öffentlichen, von der gesamten Künstlerschaft veranstalteten Festlichkeiten. Ich habe zwei davon besucht, ein Feldlager aus der Zeit des 30-jährigen Krieges, das in dem Walde von Großhesselohe sich abspielte und ein köstliches Bild der bewaffneten Macht vor 400 Jahren gab. Das zweite war ein Maskenfest in München, das leider infolge eines Brandunglücks einen tragischen Abschluß erfuhr. Einer der größten Säle Münchens war durch außerordentlich geschickte Ausbauten in eine Art Jahrmarkt verwandelt, wo man alles genießen konnte, was an Sehenswürdigkeiten bei solcher Gelegenheit geboten zu werden pflegt. Da der Zutritt nur Herren gestattet wurde, die natürlich zum großen Teil durch Masken in Frauen jeden Kalibers verwandelt waren, so hatte die künstlerische Phantasie sich ohne jede Einschränkung bis in die wunderlichsten Auswüchse austoben können. Ich habe nicht einmal in Cöln solche tolle karnevalistische Ausgelassenheit gesehen wie bei diesem Feste, an dem die Künstler bis ins höchste Alter hinein ziemlich vollständig versammelt waren. Eine köstliche Maskenidee in vortrefflicher Ausführung ist mir in Erinnerung geblieben. Der bekannte Maler Piglheim machte einen Prinzen natürlich mit Gefolge und ließ sich von dem Komitee rundführen, selbstverständlich ganz in der Form, wie sie bei solchen prinzlichen Besuchen üblich ist. Kurze Zeit darauf erschien ein wirklicher bayerischer Prinz, der genau ebenso empfangen wurde. Zum großen Spaß der ganzen Versammlung stießen nun die beiden Züge zusammen. Der wirkliche Prinz hatte aber Humor genug, den Pseudokollegen freundlich zu begrüßen und sie setzten zusammen die Besichtigung fort. Leider wurde das Fest in ziemlich später Stunde durch ein furchtbares Unglück gestört. Etwa ein Dutzend Schüler der Akademie der Künste machten eine Eskimogruppe. Sie hatten sich für den Zweck eine besondere Hütte gebaut und Eskimoanzüge angelegt, die ganz aus Werg hergestellt waren. Leichtsinnigerweise hatte man versäumt, diesen Stoff chemisch zu imprägnieren und dadurch unbrennbar zu machen. Auch die Hütte selbst war mit brennbaren Stoffen jeder Art erfüllt. Mit ähnlichem Leichtsinn war übrigens die Mehrzahl der

Bauten errichtet und da überall geraucht wurde, so hatten wir Chemiker sofort das Gefühl der höchsten Feuersgefahr, und ich erinnere mich, daß wir uns ziemlich frühzeitig in einen geschützten Bierkeller zurückzogen und nur von Zeit zu Zeit wieder das Treiben im Hauptsaale uns anschauten. Plötzlich hieß es, die Eskimos sind am Brennen, und in der Tat sahen wir diese armen Menschen brennend herumlaufen. Sie wurden zwar außerhalb ihrer Hütte ziemlich rasch gelöscht, und einige davon glaubten sich nach der überstandenen Gefahr durch einen reichlichen Trunk Bier entschädigen zu sollen, aber die Verwundungen waren doch so schwer, daß sie meines Wissens in den nächsten Tagen alle gestorben sind. Glücklicherweise konnte das Feuer in der Eskimohütte gelöscht werden. Hätte es um sich gegriffen und die benachbarten sehr brennbaren Gegenstände erfaßt, so wären viele Hunderte von Menschen umgekommen; denn die Zugänge zum Saal waren sämtlich durch die Ausbauten so beengt, daß die große Menschenzahl unmöglich sich hätte retten können.

Bei allen solchen Festlichkeiten sollten die Aufsichtsbehörden in rigoroser Weise stets verlangen, daß alle zur Bekämpfung einer etwaigen Feuersgefahr nötigen Maßregeln getroffen sind. Das Publikum ist geneigt, in derartigen Anordnungen der Behörde Willkür und kleinliche Belästigung zu erblicken; aber wer einmal ein Brandunglück wie damals auf dem Maskenfest in München erlebt hat, und wer wie wir Chemiker die rasche Entwicklung eines Brandes kennt, der wird meiner Ansicht gerne beipflichten.

Neben der bildenden Kunst spielten in München auch Musik und Theater eine große Rolle, und ich muß gestehen, daß ich viel Genuß von beiden gehabt habe.

Das Klavierspiel hatte ich leider schon in Straßburg vernachlässigt, und in München wagte ich nicht, wieder anzufangen, weil in unserem Hause ein Professor der Musik wohnte, der als Virtuose galt. Dazu kam das Bewußtsein, daß man in großen Mietshäusern durch mittelmäßiges Spiel vielen Leuten zur rechten Plage werden kann. Ich zog es deshalb vor, gute Musik von Zeit zu Zeit anzuhören durch den Besuch der ausgezeichneten Odeonkonzerte oder der königlichen Oper. Diese verfügte damals über vorzügliche Kräfte. Besonders gepflegt wurde die Wagnersche Musik, und obschon die alten klassischen Werke von Mozart, Beethoven, Maria von Weber, Lortzing usw. meinem Verständnis näher lagen, so haben mir doch die großen Opern Wagners z. B. der Ring der Nibelungen in der vortrefflichen Münchener Darstellung gewaltigen Eindruck gemacht. Dieser steigerte sich noch bei einem Besuch in Bayreuth, den ich mit mehreren Freunden später von Erlangen aus unternahm, und wo wir einer Aufführung des Parsival beiwohnten.

In München glänzte damals in den Wagner-Opern das Ehepaar Vogel am meisten, aber auch den alten Kindermann mit seinem unverwüstlichen kräftigen Bariton, der mit dem Münchener Publikum wie mit alten Bekannten verkehrte, wird kein Zuhörer vergessen haben.

Noch mehr Genuß als von der Oper habe ich gehabt vom guten Schauspiel, wie es ebenfalls in München gepflegt wurde. Es bestand an der Münchener Bühne die gute Gewohnheit, von Zeit zu Zeit eine Reihe von Dichtungen desselben Meisters zu ermäßigten Preisen aufzuführen. Eine solche Serie, welche die Königsdramen Shakespeares umfaßte, war in ihrer Wirkung so gewaltig für uns junge Leute, daß darüber die Experimentalarbeit im Laboratorium für mehrere Wochen eine ernstere Störung erhielt. Ich wurde von manchen Szenen so ergriffen, daß mir die hellen Tränen über das Gesicht flossen, und daß ich die Umgebung gänzlich vergaß. Diese starke Empfindung für dramatische Reize ist mir bis ins reifere Alter geblieben, und meine spätere Frau hat mir mehr als einmal gesagt, mein Mangel an Selbstbeherrschung im Theater sei für sie direkt unbequem.

Bei den Besuchen von Theater, Konzerten und ähnlichen Lokalen, wo die Menschen dicht zusammen sitzen, kann es dem Chemiker leicht passieren, daß er Anstoß erregt durch die schlechten Gerüche, die sich in seinen Kleidern oder auch in Haar und Haut festgesetzt haben. Auf den billigen Plätzen, die wir als junge Leute besuchten, ist mir das nie verübelt worden, aber einen Parkettplatz in der Münchener Oper habe ich einmal aufgeben müssen, weil die Nachbarschaft zu stark durch die von mir ausströmenden Dünste belästigt wurde. Auch in Privatgesellschaft sollte der Chemiker auf diese Möglichkeit achten und falls er gerade mit starkriechenden Stoffen zu tun hat, auf die Reinigung des Körpers und des Anzugs besondere Sorgfalt verwenden.

Leider kam der Sport bei der Münchener Lebensweise zu kurz. Wir hatten dazu keine Zeit und waren auch abends zu müde. Einen gewissen Ausgleich boten allerdings die Herbstferien, die ich teilweise stets in Euskirchen zubrachte und zusammen mit meinem Vater fleißig für die Feldjagd ausnutzte.

Im geselligen Verkehr, soweit er sich in der Familie abspielte, habe ich niemals viel geleistet, teils aus Bequemlichkeit, teils wegen einer gewissen Ungeschicklichkeit im Verkehr mit Damen. Gelegenheit dazu war übrigens in München genug gegeben; denn Frau Baeyer führte ein großes Haus, veranstaltete von Zeit zu Zeit hübsche Feste und gab sich die größte Mühe, uns junge Chemiker dazu heranzuziehen. Sie spielte bis zu gewissem Grade überhaupt die Mutter des Laboratoriums und war stets zur Hilfe bereit, wenn jemand von uns durch Krankheit oder andere Ursachen in Not geriet. Im Hause bei ihr habe ich auch eine große Anzahl interessanter Menschen kennen gelernt. Vor allen Dingen lud sie uns immer zu Tisch, wenn auswärtige Chemiker zu Besuch kamen. Nach Art hilfreicher und tatkräftiger Damen hätte Frau Baeyer uns auch gar zu gerne verheiratet, worauf sie um so mehr glaubte rechnen zu dürfen, als sie einen Kreis von hübschen und liebenswürdigen jungen Damen um sich zu versammeln wußte. Soweit ich unterrichtet bin, ist es ihr aber nicht gelungen, außer der eigenen Tochter eine dieser Fräuleins an einen chemischen Mann zu bringen. Die junge um Frau Baeyer versammelte Chemie war damals durchgängig unbeweibt, und es sind auch auffallend viele von ihnen in diesem Zustand geblieben, nicht zum Nutzen ihrer eigenen Person oder der Wissenschaft.

Meine Arbeiten über die Hydrazine und das Rosanilin hatten auswärts Anerkennung gefunden. Das zeigte sich zuerst in einem Ruf, den ich im Frühjahr 1880 an das Polytechnikum zu Aachen erhielt. Landolt war von dort an die Landwirtschaftliche Hochschule in Berlin berufen worden und an seiner Stelle sollten zwei Ordinarien für anorganische und organische Chemie ernannt werden.

Die anorganische Lehrstelle mit der großen Experimentalvorlesung hatte man nach dem Beschlusse des Aachener Lehrerkollegiums dem langjährigen Assistenten des Instituts Professor Classen zugedacht und die Professur für organische wurde mir vom Kultusministerium in Berlin angeboten. Um mich genau über die Stelle zu unterrichten, reiste ich nach Aachen hin umd verhandelte mit den Fachgenossen Professor Stahlschmidt und Professor Classen, sowie mit dem Direktor des Polytechnikums Herrn von Kaven. Selbstverständlich besuchte ich auch Landolt, der mich im Bett empfing, weil er an starkem Podagra litt. Für das chemische Institut war gerade ein prächtiger Neubau errichtet worden, weil die Aachen-Münchener Feuerversicherungsgesellschaft große Mittel zur Verfügung gestellt hatte. Aber die Betriebsmittel waren verhältnismäßig gering. Man wollte mir in dieser Beziehung in Aachen keine Sicherheit gewähren, und die ganze Atmosphäre an dem Polytechnikum erschien mir so wenig anziehend, daß ich trotz des lebhaften Wunsches meiner Mutter, mich wieder in der Nähe zu haben, den Ruf kurzweg ablehnte. Nun erlebte ich das Höchstsonderbare, daß nach etwa sechs Wochen der Ruf von Berlin aus erneuert wurde und daß der betreffende Referent im Kultusministerium Dr. Wehrenpfennig mir schrieb, ich habe gewiß die Absicht des Ministeriums nicht richtig verstanden. Wie ich später erfuhr, hatte er inzwischen mit seinem Freunde Dr. von Brüning aus Frankfurt a. M. über meine Berufung gesprochen und sich beklagt, daß ich den schönen Ruf in so kühler Weise abgelehnt habe. Als dieser ihm aber erwiderte, daß ich größere Ansprüche machen könne, als man mir geboten habe, entschloß man sich zu einem zweiten Versuch, mich zu gewinnen. Ich wurde dann auch in Berlin sowohl von Wehrenpfennig wie von Landolt, der inzwischen übergesiedelt war, in der liebenswürdigsten Weise empfangen und man versprach, allen meinen Wünschen entgegenzukommen. Aber die Abneigung gegen Aachen war bei mir einmal vorhanden, und da auch Baeyer mir nicht im geringsten zur Übersiedlung an das Polytechnikum riet, so habe ich zum zweiten Male abgelehnt und diesen Entschluß auch keinen Tag bereut. Statt meiner ist dann Professor Michaelis dorthin gekommen.

Zwei Jahre später wurde die chemische Professur an der Universität Erlangen wieder frei, weil Volhard als Nachfolger von Heinz nach Halle ging. Ich erhielt den Ruf und nahm ihn an. Der Ort hatte zwar wenig Anziehendes, aber es handelte sich doch um ein Ordinariat an einer Universität und das Institut war durch einen Neubau verhältnismäßig gut ausgestattet.

Ich bin bei einem Erkundigungsbesuch in Erlangen auch von den Kollegen sehr freundlich empfangen worden, und Volhard gab sich alle Mühe, mir die Vorteile der neuen Stellung klar zu machen.

Bevor man mich in München entließ, mußte ich natürlich einige Festlichkeiten über mich ergehen lassen. So veranstaltete die Chemische Gesellschaft, deren Präsident ich war, einen schlichten Abschiedsabend, für den Königs das früher erwähnte Guanolied dichtete. Aber auch die Studierenden wollten sich nicht die Gelegenheit entgehen lassen, einen Kommers abzuhalten. Nach den Erfahrungen gelegentlich des Rufes nach Aachen, wo ich von betrunkenen Chemikern beinahe Prügel erhalten hatte, war ich nicht sehr geneigt, mich dieser Möglichkeit nochmals auszusetzen und erst die Drohung, daß ich durch Ablehnung die chemische Jugend, mit der ich auf sehr gutem Fuße stand, vor den Kopf stoßen würde, machte mich williger. So fand denn wirklich ein Kommers auf einem Bierkeller statt, der für mich ohne unbequemen Zwischenfall verlief, weil ich mich frühzeitig empfahl. Aber am nächsten Morgen mußte ich doch erfahren, daß man einige junge Leute, die mit mir nach Erlangen gehen wollten, feierlich aus dem Lokal hinausgeworfen hatte.

Ich zog nun anfangs April an die fränkische Hochschule und der Zufall wollte, daß in Nürnberg in dasselbe Abteil des Schnellzuges, wo ich saß, Fräulein Agnes Gerlach, meine spätere Frau, mit ihrem Vater einstieg, und wir also zusammen in Erlangen ankamen.

Die Einquartierung in Erlangen war nicht ganz leicht. Zunächst fehlte es an einer passenden Wohnung, und auch die Gasthäuser der armen Stadt boten kein bequemes Unterkommen. Ich habe mich deshalb zuerst in dem geräumigen Sprechzimmer des chemischen Instituts niedergelassen, wo ein sehr brauchbarer Laboratoriumsdiener namens Griesinger für meine Sachen sorgte. Mit der Beköstigung war ich allerdings auf das Gasthaus angewiesen, wo ein größerer Kreis von Professoren und Privatdozenten sich zu einer Tischrunde vereinigt hatte. Da ich inzwischen aber zehn Jahre Gasthausleben hinter mir hatte, so fühlte ich das Bedürfnis, einen eigenen Haushalt zu beginnen. Zu dem Zweck hatte ich mir schon in München die Möbel bestellt. Um mich dabei vor Übervorteilung zu schützen, war meine Schwester Emma mit ihrer Tochter Hedwig eigens von Rheydt nach München gekommen und von Frau Baeyer, bei der sie Besuch machten, sofort als Logiergäste aufgenommen worden. Das Wohnungsprovisorium in Erlangen dauerte nicht allzu lange. Nach etwa zwei Monaten trafen alle Möbel ein und mit ihnen auch meine erste Haushälterin, eine Witwe aus Westfalen, die von meiner Schwester Bertha und meiner Mutter ausgesucht worden war. So zog ich denn in die erste mit einem selbständigen Haushalt eingerichtete Wohnung in eines der wenigen vornehmen Häuser Erlangens, die früher den Beamten des markgräflichen Hofes gedient hatten. Die Wohnung enthielt einen großen Saal mit Marmorfußboden und Stuckwerk im Rococostil. Er wird später noch bei der Besprechung meiner Arbeiten zu Ehren kommen; denn ich habe darin die Riechversuche mit Mercaptan ausgeführt.

Inzwischen hatten die Arbeiten im Laboratorium und die Vorlesungen begonnen. Zum Glück waren mir von München zwei prächtige junge Männer, Knorr und Reisenegger, gefolgt, die zwar noch nicht promoviert hatten, denen ich aber ohne Bedenken Assistentenstellen übertragen konnte. Wir sind rasch Freunde geworden, und da es mir zu langweilig war, allein zuhause zu essen, so habe ich diese beiden jungen Herren eingeladen, wenigstens mittags meine Tischgenossen zu sein. Sie haben das gerne getan, weil meine Haushälterin als Köchin Vortreffliches leistete und auch an gutem Wein kein Mangel herrschte.

Beide hatten ihre analytische Ausbildung unter meiner Leitung in München erhalten, und Knorr war auch schon mit der Anfertigung einer Dissertation über das Piperylhydrazin beschäftigt, als wir in Erlangen einzogen. Reisenegger hat bald nachher von mir als Thema für die Dissertation „Die Reaktion zwischen Phenylhydrazin und den Ketonen" erhalten. Beide wurden in Erlangen mit gutem Erfolg promoviert. Reisenegger ist noch einige Zeit als Assistent geblieben, ging aber dann in die Industrie und zwar zu Meister Lucius & Brüning in Höchst a. M. Er stammt aus Oberbayern, wenn ich mich recht erinnere, aus Murnau, wollte ursprünglich Offizier werden und hatte das Kadettenhaus in München absolviert. Infolge eines körperlichen Schadens war er aber zur Chemie übergegangen. Er ist in Höchst durch alle Stufen des technischen Chemikers bis zum Mitglied des Direktoriums hindurchgegangen und jetzt im reiferen Alter vor einigen Jahren als Professor der Technologie an der technischen Hochschule zu Charlottenburg der Nachfolger von Witt geworden.

Länger ist Ludwig Knorr bei mir geblieben, da er von vornherein sich für die akademische Laufbahn entschied. Er entstammt einer Münchener wohlhabenden Kaufmannsfamilie aus dem bekannten Geschäft von Angelo Sabadini in der Kaufingerstraße und war schon in der Jugend nicht allein durch Klugheit, experimentelle Geschicklichkeit und Redegewandtheit, sondern auch durch besondere persönliche Liebenswürdigkeit ausgezeichnet. Nebenher huldigte er allem möglichen Sport, tanzte vorzüglich und spielte schon während der Münchener Zeit in der dortigen Gesellschaft z. B. auch im Hause Baeyers eine bedeutende Rolle. Unter anderem gelang es ihm trotz seiner Jugend und seiner noch unfertigen Laufbahn, das hübsche, viel umworbene Fräulein Elisabeth Piloty, Tochter des großen Historienmalers in München zu gewinnen. Diese wurde schon im Frühjahr 1884 seine Gattin. Das Ehepaar ist dann später mit mir nach Würzburg übergesiedelt, und da ich von ihnen bis heute viel herzliche Freundschaft erfahren habe, so werde ich später auf sie zurückkommen.

In Erlangen trafen wir im Laboratorium den Privatdozenten der Chemie Dr. Vongerichten, einen geborenen Pfälzer, der vorher eine Zeitlang im Baeyer'schen Laboratorium gearbeitet hatte und mir deshalb gut bekannt war. Es hat mich gefreut, ihm einen Platz in dem hübsch eingerichteten Privatlaboratorium überlassen zu können, wo er seine wichtigen Versuche über die Verwandlung des Morphins in Phenanthren ausführte und damit für die Chemie des Alkaloids eine neue Periode eröffnete. Auch zu ihm bin ich in ein recht freundschaftliches Verhältnis getreten. Aber nach etwa drei Semestern verließ er die Universität und trat gleichfalls in die chemische Fabrik von Meister Lucius & Brüning in Höchst a. M. ein, wo er zunächst im wissenschaftlichen Laboratorium Verwendung fand. Seine Anstellung hat sich für die Fabrik sehr gelohnt, denn sie ist dadurch indirekt in den Besitz des Antipyrinpatentes gelangt.

Studierende aus der früheren Zeit waren sehr wenig im Institut vorhanden, da Volhard, dem man allzu große Strenge in den Examinas nachsagte, nicht anziehend gewirkt hatte.

Die Abhaltung der großen Experimentalvorlesungen hat mir zuerst ziemlich viel Mühe gemacht; denn die zahlreichen Versuche, die dabei unentbehrlich sind, erfordern sehr genaue Vorübung, sowohl für den Professor wie für den Assistenten, und obschon das gute Buch von K. Heumann über Vorlesungsexperimente schon existierte, so habe ich es doch für nötig gehalten, noch ein besonderes derartiges Buch für meine Vorlesungen mit allen Einzelheiten anzulegen. Das ist dann mitgezogen nach Würzburg und Berlin und selbstverständlich im Laufe der Zeit immer reichhaltiger geworden. Auch im Laboratorium lag der ganze Unterricht auf meinen Schultern, da mir nur junge und wenig geübte Assistenten zur Verfügung standen. Dazu kamen noch die Prüfungen und die Fakultätsgeschäfte und außerdem selbstverständlich die eigenen Untersuchungen.

Leider waren die technischen Einrichtungen des Instituts, vor allen Dingen die Ventilation ungenügend und sowohl ich, wie meine Assistenten haben besonders bei den häufigen Arbeiten mit Chlorphosphor darunter sehr gelitten. Ich werde auf diese Ventilationsfrage bei der Besprechung des Würzburger und Berliner Neubaues ausführlich zurückkommen.

Im Laufe von einigen Semestern wurde der Besuch des Laboratoriums wieder befriedigend. In der analytischen Abteilung hatten sich ziemlich viele Apotheker, einige Mediziner und auch Chemiker eingefunden und die kleine organische Abteilung war schon nach zwei Semestern so überfüllt, daß ich einen außer Betrieb gesetzten Teil des alten Gebäudes wieder in Benutzung nehmen mußte.

Von den neu hinzugekommenen Chemikern verdienen drei besondere Erwähnung, Dr. Ernst Täuber, ein Schlesier, der mir als Privatassistent bei der Bearbeitung der Acetonbasen geholfen hat; dann Kužel, der zuletzt Unterrichtsassistent in der analytischen Abteilung war, und mit dem ich über die Hydrazine der Zimtsäure, das Indazol und Benzoylaceton gearbeitet habe; endlich Julius Tafel, der erst die Isomerie von Indazol und Isindazol untersuchte und in Würzburg an meinen Zuckerarbeiten teilnahm.

Täuber kam später an das technologische Institut zu Berlin und ist auch Mitglied des Patentamtes geworden. Kužel, ein sowohl körperlich wie geistig bevorzugter Mann, hat in der Fabrik von Meister Lucius & Brüning eine recht erfolgreiche Tätigkeit ausgeübt, ist aber später nach seiner Heimatstadt Wien zurückgekehrt und hat sich in der Elektrotechnik durch die Darstellung von Metallfäden aus kolloidalem Metall einen Namen gemacht.

Julius Tafel ist der akademischen Laufbahn treu geblieben. Er wurde in Würzburg Privatdozent und später als Nachfolger von Hantzsch ordentlicher Professor und Direktor des Instituts. Leider zwang ein Lungenleiden ihn frühzeitig, seine erfolgreichen wissenschaftlichen Arbeiten, besonders die elektrolytische Reduktion organischer Verbindungen, aufzugeben und ausschließlich seiner Gesundheit zu leben.

Von sonstigen Mitarbeitern erwähne ich noch O. Bülow, den Vorlesungsassistenten Koch, der das Trimethylendiamin und die Synthese von Harnstoffen aus solchen Diaminen bearbeitete, ferner Elbers, der die ersten Hydrazinosäuren darstellte und Hess, der an der Synthese von Indolderivaten aus Hydrazin teilnahm, dann Hegel, einen Enkel des Philosophen, jetzt Mitglied des Patentamtes, Roese und endlich Antrick, z. Zt. Direktor der chemischen Fabrik auf Aktien vormals Schering Berlin. Ein gewisse Ausnahmestellung nahmen ein C. Paal und O. Hinsberg, weil sie über Thematas arbeiteten, die nicht von mir gestellt waren, aber dabei gerne theoretischen und experimentellen Rat und Hilfe von mir annahmen. Der erste hat damals seine interessanten Versuche über das Acetonyl-Aceton und seine Verwandlung in Furanderivate ausgeführt, ist später in Erlangen und dann auch in Leipzig als Nachfolger von E. Beckmann Professor für pharmazeutische und angewandte Chemie geworden. Hinsberg kam von Göttingen und hat in Erlangen die Chinoxaline entdeckt. Seine Publikation, die aus dem Erlanger Institut datiert war, hat mir einen Angriff von G. Körner in den Abhandlungen der römischen Akademie eingetragen. Körner behauptete dort, er habe das Chinoxalin vor Hinsberg dargestellt und mir bei einem Besuche in Mailand davon mündliche Mitteilung gemacht. Ich habe auf den Vorwurf Körners, der viel zu spät zu meiner Kenntnis kam, öffentlich nicht geantwortet, wohl aber habe ich an ihn privatim geschrieben und ihn darauf aufmerksam gemacht, daß seine Klage durchaus unberechtigt sei. Ich selbst konnte mich in keiner Weise an eine solche Mitteilung Körners erinnern; aber selbst wenn diese wirklich geschehen ist, so hätte Körner mehrere Jahre Zeit gehabt, seine Beobachtungen zu publizieren. Nach einer so langen Frist aber auf eine angeblich private Äußerung hin, die sich in keiner Weise prüfen läßt, einen Prioritätsanspruch zu gründen und auch noch eine Anklage wegen Indiskretion gegen einen Fachgenossen zu erheben, ist eine sehr bedenkliche Art der Polemik, die gewiß kein vernünftiger Naturforscher anerkennen wird. Ich kann hier nur bezeugen, daß die Arbeit von Hinsberg auch nicht im geringsten durch die angebliche Körner'sche Beobachtung beeinflußt war.

Wenn ich noch zufüge, daß ich in Erlangen die Osazone der Zucker auffand und damit die Grundlage für meine weiteren Zuckerarbeiten schuf, daß ich ferner durch Behandlung der Methylharnsäuren mit Chlorphosphor die ersten Oxypurine erhielt, daß ferner L. Knorr das Antipyrin entdeckte, wobei er die schon von mir flüchtig beschriebene Reaktion zwischen Phenylhydrazin und Acetessigäther in sehr sinnreicher Weise benutzte, so wird man den Eindruck bekommen, daß wir fleißig bei der Arbeit waren. Nur die Abende blieben der Geselligkeit vorbehalten.

Die Stadt bot wenig, hier und da ein Konzert, das von musikliebenden Männern veranstaltet wurde. Wir hatten einmal bei solcher Gelegenheit das Vergnügen, Bülow mit seinem vortrefflichen Orchester in Erlangen zu hören. Dann gab es zuweilen auch eine Vorstellung in dem alten markgräflichen Theater, das jetzt der Stadt gehörte und wo eine Nürnberger Truppe zu gastieren pflegte. Wollte man mehr von diesen Genüssen haben, so mußte man nach dem nahe gelegenen Nürnberg fahren.

Die Enge der kleinen Stadt führte die Angehörigen der Universität zu innigem Zusammenschluß. Der Familienverkehr wurde eifrig gepflegt. Ich selber habe bei meinem Freunde Wilhelm Leube und seiner Gemahlin Natalie, einer Tochter des ausgezeichneten Chemikers Adolf Strecker, viel davon profitiert. Selbstverständlich verlangte Frau Leube von ihren Freunden auch, wenn es nötig war, tätige Hilfe bei geselligen Veranstaltungen, und so mußte ich bei einem Ball, den Leube als Prorektor der Universität gab, die Herstellung einer Riesenbowle übernehmen. Ich habe die Gelegenheit benutzt, um einen karnevalistischen Aufzug zu veranstalten, der auf eine Verherrlichung des Weines hinauslief. Die Hauptfigur war dabei der starke, blühende Diener des Instituts Griesinger, der als Küfermeister auftrat und die Bowle in einem stattlichen Faß auf geschmücktem Handwagen in den Ballsaal hineinfuhr. Auf dem Faß saß als Bacchus verkleidet ein hübscher zehnjähriger Knabe, der Sohn des Chirurgen Heinecke, und um das Faß verteilt saßen die drei Töchter Leubes als Nymphen. Beim Umzug der Gruppe stimmte die ganze Gesellschaft das bekannte Rheinweinlied an: „Bekränzt mit Laub den lieben, vollen Becher. . . . . " Die Bowle, die ich aus 150 Flaschen Wein komponiert hatte, fand Beifall und war in wenigen Stunden ausgetrunken.

Am häufigsten trafen wir Unverheirateten uns natürlich im Gasthaus, wo ein sehr behaglicher Ton herrschte.

Von den jungen Medizinern sind mir am nächsten getreten die beiden Assistenten Leubes Penzoldt und Fleischer. Mit dem ersten habe ich auch einige Versuche ausgeführt, besonders diejenigen über die Empfindlichkeit des Geruchsinns. Er ist jetzt ordentlicher Professor der inneren Medizin in Erlangen. Weniger glücklich hat sich das Schicksal von Fleischer gestaltet.

Unter den älteren Medizinern war Leube wohl die hervorragendste Persönlichkeit. Sein Ruf als Arzt bei Krankheiten des Verdauungstraktus führte zahlreiche Patienten, darunter manche interessante Menschen, in die Erlanger Klinik.

Gerne erinnere ich mich auch an die anderen Mediziner, den pathologischen Anatomen Zenker, den Physiologen Rosenthal, den Chirurgen Heinecke, den Frauenarzt Zweifel und vor allem an den Senior der Fakultät, den Anatom J. v. Gerlach, meinen späteren lieben Schwiegervater. Er spielte in der medizinischen Fakultät eine große Rolle nicht allein wegen seines wissenschaftlichen Ansehens und seiner Verdienste als Lehrer, sondern auch wegen seiner liebenswürdigen Persönlichkeit und seiner Bemühungen um das Wohl und Ansehen der Fakultät.

Uns Chemikern recht nahe stand der außerordentliche Professor der Pharmakologie Filehne, der die fieberstillende Wirkung des Antipyrins zuerst feststellte und seine Einführung in die praktische Medizin besorgte.

Von den Juristen ist Professor Marquardsen als Politiker in der Öffentlichkeit am bekanntesten geworden. Er gehörte viele Jahre sowohl zum Reichstag, wie zum bayerischen Landtag. Ich habe später wiederholt in Berlin die Ehre und das Vergnügen gehabt, den alten erfahrenen, klugen und lebenslustigen Herrn in meinem Hause zu sehen. Solange meine Frau lebte, brachte er auch zuweilen seine Familie mit, die in Erlangen mit Gerlachs sehr befreundet war. Der zweite Jurist, mit dem ich öfter in Berührung kam, war der Schwabe Hoelder, der später nach Leipzig gekommen ist.

Die Theologie war in Erlangen nur durch eine evangelische Fakultät vertreten, und diese betonte mit größter Entschiedenheit ihren rein lutherischen Charakter, so daß sie einem Vertreter der pfälzischen reformierten Gemeinden die Mitgliedschaft versagte. Die Fakultät genoß in der kirchlichen Welt einen guten Ruf. Zeugnis davon gab die große Anzahl junger Theologen aus Norddeutschland, die hier wenigstens einen Teil ihrer Studien absolvierten und gleichzeitig von der sehr billigen, aber auch sehr einfachen Lebensweise in der süddeutschen kleinen Stadt Nutzen zogen. Die Gewohnheit junger Theologen, sich frühzeitig zu verloben, wirkte anziehend auf heiratslustige Mädchen, und so war Erlangen das Refugium von mehr als 100 Pfarrerswitwen geworden, die hofften hier ihre Töchter an den Mann zu bringen. Um sich darüber lustig zu machen, hatte eine lose Verbindung von alten, etwas verbummelten Korpsstudenten, die glaubten in Erlangen leichter das Examen bestehen zu können, sich den Namen „Pfarrerstöchter" zugelegt.

Die philosophische Fakultät war im Gegensatz zu Würzburg und München nicht geteilt und ließ sich ohne sichtbares Widerstreben von ihrem Senior, dem Historiker Hegel, Sohn des Philosophen, in wichtigen Dingen gerne leiten. Von den Naturforschern erwähne ich den Physiker Lommel, einen verständigen und behaglich veranlagten Pfälzer, ferner den Botaniker Rees, der unter de Bary eine vortreffliche Arbeit über die Hefen ausgeführt hatte, aber in Erlangen sich auf die Lehrtätigkeit und auf gesellige Bemühungen beschränkte, dann den talentvollen und witzigen Zoologen Selenka, der später mit seiner

Frau große Reisen auf Java ausführte, den sehr verdienten Mathematiker Noether, seinen äußerst komischen Spezialkollegen Jordan und endlich den Professor der Pharmazie und angewandten Chemie Albert Hilger. Dieser stand mir natürlich am nächsten, und wir haben auch zusammen ein chemisches Kolloquium eingerichtet. Hilger besaß damals den Ehrgeiz, sich mit rein chemischen Problemen zu befassen und hatte für den Zweck auch eine neue Auflage des bekannten Werkes von Husemann über Pflanzenstoffe herausgegeben. Da ich aber bald zu der Überzeugung kam, daß für diese Dinge seine Begabung und Ausbildung nicht ausreichten, so riet ich ihm, sich mehr auf praktische Dinge, namentlich auf die Nahrungsmittelanalysen zu werfen. Er hat das auch getan und besten Erfolg gehabt; denn seinen Bemühungen ist es wohl mit zuzuschreiben, daß in Bayern die Nahrungsmitteluntersuchung den Instituten für angewandte Chemie an den drei Landesuniversitäten übertragen wurde und daß dadurch die Kontrolle der Nahrungsmittel in Bayern viel früher und besser geordnet war, als in Norddeutschland, besonders in Preußen. Hilger war musikalisch und sorgte mit seiner klugen Frau, einer Holländerin, für die Pflege des musikalischen Lebens in Erlangen. Er ist später an die Universität München gekommen, wo er in der Nähe des Baeyer'schen Laboratoriums einen Neubau für pharmazeutische und angewandte Chemie errichtete.

Die Studenten spielten in Erlangen natürlich die Hauptrolle, weil ein erheblicher Teil der Einwohner von ihnen lebte. Auch das Verbindungswesen blühte in ungewöhnlicher Weise, und damit im Zusammenhang stand das Duellwesen, das trotz des gesetzlichen Verbots von jedermann als etwas Selbstverständliches und Nötiges angesehen wurde. Ja, ich selbst habe mich mit Penzoldt, Knorr und einigen anderen jungen Leuten noch an einem Kursus von Säbelfechten beteiligt, natürlich nur aus Freude an körperlicher Übung, während wir später zu unserer Überraschung hörten, daß es für Knorr eine Vorübung zu einer Säbelmensur war, die er mit Dr. Friedländer in München auszufechten hatte. Da ich außerdem wegen der Freude am Schwimmen die Universitätsbadeanstalt öfters besuchte, so widerfuhr mir die unerwartete Ehre, daß ich vom Senat zum Mitglied der Fecht- und Badekommission ernannt wurde und als solches tätigen Anteil an der Auswahl eines neuen Fechtmeisters nehmen konnte.

Im Jahre 1883 fand in der Direktion der Badischen Anilin- und Sodafabrik eine Personalveränderung statt. Vor allen Dingen wollte Dr. Heinrich Caro, der damalige Leiter des wissenschaftlichen Laboratoriums, in den Ruhestand bzw. Aufsichtsrat der Fabrik eintreten. Seinem Einfluß war es wohl zuzuschreiben, daß der Hauptaktionär der Fabrik und Vorsitzender des Aufsichtsrates Herr Sigl aus Stuttgart mir den Vorschlag machte, Nachfolger von Caro zu werden. Obschon diese Stellung materiell sehr viel mehr einbrachte, wie jede Professur in Deutschland, so war mir doch die akademische Tätigkeit mit der vollen Freiheit wissenschaftlicher Arbeit sympathischer. Ich lehnte deshalb ab, nahm aber eine Einladung zu mehrwöchentlichem Besuch der Fabrik gerne an, teils aus Interesse für die Industrie der Teerfarben, teils in der Hoffnung, mir Rohmaterialien für meine Untersuchungen in großer Menge bereiten zu können. So bin ich denn im August 1883 nach Ludwigshafen-Mannheim gezogen. Der 14-tägige Aufenthalt war eine glückliche Kombination von eifriger Arbeit und fröhlicher Unterhaltung. Zunächst wurde ich durch die ganze Fabrik geführt und in jeder Abteilung von dem betreffenden Leiter in vertraulicher, aber sehr weitgehender Weise über die Einzelheiten der Betriebe unterrichtet. Das war eine Begünstigung, die nur selten wissenschaftlichen Chemikern zuteil wird, und ich verdankte sie wohl dem Wunsche der Direktion, dauernd mit mir in Verbindung zu bleiben. Gleichzeitig begannen im technischen Laboratorium größere Versuche zur Methylierung von Harnsäure nach der Vorschrift, die ich dafür in Erlangen ausgearbeitet hatte.

Als Rohmaterial hatte ich mir mehrere Kilo Schlangenexcremente aus Amsterdam durch Vermittlung von Professor Forster verschafft. Auf meine Erkundigung, weshalb das Material so teuer sei, erhielt ich die überraschende Antwort, daß es in Holland für die Herstellung von medizinischen Geheimmitteln verwendet werde und deshalb einen richtigen Marktpreis besitze. Die Überwachung der Methylierungsversuche, die wohl 10 Tage in Anspruch nahmen, hatte mein früherer Privatassistent Dr. Boesler übernommen, so daß ich mich darum kaum zu kümmern brauchte. Anders war es mit einem zweiten Präparat, der Orthoaminozimtsäure, die ich aus einem in der Fabrik vorrätigen Material, der Orthonitrozimtsäure, durch Reduktion mit Eisenvitriol und Ammoniak nach Tiemann darstellte. Die Operation mußte in stark verdünnter Lösung vorgenommen werden und erforderte die Filtration eines dicken Schlammes von Eisenhydroxyd. Zu dem Zweck wurde mir eine richtige Betriebsapparatur der Azofarbenfabrik zur Verfügung gestellt. Bei dieser Arbeit, die ich selbst übernehmen mußte, war mir das strenge Rauchverbot innerhalb des Fabrikgebäudes recht schmerzlich, aber wenn der alte Engelhorn, Mitglied der Direktion, zum Plaudern mich besuchte, dann zündeten wir uns beide heimlich eine Zigarre an. Einmal erwischte uns dabei der Betriebsführer Dr. Burkhardt, ein alter Münchener Bekannter von mir. Es gab nun ein großes Donnerwetter und wir wurden mit unseren Zigarren rücksichtslos vor die Türe gesetzt.

Den Nachmittag benutzte ich entweder zu kleinen Ausflügen oder zur Ausübung der Hühnerjagd in der frucht- und wildreichen Umgebung von Ludwigshafen, wohin mich das Direktionsmitglied Dr. Karl Klemm in freundschaftlicher Weise mitnahm. Die Abende verbrachte ich regelmäßig im großen chemischen Kreise im Pfälzerhof zu Mannheim, der sich durch den Ausschank vortrefflichen Pfälzer Weines auszeichnete. Die Herren aus der Technik waren dankbar dafür, daß ich ihnen neben mancherlei Schnurren auch über die Fortschritte der Wissenschaft in breiterer Form als sie sonst es erfahren konnten, Auskunft gab, und es wurde scherzhaft der Vorschlag gemacht, mich dauernd als „Vortragenden Rat" der Fabrik anzugliedern. In der Tat machte mir zum Schluß meines Aufenthaltes das Direktorium den Vorschlag, gegen ein mittleres Jahresgehalt in ein Vertragsverhältnis zur Fabrik zu treten, wobei ich nur die Verpflichtung übernehmen sollte, bei Erfindungen auf dem Gebiete der Teerfarben und Heilmittel der Fabrik das Verkaufsrecht einzuräumen. Dasselbe Angebot richtete man an Victor Meyer und Adolf von Baeyer. Es ist aber nicht verwirklicht worden, weil bald nachher ein großer Personenwechsel in der Direktion der Fabrik stattfand.

Nach Erledigung meiner Arbeiten reiste ich von Mannheim nach Euskirchen, um meinem Vater bei der Hühnerjagd zu helfen. Hier hatte ich das Vergnügen, den Besuch von Victor Meyer zu empfangen. Er hatte etwa $^1/_2$ Jahr vorher im Baeyer'schen Hause den Wunsch geäußert, auch die Jagd kennen zu lernen, da dieser Sport ihm bisher fremd geblieben sei. Daraufhin lud ich ihn ein, während der Herbstferien zu uns an den Niederrhein zu kommen. Seine Zusage löste er nun ein. Leider hatte er kurz vorher anstrengende Hochgebirgstouren im Gebiete des Bernina gemacht und kam nun direkt aus dieser frischen Höhe ziemlich abgehetzt nach der warmen Ebene. Infolgedessen reichten seine körperlichen Kräfte nicht mehr aus, mit Genuß an der immerhin etwas anstrengenden Hühnerjagd teilzunehmen. Während mein Vater und ich ohne Ermüdung 6 bis 8 Stunden durch die Kartoffeläcker stiefeln konnten, lag er gewöhnlich nach 2 Stunden ermüdet hinter einem Busche. Leichter und interessanter war ihm deshalb eine kleine Jagd, die ich für ihn im Flamersheimer Walde veranstaltete. Er hatte dabei das auffallende Glück, im Laufe von etwa 2 Stunden von einem Rehbock und 5 Wildsauen angelaufen zu werden. Allerdings fehlte ihm die Übung, um eins der Tiere zu erlegen. Bei der Rüstung des Mittagsmahles, das unser Förster besorgte, mußten sich alle Mitglieder des kleinen Kreises am Kartoffelschälen beteiligen mit der Maßgabe, daß jeder so viel zu schälen habe, als er beabsichtige zu essen. Über diese eigenartige Sitte hat Meyer besonders gelacht und dazu das

Geständnis abgelegt, daß er bisher nie in seinem Leben eine Kartoffel geschält habe.

Der Tag war wenig anstrengend, und diese Art der Jagd hat ihm besser gefallen, als das Ablaufen des freien Feldes. Da er aber offenbar jetzt seine Erfahrungen auf der Jagd für ausreichend hielt, so schlug ich ihm einen Ausflug nach der vulkanischen Eifel vor. Er war damit gerne einverstanden, und wir sind nun teils zu Wagen, teils zu Fuß über Rheinbach nach dem Ahrthal, von dort abwärts an den Rhein und dann durch das landschaftlich hübsche und geologisch sehr interessante Brohlthal nach dem lieblichen Laacher See, einem der schönsten Maare der vulkanischen Eifel gereist. Das hat Meyer natürlich in hohem Grade interessiert, und um ihm noch eine der charakteristischen Kohlensäurequellen dieser Gegend zu zeigen, führte ich ihn am nächsten Tage nach Burgbrohl zu dem aus der Münchener Zeit mir wohl bekannten Vetter Dr. Hans Andreae, der hier nach einer stürmischen Studienzeit als Fabrikant und Familienvater gelandet war. Er erzeugte hauptsächlich reine Alkalibicarbonate und benutzte dafür die natürliche Kohlensäure, die als sogen. Mofette in mächtigem Gasstrom aus der Erde kam und für die Zwecke der Fabrik durch ein Metallrohr gefaßt war. Die Quelle dient jetzt zur Herstellung von flüssiger Kohlensäure, die seitdem ein bedeutender Handelsartikel geworden ist.

Selbstverständlich statteten wir auch dem am See gelegenen schönen Jesuitenkloster Marialaach unseren Besuch ab und kehrten dann, immer bei herrlichstem Wetter, wieder nach Euskirchen zurück, wo Meyer von meinem Vater, mit dem er sich rasch angefreundet hatte, Abschied nehmen wollte.

Inzwischen war von Wilhelm Königs eine Einladung an uns eingelaufen, ihn in Cöln zu besuchen, wo er Meyer die vielen Schönheiten der alten Handelsstadt zeigen wollte. Sie wurde angenommen. Auf der Fahrt empfing uns Königs schon in Kierberg bei Brühl mit der Einladung, Station in dem Landhaus seines Bruders, des Cölner Bankiers zu machen. Den Auftrag dazu hatte er von seiner Schwägerin, einer geistig sehr angeregten Frau, erhalten, weil sie die beiden Professoren kennen lernen wollte. Mit der Lebendigkeit der Rheinländerin wußte sie uns bei einem lustigen Frühstück ziemlich ausführlich über unsere wissenschaftlichen Ziele auszufragen. Sie belohnte uns dafür durch reichliche Bewirtung und durch das Vorzeigen ihrer eigenen Kunstwerke, einer Reihe von leidlich gemalten Ölbildern. Bei dieser Gelegenheit habe ich zum ersten Mal meinen späteren Schüler Ernst Königs, der jetzt Privatdozent in Breslau ist, als Knaben gesehen.

Meyer und ich haben uns dann getrennt, und er ist über Cöln nach Berlin zum Besuch seiner Eltern gefahren.

Zuvor hatten wir in Euskirchen auf seinen Vorschlag Duzbrüderschaft geschlossen, und ich muß gestehen, daß er mir dauernd ein lieber Freund geblieben ist. Bei Meyer war körperliche Anmut mit ungewöhnlicher geistiger Begabung in glücklicher Weise vereinigt. Dazu kam eine natürliche Liebenswürdigkeit des Wesens und eine große Geschicklichkeit, sich der Umgebung anzupassen, so daß ihm die Sympathien der Menschen rasch zuteil wurden.

Sein Bruder Richard hat ihm eine ausführliche Lebensbeschreibung gewidmet und ein weiteres Denkmal durch die Herausgabe seines Briefwechsels gesetzt. Aber trotzdem erscheint es mir nicht allein gerechtfertigt, sondern wie eine Art von Freundespflicht, auch hier eine kurze Charakteristik von ihm zu geben. Er war ein rascher Denker und verfügte infolge seiner Belesenheit und seines ausgezeichneten Gedächtnisses über ebenso gründliche wie ausgedehnte Kenntnisse. Drum schossen bei ihm die Ideen wie ein frischer und unversiegbarer Sprudel hervor, ohne daß er dabei die gesunde Kritik verloren hätte. So erklären sich auch seine außerordentlichen Erfolge in der Experimentalchemie, wo schöpferische Phantasie mit nüchterner Auswahl der lösbaren Probleme und der einfachsten Versuchsbedingungen verbunden sein muß. Sehr interessant war es, ihn über Fachgenossen reden zu hören, deren Vorzüge er gerne anerkannte und deren Schwächen er mit Freimut, aber ohne jede Bosheit, mehr im humoristischen Sinne beleuchtete.

Nebenher steckte in ihm ein gutes Stück Künstler mit aufrichtiger Freude an Musik, Deklamation, Schauspiel und Dichtkunst im weitesten Sinne. Das alles brach bei ihm spontan und mit natürlicher Anmut von Zeit zu Zeit hervor, so daß es, chemisch gesprochen, wie eine Transmutation vom Naturforscher zum Künstler aussah. Leider war damals sein Nervensystem schon durch übermäßige Arbeit, vielleicht auch durch zu reichlichen Genuß der Lebensfreuden, erschüttert, so daß er 1 Jahr später in Zürich zusammenbrach und Urlaub nehmen mußte. Es war um dieselbe Zeit, als ich aus anderen Gründen die Laboratoriumstätigkeit aufgeben mußte. Glücklicherweise haben wir beide eine Art von Renaissance erlebt, die allerdings bei Meyer nur etwa 12 Jahre dauerte.

Im Oktober 83 trafen wir für kurze Zeit in München bei Baeyer zusammen und erfuhren hier die freudige Überraschung, daß uns Baeyer in einer behaglichen Plauderstunde gleichfalls die Duzbrüderschaft anbot. Wir beide verehrten in ihm den ausgezeichneten und lieben Lehrer und hatten nun das Recht erhalten, ihn in besonders trauter Weise „Freund" nennen zu dürfen. Soviel ich weiß, sind wir die einzigen Chemiker geblieben, die sich dieses Vorrechts rühmen durften.

In unserem Junggesellenkreise zu Erlangen gab es kein interessanteres Ereignis als die Verlobung einzelner Mitglieder. Im Winter 1883/84 erlebten wir das dreimal, bei Ludwig Knorr, Leo Gerlach und dem schon 40-jährigen Mediziner Kieselbach. Sie wurden natürlich feierlich aus unserem Kreise entlassen und dafür haben wir an den Hochzeiten teilgenommen. Die erste von Leo Gerlach fand in den Osterferien 84 in Nürnberg statt, weil die Braut der dort altangesessenen Familie Seitz angehörte. Sie wurde mit dem großen, etwas steifen Pomp der alten Kunststadt abgehalten. Die zweite Hochzeit war in den äußeren Verhältnissen, besonders in der Zahl der Teilnehmer, bescheidener, aber mit feinem künstlerischen Geschmack hergerichtet. Sie fand statt in dem Hause des Direktors der Akademie der Künste Piloty zu München. Bei dem Festmahl war die Braut so gesetzt, daß das prächtige blonde Haar, welches dem Vater als Modell bei der Idealfigur der Thusnelda in dem bekannten Bild „Der Triumphzug des Germanikus" gedient hatte, allein von der Sonne beleuchtet und deshalb von einer Art Glorienschein umgeben war. Meine Nachbarinnen bei dieser Hochzeit waren die durch Schönheit ausgezeichnete Schwester der Braut Johanna, die spätere Frau von Hefner-Alteneck, und die gewandte redefertige Baroneß L. von Hornstein, die spätere zweite Frau von Lenbach.

Wegen meiner freundschaftlichen Beziehungen zu Knorr mußte ich eine Rede auf das Brautpaar halten und die Glückwünsche der jungen Erlanger Gesellschaft mit einem stattlichen Album von Photographien überbringen. Ich hatte mir die Rede natürlich vorher überlegt und eine Einleitung ausgedacht, die an die Namen Knorr und Piloty anknüpfte. Es war ein Schiffsvergleich, und nun wollte es der Zufall, daß gerade vor mir auf der Tafel ein prächtiges, in Silber gehaltenes Segelschiff stand, von dem ich jetzt natürlich ausging. Dadurch bekam die Rede einen ganz improvisierten Zug und am Schluß erklärte mir der Hochzeitsvater, dem seine Rede recht sauer geworden war, daß wir Professoren den Künstlern im Schwätzen doch über seien.

Von diesen beiden Hochzeiten kam ich mit einem Katarrh, der durch das 2-jährige Arbeiten mit Chlorphosphor in dem schlecht ventilierten Privatlaboratorium vorbereitet und durch eine akute Erkältung verstärkt war, nach Hause zurück, reiste dann aber bald, ohne mich darum zu kümmern, zum Besuch meines Vaters und Schwagers nach Uerdingen. Hier habe ich mir wahrscheinlich auf der Jagd eine kleine Verletzung des Darms zugezogen, zu deren Beseitigung ich mich an den Chirurgen Professor Bardenheuer im Cölner Bürgerhospital wandte. Ich wurde dort operiert und mußte 14 Tage zu Bett liegen. Leider stellte sich ziemlich hohes Fieber ein und infolge dieser ungünstigen Umstände entwickelte sich mein Bronchialkatarrh zu einem tüchtigen Husten. Aus dem Spital entlassen, bin ich statt nach dem Süden törichterweise nach Euskirchen gegangen, habe dort an der Jagd teilgenommen und mich neuen Erkältungen ausgesetzt. Jetzt nutzte auch ein 14-tägiger Aufenthalt in Wiesbaden, wo abends immer ziemlich stark gekneipt wurde, nichts und so hat der Katarrh allmählich eine chronische Form angenommen. Am meisten beschädigt waren Nase, Hals und Trachea, und über den Nasenkatarrh war ich besonders unglücklich, weil mein sonst so feiner Geruchssinn völlig aufgehoben war und ich fast 1/2 Jahr kein Geruchsempfinden gehabt habe. Daran mögen zum Teil auch die Riechversuche, auf die ich später zurückkommen werde, schuld gewesen sein. Trotz des Katarrhs habe ich im Sommer Vorlesungen und Praktikum in Erlangen abgehalten, weil ich glaubte, daß in den Herbstferien die Krankheit geheilt werden könnte. Aber ich hatte noch immer nicht die richtige Lebensweise angenommen, denn das Rauchen, dem ich leidenschaftlich ergeben war, konnte ich nicht lassen, und im Weintrinken habe ich auch vielleicht damals mehr geleistet als gut war. Zudem ließ ich mich im August von Fleischer und Penzoldt überreden, mit nach Pontresina im Engadin zu gehen. Die Reise dahin fing schon mit einem Wagenunglück an, das recht böse Folgen hätte haben können. Fleischer und ich hatten nämlich, um die überfüllte Post zu vermeiden, in Chur einen Privatwagen gemietet, ohne über die Eigenschaften der Pferde und des Kutschers uns zu unterrichten. Nach einigen Stunden leidlicher Fahrt begegneten wir einem italienischen Orgeldreher, der seinen Leierkasten mit einer buntgefärbten Decke überzogen hatte. Davor scheute das eine Pferd, der Kutscher verlor die Herrschaft über die Tiere, und wir stürzten, nachdem das schlechte Geländer durchbrochen war, von der Straße etwa 5 m bergab, glücklicherweise auf eine Wiese. Ich hatte das Unglück kommen sehen, war aufgestanden und wollte aus dem Wagen herausspringen. Es war aber zu spät und ich flog in weitem Bogen aus dem Gefährt heraus in die Wiese hinein. Ich bin niemals in meinem Leben vom Boden so rasch wieder aufgesprungen, weil ich fürchtete, daß der Wagen nachkommen würde. Der war aber inzwischen ganz umgeschlagen und stark beschädigt liegen geblieben. Auch Fleischer war herausgeflogen und hatte sich einen Arm ziemlich stark verstaucht. Merkwürdigerweise blieben die Schuldigen, d. h. der Kutscher und die Pferde, ganz unverletzt. Der Absturz war auf etwa 500 m Entfernung von den Gästen eines kleinen Schwefelbades Alvaneu beobachtet worden, und als wir dort einkehrten, um uns durch ein Mittagsmahl von dem Schrecken zu erholen, wollte die Tischgesellschaft es nicht glauben, daß Männer, die soeben einer wirklichen Lebensgefahr entgangen waren, Lust zum Essen haben könnten. Ich hatte nun die Freude am Wagenfahren verloren. Wir gaben deshalb unsere Koffer auf die Post und machten den Rest des Weges nach Pontresina zu Fuß. Hier bin ich nur einige Wochen geblieben, weil der Aufenthalt in der trockenen und abends kalten Luft, das Unternehmen verschiedener kleiner Gletschertouren und das abendliche stundenlange Verweilen in einer rauchigen Bierkneipe meinem Katarrh nur schädlich waren. Ich zog es deshalb vor, nach dem niedriger gelegenen Kurort Flims in Graubünden zu gehen, wo ich mit Freund Königs zusammentraf und einige vergnügte Wochen verbrachte. Hier habe ich auf eigentümliche Art den Präsidenten des Schweizer Schulrats Kappler kennen gelernt. In dem mit dem Hotel verbundenen Bierhause war nämlich allabendlich eine Gesellschaft von älteren Schweizer Herren versammelt, die sich mit dem in der Schweiz üblichen Kartenspiel „Jas" vergnügten. In diesem kleinen Kreise zeichnete sich durch Lebhaftigkeit, originelles Äußere und kräftige Witze ein alter Herr so sehr aus, daß wir uns nach seinem Namen erkundigten. Es war Herr Kappler, den alle jungen Dozenten der Naturwissenschaften in Deutschland dem Ruf nach kannten. Unsere Neugierde war dem alten Herrn verraten worden. Er hat sich dann auch erkundigt, und als wir am nächsten Tage bei Tisch saßen, schickte er den Kellner zu mir mit der Frage, ob ich der Otto oder der Emil wäre; denn er war über die jungen Naturforscher in Deutschland ausgezeichnet unterrichtet. Wir sind dann in persönliche Berührung gekommen, und er sprach sofort den Wunsch aus, daß ich die Professur der Chemie am Polytechnikum in Zürich übernehmen möchte, da Victor Meyer am Ende des nächsten Wintersemesters nach Göttingen übersiedeln werde. Als ich ihm erwiderte, daß ich augenblicklich leidend sei und erst meinen Katarrh kurieren müsse, wollte er mit Rücksicht auf mein gesundes Aussehen nichts davon wissen und wiederholte mehrere Wochen später das Angebot brieflich, nachdem er seinen Kollegen im Schulrat Bericht erstattet hatte. Der Ruf war sehr verlockend, da ein prächtiges neues Institut für Chemie gebaut werden sollte, wozu die Pläne von Victor Meyer und Lunge in Verbindung mit einem ausgezeichneten Baumeister schon fertiggestellt waren. Auch hätte es für mich einen Reiz gehabt, der Nachfolger Meyers zu werden, aber ich war doch zu unsicher, ob ich bei meinem Gesundheitszustand den Anstrengungen der Züricher Professur gewachsen sein würde, denn Meyer war doch zuletzt auch zusammengebrochen und hatte lebhafte Klage über das aufreibende Leben und Treiben in Zürich geführt. So lehnte ich denn wieder ab.

Auf der Rückreise von Flims drohte mir wieder die Gefahr eines Wagenunglücks, denn als Königs und ich von dort in einem Zweispänner nach Chur fuhren, stürzten unmittelbar vor dem Hotel auf dem glatten Pflaster beide Pferde zur Erde. Glücklicherweise blieben wir im Wagen unversehrt, aber meine Abneigung gegen Wagenfahrten ist durch den Vorfall noch verstärkt worden.

Den Rest der Ferien verbrachte ich in Südtirol, Brixen und Meran, wo aber infolge der Hitze und des Staubes der Bronchialkatarrh auch nicht völlig ausheilte. Die Folge davon war, daß ich bei einem kurzen Aufenthalt in München mir sofort einen neuen akuten Katarrh zuzog und kränker nach Erlangen zurückkehrte, als ich es im August verlassen hatte. Ich kam deshalb zu der Überzeugung, daß eine längere ernsthafte Kur nötig sei und nahm langen Urlaub, der mir vom Ministerium in München in der freundlichsten Weise gewährt wurde. Damit aber das Institut nicht ganz verwahrlost bleiben sollte, so schlug ich der Fakultät vor, meinen Vetter Otto Fischer, der in München Privatdozent der Chemie war, als provisorischen Vertreter für die Zeit der Krankheit anzunehmen. Das ist auch geschehen, und hat dann zur Folge gehabt, daß er ein Jahr später, als ich nach Würzburg übersiedelte, definitiv mein Nachfolger wurde. Der Vetter hat nicht allein meine Vorlesungen und die Leitung des Laboratoriums, sondern sogar die Wohnung einschließlich der Haushälterin für die Zeit meines Urlaubs übernommen. Ich bin noch bis Ende November in Erlangen geblieben, um ihn in alle Geschäfte einzuführen, und während dieser Zeit hatte ich das besondere Vergnügen, Herrn Kappler aus Zürich nochmals zu sehen. Nach meiner Ablehnung hatte er sich trotz seines hohen Alters und seiner schlechten Augen entschlossen, eine Rundfahrt durch Deutschland zu machen, um die jungen Dozenten der Chemie kennen zu lernen. Begleitet von seiner Tochter erschien er auch in Erlangen, um die Bekanntschaft meines Vetters zu machen und seine Vorlesungen zu besuchen. Ich lud ihn zu Tisch, und als wir vergnüglich getafelt hatten und er mich in guter Laune glaubte, machte er einen letzten Versuch, mich zu gewinnen. Er behauptete dabei, daß ich gar nicht so krank sei, er riskiere es ruhig mit mir und dann setzte er mit erstaunlicher Beredsamkeit die Vorzüge von Zürich auseinander, wobei er besonders die Annehmlichkeiten betonte, welche dort einem Junggesellen durch die Freiheit der Sitten geboten seien. Da er aber bald einsehen mußte, daß er mit mir kein Geschäft machen könne, so beschränkte er sich schließlich darauf, mir eine Reihe von interessanten Begebenheiten aus seinem Leben, vermischt mit köstlichen Schnurren, zu erzählen. Er war ein vortrefflich unterrichteter, sehr kluger Mann, mit allen guten Eigenschaften des Schweizers ausgestattet, der keine Mühe scheute, seinem geliebten Polytechnikum die bestmöglichen Lehrkräfte zuzuführen. Die große Blüte dieser Schule ist damals sicherlich zum erheblichen

Teil das Werk von Kappler gewesen. Als ich ihm 4 Jahre später von Würzburg meine Verlobung anzeigte, schrieb er mir einen ebenso liebenswürdigen wie interessanten Brief. Das einzige, was ihm leid tue, sei, daß die Verlobung nicht in Zürich erfolgte. Dann kam eine lange Auseinandersetzung über die Bemühungen der Schweiz, auf dem weiten Gebiete des Unterrichts verhältnismäßig mehr zu leisten, als die europäischen Großstaaten, wo so viel geistige Kräfte durch die Politik und das Militär in Anspruch genommen seien.

Als er in Erlangen von mir Abschied nahm, sagte er: da ich nicht zu haben sei, so werde er sich jetzt um keinen fremden Rat mehr kümmern und einfach seiner Nase nachlaufen, um einen möglichst guten Nachfolger für Meyer zu gewinnen. Seine Wahl ist dann auf Hantzsch gefallen.

Anfangs Dezember verließ ich Erlangen, und da mir der dauernde Aufenthalt in den Kurorten zuwider geworden war, so ging ich zuerst zu meinem Schwager Arthur Dilthey in Rheydt, der ein behaglich eingerichtetes, mit Zentralheizung versehenes Haus besaß und mich ebenso liebenswürdig wie scherzhaft zur Kur in der Winterfrische zu Rheydt eingeladen hatte. Hier habe ich drei vergnügte Monate zugebracht. Der Tag wurde zu größeren Spaziergängen benutzt, und abends spielte ich, angeblich um die Stimme zu schonen, mit dem Schwager und August Fischer Skat. Dabei passierten aber so komische Dummheiten, daß wir aus dem Lachen nicht herauskamen, und da auch fleißig Wein getrunken wurde, so war es nicht gerade die Kur, die für die Heilung des Katarrhs nötig gewesen wäre. Aber er wurde auch nicht schlimmer und meine Gemütsverfassung hatte sich in dem lustigen rheinischen Kreise außerordentlich gebessert. Selbstverständlich kam ich auch mit meinen anderen Schwägern und den Schwestern häufig zusammen, und die alte Tante „Lisettchen" ließ es sich ebenfalls nicht nehmen, den Neffen von Zeit zu Zeit einzuladen.

In der Weihnachtswoche kam es bei meinem Schwager zu einem kleinen Brand, der recht üble Folgen hätte haben können. Der in der trockenen Luft der Zentralheizung ganz ausgedörrte Christbaum wurde auf Wunsch der Kinder nochmals angezündet. Dabei fingen die harzreichen Nadeln Feuer und in kurzer Zeit war der ganze Baum am Brennen. Mir selbst war das kein ungewohntes Schauspiel, da man im Laboratorium ja öfters solche raschen Brände erlebt. Aber auf die Familie meines Schwagers, besonders auf die Kinder machte es einen ganz lähmenden Eindruck, und die Erzieherin der Kinder war so außer Fassung, daß sie direkt in das Feuer hineinlaufen wollte, um zu löschen. Mir blieb nichts anderes übrig, als sie und die Kinder mit einiger Gewalt vor die Tür zu setzen, und dann das Kommando auszugeben: „Ruhig ausbrennen lassen." Das war in einer halben Stunde geschehen, einige Vorhänge waren mitverbrannt, einige Bilder und Teppiche beschädigt. Dem Umgreifen des Feuers hatten wir mit ein paar Eimern Wasser gewehrt und die Sache war erledigt. Ich habe seitdem immer davor gewarnt, Christbäume in Häusern mit Zentralheizung nach mehrtägigem Stehen nochmals anzuzünden.

Bei der Weihnachtsbescherung passierte eine schnurrige Geschichte, die bezeichnend ist für den Kunstsinn der Kinder. Mein Schwager hatte als Geschenk einen Gipsabguß der Büste der Venus von Milo bekommen, die von den Kindern wegen der abgeschnittenen Glieder nicht sehr freundlich kritisiert wurde. Plötzlich erhebt die kleine Else die Frage: „Was hat die da für Buckel auf der Brust?", worauf der noch jüngere sechsjährige Alfred ihr antwortete: „Wie dumm, Else, das sind doch Furunkel."

Im Februar 1885 entschloß ich mich, das Frühjahr am Mittelmeer zu verleben, und mein Freund Victor Meyer, der inzwischen ebenfalls nervös erkrankt war und an heftiger Neuralgie litt, riet mir nach Ajaccio auf Corsica in den Schweizer Hof der Frau Dr. Müller zu gehen. Niemals ist mir ein besserer Rat bezüglich eines Kurortes erteilt worden. Die Reise ging über Paris und Marseille. Als ich in Paris abends einen kleinen Spaziergang auf dem Boulevard machte, verlor ich meine goldene Uhr und merkte es erst, als ich etwa 20 Schritte weiter gegangen war. Ich kehrte natürlich sofort um und hatte auch das Glück, die Uhr noch auf dem Boden zu finden, obschon eine ganze Reihe von Menschen die Stelle passiert hatten.

In Marseille kam ich mit einem großen Kreditbrief eines Cölner Bankhauses an, der auf eine ziemlich hohe Summe ausgestellt war, weil ich die Absicht hatte, an den Aufenthalt in Corsica eine Seereise nach Brasilien anzuschließen. Als ich mit diesem Brief in dem Geschäftshause, auf das es ausgestellt war, einige 1000 Frcs. abheben wollte, erklärte man mir, die Kasse enthielte so viel nicht. Es gab ein großes Gelächter, und ich mußte warten, bis das Geld von der Bank herbeigeschafft war. Marseille, das ich schon kannte, hat mir wegen seiner prächtigen Lage und der hübschen Hafenbauten immer von neuem sehr gefallen.

Die Seefahrt nach Ajaccio, die etwa 20 Stunden dauerte, ist mir noch in freundlicher Erinnerung. Die Franzosen waren bei der Tafel von ausgezeichneter Höflichkeit. Auf dem Schiff befanden sich auch eine Reihe von Sträflingen, die in Corsica eine längere Gefängnisstrafe abbüßen sollten. Sie unterhielten sich gegenseitig in der lauen, sternenklaren Nacht bei ganz ruhiger See mit Gesang und übermütigen Scherzen. Von dieser Fröhlichkeit wurde die ganze übrige Gesellschaft beeinflußt, und bei mir kam das angenehme Gefühl hinzu, daß die milde feuchte Luft des Mittelmeers für die erkrankten Schleimhäute meiner Atmungswege das richtige Kurmittel sei.

Früh morgens sahen wir schon die Schneeberge Corsicas winken und bei der Einfahrt in Ajaccio bot sich uns ein so prächtiger Anblick, wie ihn nicht viele Orte des Mittelmeeres gewähren.

Im Gasthof wurde man besonders freundlich empfangen, da der Besuch infolge der Choleraepidemie des vergangenen Sommers noch recht schwach war.

Corsica hat in den Monaten März und April so gute klimatische Verhältnisse, wie wenige Orte am Mittelmeer, sehr viel Sonne und mittags die kühle frische Seebrise, abends bei Sonnenuntergang nochmals eine kurze Periode der Abkühlung und dann gleichmäßige Temperatur bis tief in die Nacht hinein. Da man außerhalb der Stadt wohnte, blieb man auch von jedem Staub verschont, der an der Riviera die Menschen so stark belästigt; denn es gab in Corsica wohl Straßen, aber keine Fuhrwerke darauf. Unter diesen günstigen Bedingungen ist mein Bronchialkatarrh in 8 Wochen geheilt. Es blieb aber eine Neigung zu akuten Rückfällen, die mich noch einige Jahre nötigte, Erkältungen und auch die schädlichen Gase des Laboratoriums so weit wie möglich zu vermeiden. Erst nach 33 Jahren bin ich wieder von einem influenzaartigen Bronchialkatarrh überfallen worden, der nicht heilen wollte, und nach sechs Wochen zu einer Lungenentzündung führte. Diese doppelte Erkrankung war dann der Grund, im April und Mai 1918 zur Erholung sechs Wochen in Locarno am Lago Maggiore zuzubringen, und die unfreiwillige Muße habe ich benutzt, um den ersten Teil dieser Erinnerungen niederzuschreiben.

Corsica ist ein wildes Gebirgsland, dessen höchste Spitzen, der Monte d'Oro und Monte Rotondo, ewigen Schnee und sogar kleine Gletscher tragen. Der Hauptgebirgsstock besteht aus Granit, der hier durch den am Mittelmeer überwiegenden Kalk durchbrochen ist. Von der Wildheit der Natur ist etwas auf die Einwohner übergegangen; denn sie sind bekannt als kühne Krieger und rühmen sich gerne als die Landsleute von Napoleone Buonaparte. Seit Jahrhunderten üben sie die Blutrache, und noch zu meiner Zeit gab es ein Dorf in entlegener Gegend, das fast ausschließlich von solchen Mördern bewohnt war, die sich hierher geflüchtet hatten und der französischen Gendarmerie nicht selten bewaffneten Widerstand entgegensetzten. Die barbarischen Sitten des Landes sind in dem bekannten Buche von Gregorovius genau geschildert, und wer sie in anmutiger Schilderung kennen lernen will, der lese die vortrefflichen Novelle von Prosper Merimée „Colomba". Die Wahrung der Familienehre erschien den Corsen seit Jahrhunderten als erste Pflicht. Wir haben davon ein treffliches Beispiel miterlebt. In einem benachbarten Hotel hatte ein Rechtsanwalt aus Zürich versucht, ein corsisches Dienstmädchen zu gewinnen. Der Erfolg war negativ. Aber nach kurzer Zeit erschienen zwei Verwandte des Mädchens, Bauern aus der Umgegend, natürlich bis an die Zähne bewaffnet, und ersuchten den Rechtsanwalt um eine Unterredung. Dieser war klug genug, zu erklären, er habe dem Mädchen einen ehrlich gemeinten Heiratsantrag gemacht. Das wirkte beruhigend und die beiden Männer entfernten sich, nachdem sie noch den dringenden Rat erteilt hatten, er möge das Mädchen nicht weiter belästigen. Nach der Aussage von Sachverständigen würden diese Männer den Rechtsanwalt, falls er nicht eine der Familienehre genügende Erklärung abgegeben hätte, im Hotel niedergeschossen haben.

Geschossen wurde überhaupt in Corsica viel zu viel; denn die Jagd war frei und jeder erwachsene Mann hielt sich für verpflichtet, davon Nutzen zu ziehen. Bei weiteren Spaziergängen mußte man in der Tat einige Vorsicht gebrauchen, um nicht von leichtsinnigen Jägern angeschossen zu werden. Für die friedliche Arbeit scheinen die Corsen wenig geschaffen zu sein; denn die Landeskultur war durchweg vernachlässigt, und wo fleißige Arbeiter am Werk waren, konnte man sicher sein, daß es geworbene Italiener vom Festlande waren.

Da mein Katarrh geheilt war und mir die Nachricht zuging, daß durch den Tod von Kolbe demnächst eine Verschiebung der Professuren der Chemie stattfinden werde, so gab ich die Reise nach Südamerika auf und kehrte über die Schweiz nach Erlangen zurück. Auf der Durchreise habe ich in Genf Carl Graebe besucht, der mir das neue, mit großem Geldaufwand gebaute Universitätslaboratorium zeigte. Wir sind später noch öfter während der Ferien in der Schweiz oder im Schwarzwald zusammengetroffen, und ich werde noch mehr über diesen vortrefflichen Mann zu sagen haben.

Da mein Urlaub bis zum Herbst lief, so habe ich mich nur einige Wochen in Erlangen aufgehalten, um meinen Vetter Otto und die anderen Freunde zu begrüßen, und bin dann zum längeren Aufenthalt nach Badenweiler in den Schwarzwald gegangen. Auf der Hinreise machte ich kurze Rast in Frankfurt und besuchte die Höchster Farbwerke, den dort tätigen Freund Vongerichten und vor allem auch Herrn Dr. Lucius, Mitinhaber der Firma, der mir vorher einen Besuch in Erlangen abgestattet hatte. Ich wurde nicht allein in der Fabrik, sondern auch in der Familie Lucius auf das Freundlichste aufgenommen und bin mit dem Hausherrn bis in die Berliner Zeit hinein in Verkehr geblieben.

In Badenweiler mit seiner prächtigen Umgebung habe ich das Schlaraffenleben von Corsica etwa noch 5 Wochen fortgesetzt, bis eines Tages eine Depesche von dem Würzburger Professor Semper eintraf, der mich zu einer Unterredung nach Heidelberg einlud. Es war mir bereits bekannt, daß J. Wislicenus als Nachfolger von Kolbe nach Leipzig berufen war, und daß die Würzburger Professur für Chemie neu besetzt werden mußte. Daß man an mich denken würde, hielt ich für unwahrscheinlich, da ich wegen Krankheit mich in Urlaub befand und solche Dinge gerüchtweise immer stark übertrieben werden. In der Tat hatte auch die Fakultät in Würzburg auf mich verzichtet und O. Wallach dem Ministerium zu München vorgeschlagen. Ehe aber dort die Entscheidung fiel, hatte Fräulein Bertha Strecker aus München, die Schwester von Frau Leube, in Erlangen erfahren, daß ich wieder gesund sei und diese Nachricht nach Würzburg überbracht. Das veranlaßte Professor Semper, Mitglied der philosophischen Fakultät, die Möglichkeit meiner Berufung wieder in Erwägung zu ziehen, und zu dem Zweck die Zusammenkunft mit mir zu veranstalten. Sie fand statt im Hotel Schlieder zu Heidelberg, und, wie ich bald merkte, lief sie hinaus auf eine Prüfung meines Gesundheitszustandes, wozu sich offenbar Semper als Zoologe besonders geeignet hielt. Als später die Sache in Würzburg ruchbar wurde, erzählten sich die Leute dort, man habe mich von einem Tierarzt untersuchen lassen. Genug, Semper machte mir den Vorschlag, einen Spaziergang zum Schloß zu unternehmen. Obschon er viel älter war als ich, schlug er absichtlich einen raschen Schritt an, so daß er ganz atemlos oben ankam, während ich, an das Bergsteigen damals gewöhnt, mich bei dem Tempo sehr behaglich fühlte. Dann kam die zweite Probe, Semper schlug vor, eine Flasche Sekt zu trinken. Auch das war mir nicht unsympathisch, da der Genuß von Wein zu meinen Gewohnheiten gehörte. Der Erfolg dieses Frühstücks war dann auch, wie man erwarten konnte, eine leichte Betrunkenheit des älteren Herrn ohne Mitleidenschaft des jüngeren Kollegen. Das Examen war bestanden. Semper reiste nach Würzburg zurück, erklärte seinen Fakultätsgenossen, „der Fischer ist ein ganz starker, leistungsfähiger Mann, der uns alle überleben wird", worin er auch recht behalten hat. Infolgedessen ging ein neuer Vorschlag der Fakultät nach München und etwa einen Monat später erhielt ich wirklich vom Ministerium den Ruf nach Würzburg. Um Wallach hat es mir leid getan, denn er war über den ersten Vorschlag der Fakultät unterrichtet. Aber ich konnte deshalb nicht zurücktreten, und da mir der Wechsel nach Würzburg außerordentlich sympathisch war, so nahm ich das Angebot des Ministeriums gerne an.

Zuvor hatte ich einen mehrwöchigen Aufenthalt in Homburg v. d. Höhe genommen, von dem mir zwei Ereignisse in Erinnerung geblieben sind.

Zunächst wieder ein Wagenunglück. Ich besuchte die mir von Lebzeiten ihres Gemahls her wohlbekannte Frau Dr. von Brüning, geb. Spindler aus Berlin auf ihrem prächtigen Landsitz bei Homburg und sie lud mich zu einer Wagenfahrt ein. Ich suchte das abzulehnen mit der Motivierung, daß das Wagenfahren für mich zu viele Gefahren bringe. Sie ließ diese Bedenken aber nicht gelten, weil Pferde und Kutscher bei ihr durch langjährige Dienste bewährt seien. Wie fuhren also in Begleitung ihres Sohnes nach der nahegelegenen Saalburg im Taunus. Als wir uns glücklich wieder auf dem Rückweg und auf der ebenen Landstraße befanden, stieß plötzlich der Kutscher einen Schrei aus und fiel bewußtlos vom Bock. Die Zügel schleiften natürlich an der Erde. Glücklicherweise trabten die alten Pferde ruhig weiter, bis der junge Brüning vom Wagen auf den Bock klettern und dort die Zügel fassen konnte. So ging die Sache ohne Schaden für uns vorüber. Der kranke Kutscher wurde nun in den Wagen gesetzt und nach Hause gefahren. Wir folgten zu Fuß und Frau von Brüning war doch etwas erschüttert durch das prompte Eintreffen meiner Unglücksprophezeiung. Sie hat daraus die Konsequenz gezogen, in Zukunft neben den Kutscher stets noch einen Diener auf dem Bock sitzen zu lassen.

Zufälligerweise hatte auch die Familie Meister in Homburg eine Sommerwohnung bezogen und auf meinen Besuch hin wurde ich zu einer Abendgesellschaft zugezogen, an der die Fürstin Bismarck und ihr Sohn Herbert teilnahmen. Es war mir natürlich sehr interessant, die nächsten Angehörigen des großen Reichskanzlers kennen zu lernen. Sie gaben sich als einfache natürliche Menschen. Besonders galt das von der Fürstin, die in fast burschikoser Weise von ihrer eigenen Person redete.

Frau Meister war die Schwester von Frau Lucius und beide waren wieder die Töchter eines Kunstmalers aus Frankfurt a. M., in dessen Haus Bismarck verkehrt hatte, als er preußischer Gesandter beim Bundesrat in Frankfurt war. Daher stammte die Freundschaft zwischen den beiden Damen und der Familie Bismarck, und das hatte mir die unerwartete Ehre eingetragen, mit der Fürstin und ihrem Sohn zusammenzutreffen.

Inzwischen war der Würzburger Ruf eingetroffen. Ich ging deshalb nach Erlangen, um den Umzug vorzubereiten und machte bald nachher einen Besuch bei J. Wislicenus in Würzburg, um mich bei ihm über die dortigen Verhältnisse, insbesondere auch über die Einrichtungen des Instituts zu unterrichten. Wir kannten uns schon von den Naturforscherversammlungen und von festlichen Veranstaltungen der Universitäten Würzburg und Erlangen.

Er war eine sympathische Persönlichkeit, von sehr würdiger äußerer Erscheinung und liebenswürdigem Wesen. In seiner kinderreichen

Familie herrschte er wie ein Patriarch. Von seinen Schülern wurde er allgemein verehrt als wohlwollender Lehrer und vornehmer Charakter. Die Universität Würzburg hatte ihn zweimal zum Rektor gewählt, und als solcher hat er das 300jährige Jubiläumsfest der Hochschule in mustergültiger Weise geleitet. Sein Lebenswerk ist in einer von Beckmann verfaßten Biographie geschildert und niemand wird die Verdienste leugnen wollen, die Wislicenus durch seine Arbeiten über Milchsäure, die Acetessigester-Synthese und namentlich später durch die vielen Anregungen auf stereo-chemischem Gebiete erworben hat. Aber als Chemiker und Naturforscher repräsentierte er doch einen ganz anderen Typ wie Baeyer, Hofmann oder Liebig und Wöhler. Ihm kam es mehr darauf an, vorgefaßte theoretische Ansichten durch ein Experiment zu prüfen, als empirisch den Erscheinungen zu folgen und auf unerwartete Vorgänge zu fahnden. Untersuchungen, wo große experimentelle Schwierigkeiten zu überwinden waren, hätte er wahrscheinlich nicht anstellen können. Dem entsprach auch die ziemlich dürftige Einrichtung des Würzburger Instituts, das von Scherer 20 Jahre früher erbaut worden war, ohne an die Bedürfnisse unserer rasch fortschreitenden Wissenschaft zu denken. Das erste, was ich in Würzburg tat, war der Vorschlag, die ganz ungenügende Ventilation durch Anlage neuer Abzüge zu verbessern. Wislicenus hat mich dabei in liebenswürdigster Weise unterstützt und bei dem Verwaltungsdirektor der Universität, Professor Risch, der aus den Einkünften der Hochschule ziemlich große Mittel zur Verfügung hatte, meine Forderung warm befürwortet. Die Verhandlungen endigten denn auch mit einer Bewilligung von etwa 6000 Mk. Dafür verlangte Risch von mir das feierliche Versprechen, daß ich niemals mehr die Mittel der Universität in Anspruch nehmen würde. Ich gab dasselbe lachend mit der Bemerkung, daß ich es bei der nächsten Gelegenheit brechen würde. Das ist schon nach einigen Jahren geschehen, und ich muß zu Ehren des Kollegen Risch zufügen, daß er für meine Wünsche immer eine offene Hand gehabt hat. Die kleinen Neueinrichtungen im Institut wurden während der Ferien ausgeführt, so daß Ende Oktober alles für die Arbeit bereit war. Für die Anlage der Abzüge hatte ich ein sehr einfaches System ausgedacht. Tonröhren, wie man sie für Aborte verwendet, wurden an die Wände gelegt, mit eisernen Schellen befestigt, über Dach geführt und dort durch einen zweckmäßigen Aufsatz gegen die schädliche Wirkung des Windes geschützt. Unten, im Arbeitsraum war ein Kniestück eingesetzt, um das Herabfallen des Schmutzes aus den Röhren zu verhindern. Um diese untere Öffnung des Rohres wurde dann der eigentliche Abzug in Holz und Glas gebaut. Eine Lockflamme in dem Rohr besorgt den nötigen Luftzug. Diese einfache

Form des Abzuges läßt sich in jedem Gebäude anbringen und ist deshalb höchst empfehlenswert, wo es sich darum handelt, provisorische Laboratorien in fertigen Häusern einzurichten.

Von Erlangen waren mit übergesiedelt J. Tafel und die Familie Knorr, die jetzt außer dem jungen Ehepaar noch einen Sohn zählte. Da sich für letztere keine passende Wohnung fand, so machte es mir ein Vergnügen, sie in die große Dienstwohnung des Instituts aufzunehmen, bis sie im Frühjahr ein Quartier außerhalb fand.

Die alten Assistenten des Instituts waren J. Wislicenus nach Leipzig gefolgt. Dagegen hatte er mir seinen Sohn Wilhelm zurückgelassen, dem ich gerne eine Assistentenstelle anvertraute, und der mir auch ein sehr lieber Freund geworden ist. Er konnte ebenfalls noch im Institut Quartier nehmen.

Knorr, der schon in Erlangen sich habilitiert hatte, wurde auf meinen Antrag ohne weiteres von der Würzburger Fakultät als Privatdozent übernommen und von mir mit der Leitung der analytischen Abteilung im Institut betraut. Für diese war der einzige große Arbeitssaal des Hauses reserviert, während die organische Abteilung sich mit einem kleinen Anbau und dem darunter befindlichen Keller begnügen mußte. Alles das war recht dürftig und auch recht unpraktisch angelegt. Der Hörsaal war ziemlich geräumig, weil man ihn von vornherein auf die große Zahl von Medizinern berechnet hatte, aber auch seine Einrichtung ließ viel zu wünschen übrig.

Mein Privatlaboratorium bestand aus einem einzigen einfachen Wohnzimmer, und als Raum für Wägungen und optische Untersuchungen mußten wir das danebenliegende Sprechzimmer benutzen. Trotzdem sind hier die experimentell recht schwierigen Zuckerarbeiten ausgeführt worden. Zunächst begann ich aber mit der Ausbildung der schon in Erlangen begonnenen Synthese von Indolderivaten aus Phenylhydrazonen. Daran nahmen auch die meisten Doktoranden teil, wodurch das Institut gleich in einen recht schlechten „Geruch" kam; denn der Hauptstänker, das Skatol, das wir in recht stattlichen Mengen darstellen konnten, wurde von den Praktikanten in die Gast- und Wohnhäuser der Stadt hineingetragen und gab zu mancherlei Klagen Anlaß. Von der Haftbarkeit seines Geruches habe ich selbst ein drastisches Beispiel erlebt. In den Osterferien 1886 unternahm ich nämlich eine zweite Reise nach Corsica und führte dabei die Lodenjoppe mit, die ich während des Winters im Laboratorium getragen hatte und die mir bei kleinen Bergtouren auf der Insel dienen sollte. Als mein Koffer an der französischen Grenze vor der Zollbehörde geöffnet wurde, verweigerte der Beamte mit einer Gebärde der Entrüstung die weitere Durchsicht und verlangte schleunige Schließung des Koffers; denn aus diesem war ein starker Skatolgeruch aufgestiegen und hatte offenbar bei dem Beamten den Eindruck hervorgerufen, daß in dem Koffer sehr stark mit Kot beschmutzte Wäsche enthalten sein müßte. Noch schlimmer ging es mir in dem Schweizer Hof in Ajaccio. Ich hatte dort ahnungslos meine Kleider im Zimmer aufgehängt. Aber nach einigen Tagen erschien die Wirtin und richtete an mich die ängstliche Frage: „Um Gotteswillen, was fangen Sie auf Ihrem Zimmer an, die Nachbarschaft beschwert sich über den schlechten Geruch, der von dort kommt?" Jetzt wußte ich sofort, wo die Schuld lag, und die fatale Joppe wurde nun 14 Tage ins Freie in die corsische Sonne gehängt. Als ich sie aber wieder einpackte und nach Deutschland zurückkehrte, war der Geruch zwar stark vermindert, aber noch keineswegs verschwunden.

In Würzburg hatte ich von vornherein das Glück, daß mir eine Reihe tüchtiger junger Doktoranden zuzogen. Ich erwähne zunächst A. Schlieper, Sohn des Kattundruckers Adolf Schlieper in Elberfeld, der 30 Jahre vorher durch seine Beobachtungen in der Harnsäuregruppe Adolf Baeyer Veranlassung gab, sich mit diesen Stoffen zu beschäftigen. Der Sohn ist sein Nachfolger geworden und steht, wenn ich nicht irre, noch an der Spitze des Geschäftes Baum & Co. Eine ebenso rühmliche Laufbahn hat C. Steche gemacht, der zur selben Zeit nach Würzburg kam. Er ist jetzt der Leiter der großen Riechstofffirma Heine & Co. in Leipzig. Dann Ahrheidt, später Beamter der Badischen Anilin- & Sodafabrik. Friedrich Ach aus Würzburg, ein ebenso begabter wie fleißiger Chemiker, der nebenher auch ein eifriger Corpsstudent war und mit dazu beitrug, daß seine Corpsbrüder die Vorlesungen nicht vernachlässigten. Er ist nach Beendigung der Studien in die Alkaloidfabrik von C. F. Böhringer & Söhne in Waldhof bei Mannheim eingetreten und hat dort als Leiter des wissenschaftlichen Laboratoriums großen Anteil an der technischen Ausarbeitung der Kaffeinsynthese genommen. Darüber werde ich aus der Berliner Zeit noch berichten. Leider ist er frühzeitig an Psoriasis und Nephritis zugrunde gegangen.

Ferner Jacob Meyer, der später selbständig das Tannigen entdeckte, jetzt seit einer Reihe von Jahren im Berliner Institut technische Probleme verschiedenster Art bearbeitet und mir bei der Verwaltung des Instituts eine wesentliche Hilfe leistet.

Im Sommer 86 wandte ich mich wieder den schon in Erlangen entdeckten Osazonen der Zucker zu und es gelang mir, mit Unterstützung von W. Wislicenus, der damals mein Privatassistent war, das Isoglucosamin zu gewinnen. Bald darauf, nachdem die Indolarbeiten abgeschlossen waren, begannen die Synthesen in der Zuckergruppe, die mich bis zum Ende der Würzburger Periode vorzugsweise beschäftigt haben. Die Zahl meiner Mitarbeiter ist hier so groß gewesen, daß ich sie nicht alle anführen kann, aber ich halte mich doch für verpflichtet, drei besonders zu nennen. Zuerst J. Tafel, der an der Oxydation der mehrwertigen Alkohole und der Verwandlung von Glycerose oder Acroleindibromid in den ersten synthetischen Zucker mit 6 Kohlenstoffatomen, die Acrose, teilgenommen hat. Das waren sehr mühsame und zum Teil auch recht gesundheitsschädliche Arbeiten.

Um genügende Mengen von Acrose in Form von Osazon zu gewinnen, haben wir einige Wochen zusammen in den Farbwerken Meister Lucius & Brüning zu Höchst a. M. mit den großen Hilfsmitteln der Fabrik Acrolein und sein Dibromid dargestellt. Das große Gefäß, aus dem das Acrolein destilliert wurde, befand sich im Freien, und die öfter wiederholte Operation wurde an windigen Tagen ausgeführt, um dem furchtbaren Geruch des Acroleins zu entgehen. Tafel ist einmal durch Versehen in eine Acroleinwolke hineingeraten und bekam ein so heftiges Nasenbluten, daß mir um seine Gesundheit bangte. Als etwas später die Schwierigkeiten bei der Zuckerarbeit sich immer mehr auftürmten und eine glückliche Lösung des Problems erst in weiter Ferne möglich schien, hat Tafel, der für die beabsichtigte wissenschaftliche Laufbahn auch selbständige Resultate nötig hatte, sich anderen Aufgaben zugewendet. An seine Stelle sind dann eine Reihe von anderen Herren getreten, unter denen ich besonders Josef Hirschberger, jetzt in Brooklyn, und Heller, jetzt in Leipzig, nennen muß. Bei der speziellen Synthese von kohlenstoffreichen Zuckern durch das Blausäure-Verfahren haben besonders der Engländer Passmore, der jetzt angesehener Handelschemiker in London ist, und Lorenz Ach aus Würzburg, von dem später noch die Rede sein wird, teilgenommen.

Unabhängig von mir waren die älteren Assistenten mit eigenen Problemen beschäftigt. Knorr hatte in Erlangen schon den großen praktischen Erfolg mit dem Antipyrin gehabt und wissenschaftlich nicht allein die Pyrazole, sondern auch die eleganten Synthesen von Pyrrolderivaten aus 1,4-Diketonen entdeckt. Diese Funde sind in Würzburg in weitestem Maße ausgenutzt worden. Dazu gesellte sich die Synthese des Morpholins, von dem der Entdecker annahm, daß es den Stickstoffring des Morphins enthalte.

Einen ebenso glücklichen Griff machte Wilhelm Wislicenus mit der Ausdehnung der Acetessigestersynthese auf andere Ester, z. B. den Oxalester. Wäre ihm L. Claisen nicht mit gleichzeitigen Beobachtungen in die Quere gekommen, so hätte er damals sicherlich das ganze große Gebiet der Kondensation von Estern untereinander oder mit Ketonen und Aldehyden erobert. Eine andere sehr niedliche Beobachtung von ihm war die Bildung von Natriumazid aus Stickoxydul und Natriumamid. Um dieselbe Zeit fand Tafel die Reduktion der Hydrazone zu Aminen mit Natriumamalgam und er hätte diese Reaktion sicherlich auch auf die Oxime ausgedehnt, wenn nicht Goldschmidt alsbald nach Tafels Publikation solche Versuche rasch angestellt und veröffentlicht hätte.

Von den Würzburgern Schülern verdient noch besonders Erwähnung Oscar Piloty. Nachdem er im Münchener Verbandsexamen keinen Erfolg gehabt hatte, kam er in etwas gedrückter Stimmung nach Würzburg, fand aber rasch das Selbstvertrauen wieder, als er bei seiner Doktorarbeit über die kohlenstoffreichen Zucker aus Rhamnose sein experimentelles Geschick beweisen konnte. Er hat dann auch mit Erfolg promoviert und nachher an der recht schwierigen Verwandlung der Zuckersäure in Glucuronsäure teilgenommen. Er durfte jetzt auch Baeyers einzige Tochter Eugenie, mit der er schon längst einig war, heiraten. Das junge Paar kam nach Würzburg und dem Ehemann passierte dabei das Unglück, daß er auf der Hochzeitsreise die militärische Kontrollversammlung vergaß und dafür mit einer kleinen militärischen Freiheitsstrafe belegt wurde. Zu meiner großen Freude entschloß er sich, mit mir nach Berlin überzusiedeln, um dort fast 7 Jahre als Assistent und später als selbständiger Leiter der analytischen Abteilung zu wirken. Als sich aber eine Gelegenheit bot, eine Professur an dem Münchener Laboratorium zu erhalten, zog die alte Sehnsucht nach der bayerischen Heimat sowohl ihn wie die Gemahlin unwiderstehlich dorthin, obschon er ein Vierteljahr später an dem neuen Berliner Institut die gleiche Stellung, vielleicht unter noch besseren Bedingungen, hätte haben können.

Die schönen wissenschaftlichen Leistungen Pilotys, die von einem anderen Fachgenossen in einem ausführlichen Nekrolog geschildert werden sollen, haben mein ursprüngliches Urteil über seine experimentelle Begabung durchaus bestätigt, und ich habe mich darüber um so mehr gefreut, als sein Schwiegervater diese Meinung anfangs nicht teilen wollte.

Zu Anfang des unseligen Krieges hatte Piloty das wehrpflichtige Alter schon überschritten. Trotzdem meldete er sich aus patriotischer Begeisterung als Freiwilliger, und ich konnte noch durch eine Empfehlung an die bayerische Militärbehörde seine Beförderung zum Leutnant d. R. erleichtern. An der Spitze einer Maschinengewehrabteilung ist er in der Sommeschlacht gefallen, tief bedauert nicht allein von der Familie, sondern auch von Freunden und Fachgenossen, die von ihm noch manche schöne wissenschaftliche Leistung erwarteten.

Im letzten Jahr der Würzburger Periode kam auch ein junger Däne Dr. Fogh dorthin, um eine thermochemische Arbeit, hauptsächlich über die Verbindungen der Zuckergruppe, auszuführen. Er hatte sich zuvor bei M. Berthelot in Paris mit den thermochemischen Methoden bekannt gemacht und auch eine Reihe von Instrumenten von dort mitgebracht. Nur die Verbrennungsbombe fehlte ihm, aber in den Ferien konnte er nach Paris reisen und dort die kalorischen Verbrennungsversuche ausführen. Die Arbeit ist in den „Comptes rendus" erschienen. Dr. Fogh ist ebenfalls mit mir nach Berlin gegangen und dort eine Zeitlang Assistent in der anorganischen Abteilung gewesen. Aber aus Gesundheitsrücksichten mußte er bald Urlaub nehmen und ist schließlich nach Kopenhagen zurückgekehrt, von wo er mir nur noch eine Verlobungsanzeige geschickt hat. Seitdem habe ich nichts mehr von ihm gehört.

In der großen Experimentalvorlesung über Chemie bildeten die Mediziner den größten Teil der Zuhörerschaft, dann kamen die Apotheker und erst in dritter Linie die Chemiker. Ich habe mich bemüht, dem gerecht zu werden und die eigentliche Vorlesung so einfach und populär zu halten, wie es ohne Gefährdung des wissenschaftlichen Charakters möglich ist. Aber hinterher gab ich häufig eine Ergänzung für die Fachchemiker, wobei nicht allein schwierigere theoretische Fragen, sondern auch manche speziellen experimentellen Methoden zur Sprache kamen. Soweit ich in Erfahrung bringen konnte, war diese Form des Vortrages den Studierenden recht willkommen. Leider ist sie für den Dozenten sehr anstrengend.

Der kollegiale Verkehr unterschied sich in Würzburg nicht unwesentlich von den Erlanger Sitten durch die freie, meist ganz ungeschminkte Äußerung der Meinungen und durch die Schlichtheit der Formen. Von dem Schaugepräge akademischer Würde, wie sie in Erlangen sowohl vom Senat als auch von der Fakultät geübt wurde, war in Würzburg wenig zu spüren, was mich als Rheinländer besonders sympathisch berührte.

Die philosophische Fakultät bestand wie in München aus zwei Sektionen. Unsere mathematisch-naturwissenschaftliche Abteilung zählte nur 6 Ordinarien, aber darunter waren mehrere durch hohes wissenschaftliches Ansehen ausgezeichnete Personen. Das galt besonders für den Physiker Friedrich Kohlrausch und den Botaniker Julius Sachs. Zu Kohlrausch bin ich bald in ein freundschaftliches Verhältnis getreten; denn er war ein ebenso verständiger wie wohlwollender Kollege und jederzeit bereit, vernünftige Wünsche zu unterstützen. Seine ausgezeichneten Untersuchungen über das elektrische Leitvermögen von Lösungen und seine physikalischen Konstantenbestimmungen, sowie das von ihm verfaßte vortreffliche Lehrbuch der „Praktischen Physik" brauche ich hier nicht näher zu besprechen, da sie den Naturforschern genügend bekannt sind. Ich habe mit ihm an schönen Sommertagen manchen Spaziergang unternommen und auch in seiner Familie, besonders von seiten der liebenswürdigen Gattin, freundliche Aufnahme gefunden.

Ganz anders geartet war Sachs, der berühmte Pflanzenphysiologe, der in Würzburg eine große Schule von Botanikern geschaffen und auch als wissenschaftlicher Schriftsteller durch ein großes Lehrbuch und die ausführliche Geschichte der Botanik Hervorragendes geleistet hat. Charakteristisch für seine kühne Kritik ist das vernichtende Urteil über das von Linné aufgestellte rein schematische System der Pflanzen. Er pflegte zu sagen, dieses System sei wie ein Pesthauch über die Botanik gegangen und habe die schon vorhandenen gesunden Keime eines natürlichen Systems auf lange Zeit zerstört. Zu meiner Zeit war Sachs schon ein kranker Mann mit überreiztem Nervensystem, der seine Launen in heftiger Weise auslassen konnte. Trotzdem ist es mir gelungen, mit ihm einen freundschaftlichen Verkehr zu pflegen, und ich habe bei öfteren Besuchen in seinem Institut manches gelernt, was mir für eigene Arbeiten auf physiologisch-chemischem Gebiet Nutzen brachte.

Fremder blieb mir der Mineraloge Fridolin Sandberger, ein ebenfalls verdienter Gelehrter und ein wohlwollender Mensch, aber ein Sonderling, der noch ungern das Deutsche Reich anerkannte und auch von der modernen Chemie nichts wissen wollte. Er brachte das drastisch zum Ausdruck mit der Redensart: „Chemie ist, was knallt und stinkt.“ Mindestens ebenso originell war der Zoologe Karl Semper, der eine so große Rolle bei meiner Berufung gespielt hat. Seine wissenschaftliche Blütezeit lag auch schon eine Weile zurück. Als junger Mann hatte er zu wissenschaftlichen Zwecken weite Reisen gemacht und über die Philippinen und die Palaoinseln, wo er sich jahrelang aufhielt und auch seine spätere Gemahlin kennen lernte, wertvolle Bücher geschrieben. Zu meiner Zeit beschäftigte ihn am meisten der Plan eines Neubaues für das zoologische Institut. Da die Bewilligung der Gelder in München auf Schwierigkeiten stieß, so versuchte er die sonderbarsten Mittel, um die Notwendigkeit des Baues allen Leuten klar zu machen. So hatte er im alten Institut, das im vierten Stock des Universitätsgebäudes untergebracht war, große Aquarien angelegt, die eines Tages schadhaft wurden und alle darunter gelegenen Räume unter Wasser setzten. Es blieb also nichts anderes übrig, als einen Neubau am Pleicher Ring zu errichten, und die Einweihungsfeier wurde von Semper so eindrucksvoll inszeniert, daß sie sicherlich jedem Teilnehmer in der Erinnerung geblieben ist; denn zum Schluß führte er die ganze Gesellschaft, unter der sich die Spitzen der Universitäts- und Verwaltungsbehörden sowie eine Reihe von Damen befanden, in das Warmhaus, wo Pflanzen und Tiere in bunter Abwechselung versammelt waren, und das mit einer ausgiebigen Berieselungsanlage versehen war. Als die Gesellschaft sich bei Bier und Würstchen in einer behaglichen Feststimmung befand, trat plötzlich bei geschlossenen Türen die Berieselung in Tätigkeit, bis die ganze Versammlung von einem tropischen Regen total durchnäßt war.

Einige Zeit später kam der Kultusminister Dr. Müller, der eben sein Amt angetreten hatte, zur Besichtigung der Universität nach Würzburg. Sein Besuch war auch den einzelnen naturwissenschaftlichen Instituten angekündigt, verzögerte sich aber aus irgendwelchem Grunde weit über die angesetzte Zeit. Semper, dem das Warten zu lang wurde, hatte sich ein Frühstück von Wurst und Bier kommen lassen und war eben mit dessen Vertilgung beschäftigt, als der Minister eintrat. Dieser begrüßte den schmausenden Professor mit den jovialen Worten: „Ihnen scheint es heute gut zu schmecken", worauf die prompte Antwort erfolgte: „Wenn Sie so lange wie ich vergeblich auf den Minister gewartet hätten, würden Sie auch Hunger bekommen haben." Diese freimütige, keineswegs durch Ehrerbietung ausgezeichnete Antwort gab dem Minister Anlass, sich bald wieder zu empfehlen. Das sind nur ein paar Proben von den zahlreichen Schnurren, die Semper geliefert hat. Andererseits war er ein prächtiger Mann von würdigem Äußern, voll lustiger Einfälle, aller geselligen Unterhaltung zugetan und frei von jeder gelehrten Pedanterie. Man spürte bei ihm die Abkunft aus der durch hervorragende Künstler z. B. den berühmten Architekten Gottfried Semper ausgezeichneten Familie.

Ein weiteres Original war der Mathematiker Prym, der aus einer reichen Fabrikantenfamilie zu Düren stammte. Er hatte zwar Mathematik studiert, war aber dann in ein Bankgeschäft zu Wien eingetreten, und von dort durch Kappler als Ordinarius der Mathematik an das Polytechnikum zu Zürich berufen worden. Obschon mit Glücksgütern sehr gesegnet, führte er in Würzburg mit seiner Familie ein ziemlich einfaches Leben. Mit dem Lehramt nahm er es sehr genau. Die Vorlesungen waren auf das gewissenhafteste vorbereitet, und es kam nicht selten vor, daß er säumige Studenten durch seinen Wagen aus der Wohnung abholen und in die Universität fahren ließ. Als Teilhaber an manchen industriellen Unternehmungen interessierte er sich für Chemie und hat mich später auch in Berlin mehrmals besucht.

Ein zweiter Ordinarius für Chemie war in Würzburg nicht vorhanden. Das Institut für angewandte Chemie wurde von dem außerordentlichen Professor L. Medicus geleitet, der nicht allein bei Wislicenus, sondern schon bei dessen Vorgänger A. Strecker Assistent gewesen und deshalb der Hauptträger der chemischen Tradition in Würzburg war. Große wissenschaftliche Interessen haben ihn niemals geplagt, aber er war bei den Studenten wegen seiner Jovialität recht beliebt und in unserem chemischen Kreise wegen seines köstlichen Humors und seiner freundlichen Lebensauffassung ein gern gesehenes Mitglied. Auch ich bin mit ihm aufs beste ausgekommen. Nach meinem Weggang ist er zum Ordinarius befördert worden und vor einigen Jahren auch schon zu den Vätern gegangen. Am Feldzug 1870 hatte er als bayerischer Offizier teilgenommen und war bei Orleans verwundet in Gefangenschaft geraten. Seine Schilderungen über die schlechte Behandlung und gehässige Verhöhnung von seiten der französischen Bevölkerung stimmte genau mit dem überein, was man heute über die Behandlung deutscher Kriegsgefangener in Frankreich hört, obschon der damalige Krieg so rasch zu Ende ging und die Volksleidenschaft keineswegs so aufpeitschte, wie der jetzige. Die philologisch-historische Klasse der philosophischen Fakultät zählte außer dem klassischen Philologen und Archäologen Urlich wenig hervorragende Männer. Ganz anders war es in der medizinischen Fakultät, zu deren Mitgliedern ich vielerlei Beziehungen hatte. Von Erlangen mit uns übergesiedelt, war Wilhelm Leube der Nachfolger des nach Berlin berufenen Professors Gerhardt in der inneren Klinik des Julius-Hospitals geworden; er verstand es, die große Tradition dieser Stelle in jeder Beziehung zu wahren. Neben ihm stand der Chirurg Maass, ein geschickter Operateur, anregender Lehrer und trefflicher Gesellschafter. Leider ist er in ziemlich frühem Alter gestorben. Als Senior stand an der Spitze der Fakultät der Anatom Albert Kölliker, ein geborener Schweizer. Er genoß in seiner Wissenschaft durch die Fülle wertvoller Untersuchungen großes Ansehen. Seinem Einfluß war es zuzuschreiben, daß die Anatomie äußerlich das schönste Institut hatte und in dem Unterricht der Mediziner eine überragende Rolle spielte. Nebenher war er ein ungewöhnlich schöner Mann, mit feinem Gesicht, weißem Lockenhaar, klugen dunklen Augen und einer zarten, fast frauenhaften Hautfarbe. Auf die Chemie muß er früher einen kleinen Pik gehabt haben; denn es wurde von ihm der Ausspruch kolportiert, den dümmsten seiner Söhne lasse er Chemiker werden, wozu die boshaften Würzburger den Zusatz machten, er habe tatsächlich auch den richtigen ausgesucht.

Mir ist er immer nur mit Freundlichkeit begegnet. Sein Prosektor war der außerordentliche Professor Stöhr, ein geborener Würzburger, ein sehr behaglicher und liebenswürdiger Mann, der auch Nachfolger von Kölliker geworden ist. Durch Originalität zeichnete sich aus der Polikliniker Geigel, ein Meister in der Abfassung von feinen und leicht ironischen Gutachten, deren sich die Fakultät stets bediente, wenn sie unbequeme Zumutungen des Ministeriums in München bekämpfen wollte. Er war das Haupt einer Musikbande, die aus Würzburger Professoren oder Bürgern z. B. den beiden Brüdern Stöhr und dem Pharmakologen Kunkel bestand. Während der Herbstferien hauste diese Gesellschaft zu Ammerland am Starnberger See und gab täglich kleine Konzerte, wobei das Hornblasen die Hauptrolle spielte.

Als Geburtshelfer fungierte damals noch der einst so berühmte Scanzoni, er starb aber bald und wurde durch Hofmeier ersetzt. Ganz besonders muß ich aber erwähnen den Augenarzt Julius Michel, einen lustigen Pfälzer, der vor meiner Zeit auch in Erlangen gewesen war und mit dem ich mich recht befreundet habe. Obschon er es mit seiner ärztlichen Kunst und auch der Wissenschaft sehr ernst nahm, und sowohl als Arzt wie als Gelehrter einen guten Ruf genoß, so war ihm das Reißen von losen Witzen ein richtiges Lebensbedürfnis geworden. In der Gesellschaft bildete er deshalb ein belebendes Element, und wir haben in dem gastfreien Hause Leube nicht selten stundenlang seinen Späßen mit Vergnügen zugehört. Auch in ernsten Augenblicken konnte er Witze nicht unterdrücken, und seine Vorlesung war in dieser Beziehung etwas Besonderes. Ich will nur ein Beispiel anführen: Eines Tages hatte er einen Kandidaten der Medizin namens Jerusalem zu prüfen. Das Resultat war ungenügend, und nun verkündete er der Schar der Freunde des Kandidaten das Resultat mit den Worten: „Israel weine, Jerusalem ist gefallen.“ Boshafte Leute haben allerdings hinterher behauptet, er hätte den Kandidaten nur durchfallen lassen, um diesen Witz machen zu können. Er war unverheiratet, führte aber einen trefflich geleiteten Haushalt und gab kleine Feste, bei denen ausgelassene Fröhlichkeit herrschte.

Ich hatte einmal mit einem Assistenten zusammen das kleine Unglück, daß uns ein stark gefärbtes chemisches Präparat beim Platzen eines Preßsackes ins Gesicht flog und die Augen nicht allein ganz verschmierte, sondern auch reizte. Wir hatten nichts Eiligeres zu tun, als die Universitätsklinik aufzusuchen. Freund Michel hat uns sofort sorgfältig gesäubert. Als aber die Harmlosigkeit der Sache festgestellt war, machte es ihm ein besonderes Vergnügen, in dem Buch der Poliklinik ein ausführliches Protokoll über diesen seltenen Fall niederzuschreiben, wobei er durch eine ganze Reihe von langen chemischen Namen seine wissenschaftliche Bedeutung zu illustrieren wußte. Ich habe später in Berlin, wohin er etwa 10 Jahre nach mir berufen wurde, seine Hilfe nochmals in Anspruch nehmen müssen, ebenfalls mit bestem Erfolg. Ich wurde damals von leichtem, aber anhaltendem Kopfschmerz geplagt, der sich bis zur Übelkeit steigern konnte, und die Ärzte, die ich frug, stellten ganz besorgte Diagnosen auf Zirkulationsstörungen und dergl. Da ich aber die Beobachtung machte, daß der Zustand am schlimmsten war, wenn ich tags zuvor stundenlang gelesen hatte, so kam ich schließlich auf die Vermutung, daß das von mir benutzte Augenglas, das ich mir einstmals selbst beim Optiker ausgesucht und dann länger als 20 Jahre unverändert getragen hatte, Schuld daran haben könne. Ich ging deshalb zum Freund Michel, und nachdem er die Augen sowie das Glas untersucht hatte, faßte er sein Urteil in folgende Worte zusammen: „Die Kopfschmerzen hätten Sie längst verdient, man sagt doch, Sie seien sonst nicht so dumm." Er verschrieb mir dann die richtige Brille, und die Kopfschmerzen, die mich viele Wochen geplagt hatten, waren wie weggeblasen. Niemals ist bei mir eine Kur von so glänzendem und raschem Erfolge begleitet gewesen.

Von den Juristen sind mir außer dem schon erwähnten Verwaltungsdirektor Risch der Pandektist Burkhardt und der Nationalökonom Schanz im Gedächtnis geblieben. Auch an einige Theologen erinnere ich mich gerne. Es waren katholische Herren, aber keineswegs vom Kaliber der sogen. Hetzkapläne, sondern zum Teil sehr gebildete, feinfühlige und joviale Herren. Mit dem Senior der Fakultät, den sie den Bischofsmacher nannten, bin ich öfters auf dem alten Glacis der Stadt, das so hübsche Spaziergänge bot, zusammengetroffen, und wir haben uns stundenlang in freiester Weise über Kirche, Staat und Gesellschaft unterhalten. Er war im Collegium Germanicum im Rom erzogen, hatte ein lesenswertes Buch über Kirche und Staat geschrieben, und ich habe von ihm vieles über die Institutionen der katholischen Kirche, besonders über die praktische Seelsorge gehört, was mich in höchstem Grade interessieren mußte. Ob dieser Verkehr mir in den katholischen Kreisen trotz meiner protestantischen Abstammung Vertrauen verschaffte, kann ich nicht sagen, aber ich erlebte doch eines Tages die Überraschung, daß etwa 25 katholische Theologen als eingeschriebene Hörer meine Vorlesung über anorganische Chemie besuchten. Auf meine Frage, wie die Herren dazu kämen, wurde mir die Antwort, der Bischof von Speyer habe es befohlen. Wahrscheinlich hatte dieser kirchliche Würdenträger den Wunsch, daß der in seiner Diözese tätige Klerus gewisse chemische Kenntnisse besitze, um besser mit den Arbeitern der zahlreichen chemischen Fabriken in der dortigen Gegend sich verständigen zu können. Auch der Religionslehrer des katholischen Gymnasiums war mein Zuhörer, und er erklärte mir eines Tages, er könne die Lehren der modernen Chemie in dem Religionsunterricht vortrefflich verwerten. Ich habe daraus den Schluß gezogen, daß Naturwissenschaft und Religion keine grundsätzlichen Gegner zu sein brauchen.

In Würzburg bestehen zwei wissenschaftliche Gesellschaften, an deren Arbeiten ich mich gerne beteiligt habe. Die ältere und allgemeinere führte den Namen physikalisch-medizinische Gesellschaft; ihre Mitglieder waren meistens Professoren und Dozenten. Die jüngere chemische

Gesellschaft war erst von J. Wislicenus gegründet worden. Zu ihr gehörten selbstverständlich die Professoren und Assistenten der beiden chemischen Institute, aber die Mehrzahl der Mitglieder waren doch ältere Studenten. Auch hier wurden Originalvorträge, jedoch häufiger große Referate über fremde Arbeiten gegeben und daneben spielte die Geselligkeit eine Hauptrolle. Den Höhepunkt erreichte diese bei den alljährlichen Stiftungsfesten, zu denen viel frühere Mitglieder von auswärts und auch zahlreiche Professoren der philosophischen und medizinischen Fakultät erschienen. Gewöhnlich gab es dabei außer witzigen Tischreden und Vorträgen auch ein Festspiel. Mir sind zwei in Erinnerung geblieben. In dem einen wurde das Doktorexamen eines Chemikers auf die Bühne gebracht mit sehr komischen und drastischen Ausfällen gegen die bekannte Prüfungsweise einzelner Professoren, so daß der Dekan der Fakultät, ein Philologe, unwillig den Saal verließ, während das übrige Publikum sich köstlich unterhielt. Die zweite Aufführung war eine Operette „Der Chemikado" mit den Melodien des Mikado und einem sehr witzigen Text von Dr. Reitzenstein. Die Kostüme und die Haartrachten hatte das Stadttheater geliehen. Der Mikado war mein früher erwähnter, lieber Neffe Alfred Mauritz, der damals in Würzburg Chemie studierte, und von dem ich in späteren Jahren noch viel Freundschaft erfahren habe. Er trug meinen Laboratoriumsanzug und Hut, sprach den niederrheinischen Dialekt und trat als einziger Europäer in der japanischen Gesellschaft um so mehr hervor. Ich besitze eine Photographie der Schauspieler, deren Anblick mich noch jetzt zum Lachen bringt.

Die chemische Gesellschaft hat auch nach meiner Zeit ihre Tradition zu wahren gewußt und von Dr. Reitzenstein ist noch manches hübsche Festspiel verfaßt worden. Wegen Zeitmangel habe ich leider an den Stiftungsfesten von Berlin aus nicht teilnehmen können, aber beim 25-jährigen Jubiläum hielt ich mich für verpflichtet, meine Absage in gereimte Form zu kleiden, und da es eins der wenigen Gedichte ist, die ich verfaßt habe, so will ich das Telegramm hier anführen.

Chemische Gesellschaft Würzburg.

„Zum fröhlichen Feste
Glückwünsch ich das Beste.
Blüh' immer so weiter
Und bleib auch stets heiter,
Grüß' alle Bekannte
Und chemisch Verwandte
Vom alten Giftmischer
Aemilius Fischer."

Als Antwort der Gesellschaft traf prompt ein liebenswürdiger Vers von Dr. Reitzenstein ein.

Daß in Würzburg Fröhlichkeit und Humor blühten, war kein Wunder. Die freundliche Stadt mit dem prächtigen Schlosse, dem lieblichen Flusse, den schönen Glacis-Anlagen und den rebenbekränzten Bergen, die behagliche unterfränkische Bevölkerung und die alte Tradition des Krummstabes waren wohl geeignet, die an und für sich schon heitere Stimmung der akademischen Gesellschaft zu verstärken. Der Verkehr der Professoren untereinander und auch mit den Studenten war leicht und gemütlich und nahm nur zeitweise, z. B. bei den Prüfungen eine ernstere Form an. Trotzdem herrschte unter der Studentenschaft ein guter Geist; denn es wurde im allgemeinen in Würzburg ziemlich viel gearbeitet und in dem chemischen Laboratorium konnte man sich über Mangel an Fleiß nicht beklagen. Der größere Teil der Studentenschaft bestand aus Norddeutschen. Dasselbe galt von den Professoren, und von dem Partikularismus, der der bayerischen Regierung öfters bei Berufung der Professoren vorgeworfen wurde, war in Würzburg nichts zu merken. Hatte man doch an die Universität der alten Bischofsstadt, die nur eine katholisch-theologische Fakultät besaß, als Lehrer des Kirchenrechts einen Protestanten berufen! Eine Einmischung in Berufsgeschäfte ist allerdings von der ultramontanen Kammermehrheit öfters versucht, aber vom damaligen Kultusminister Dr. Lutz meist erfolgreich zurückgewiesen worden. Auch der Familienverkehr wurde in Würzburg in anmutiger Form zwischen den verschiedenen Fakultäten gepflegt und im ersten Winter war die Erlanger Kompagnie, d. h. die Ehepaare Leube und Knorr sowie meine Wenigkeit, bei den üblichen Abendessen das Objekt einer feierlichen Begrüßung. Die Antwort darauf haben wir abwechselnd gegeben, und diese erste Tischrede ist nicht selten für die akademische Gesellschaft der kritische Maßstab, den sie an neue Mitglieder anlegt. Ich mußte bei Kohlrauschs reden und hatte mir einen launigen Toast überlegt. In ihm spielte zum Schluß die Elektrizität eine Rolle, indem ich die einzelnen Damen den damals frisch erfundenen Glühlampen verglich und für die Hausfrau die Bogenlampe reservierte. Nun glaube alle Welt, daß ich mich in diesem Bilde wie in einem Spinnetz verwickeln und höchstens durch einen brutalen Riß wieder befreien könnte. Aber glücklicherweise fiel mir ein, daß von der Bogenlampe zur Sonne rhetorisch nur ein kurzer Sprung nötig sei, und damit hatte ich das poetische Bild gewonnen, um die Hausfrau würdig zu preisen und die Zustimmung der Tischgesellschaft zu einem Hoch zu gewinnen.

Dauernden Familienverkehr habe ich aber nur bei Knorrs und ganz besonders bei dem lieben Ehepaar Leube gehabt. Die Woche mindestens einmal sind wir dort zusammengekommen. Gewöhnlich waren Michel und noch einige andere Freunde mit dabei, und wir haben bei einfachem Abendmahl überaus lustige Stunden verlebt. Leube war ein prächtiger Gesellschafter, klug, wissenschaftlich gut unterrichtet und mit den reichen Erfahrungen des erfolgreichen Arztes versehen. Er kannte eine große Anzahl von Menschen, die seinen ärztlichen Rat in Anspruch nahmen, hielt schöne Reden und machte allerliebste Gelegenheitsgedichte.

Seine liebe Frau Natalie fühlte sich mir chemisch verwandt; denn sie war, wie ich schon früher erwähnte, die Tochter von Adolf Strecker, der als Professor der Chemie in Würzburg starb. Sie selbst hatte sich in dem chemischen Institut als 18-jähriges Fräulein mit ihrem Wilhelm verlobt. Dazu kam, daß ich zufälligerweise bei mehreren chemischen Arbeiten, z. B. bei den Hydrazinen und dem Coffein der wissenschaftliche Erbe von Strecker geworden war. Kurz nach meiner Verheiratung hat das Ehepaar Leube mir die Duzfreundschaft angeboten, und wir stehen noch jetzt, wo Leubes in Stuttgart ein behagliches Alter verleben, in freundschaftlichem Briefwechsel. Die älteste Tochter Lilly hat den Gynäkologen Bumm, der jetzt ebenfalls an der Berliner Universität tätig ist, geheiratet. Die drei anderen Töchter sind die Frauen von Offizieren geworden. Leube hat mir auch in Krankheitsfällen wertvolle Dienste geleistet und meinen ältesten Sohn Hermann in frühester Jugend bei einem schweren Darmkatarrh geradezu vor dem Tode bewahrt.

Frau Leube hatte schon in Erlangen die gute Absicht, mir eine Frau zu verschaffen und glaubte das geeignete Mädchen dafür in Fräulein Agnes Gerlach in Erlangen gefunden zu haben. Aber meine Gleichgültigkeit in Sachen der Liebe und die Überhäufung mit wissenschaftlichen Problemen waren ihren Plänen nicht günstig gewesen, und schließlich trat noch als zweites Hindernis meine Erkrankung und die Befürchtung eines Rückfalles dazwischen. Aber Frauen geben so leicht ihre Lieblingsideen nicht auf, und so wußte sie das durch Liebreiz ausgezeichnete Fräulein wiederholt zu Besuchen in Würzburg zu veranlassen. Sie wurde dabei auf das kräftigste unterstützt von Frau Dr. Knorr, die sich ebenfalls mit Fräulein Gerlach angefreundet hatte. Bei einem dieser Besuche ist es dann auch wirklich zur Verlobung zwischen dem Fräulein und mir gekommen. Es war am 1. Dezember 1887, wo ich selbst 35 Jahre und meine Braut 26 Jahre alt war. Die Hochzeit fand statt in Erlangen am Sonnabend den 22. Februar 1888, kurz vor Karneval, so daß ich 4 Tage Ferien hatte, um mich in den neuen Zustand hineinzugewöhnen. Wir mußten noch einige Wochen in Würzburg bleiben und haben dann Mitte März bei Beginn der Osterferien eine vierwöchentliche Reise nach Italien gemacht. Über diese Dinge will ich aber nicht Näheres berichten, weil die Schließung des Ehebundes eine zu intime Sache ist. Ich kann nur sagen, daß meine liebe Frau ein durch körperliche Schönheit, Reinheit der Seele und Sanftmut ausgezeichnetes Wesen war. Ihre Eltern hatten sie auf den Händen getragen und dadurch vielleicht zu sehr verwöhnt; denn die Pflichten der Ehe und die Führung eines großen Haushaltes haben ihr namentlich in Berlin den ruhigen Lebensgenuß stark verkürzt und eine gewisse Gleichgültigkeit gegen ihre eigene Wohlfahrt erzeugt, die bei ihrer letzten Krankheit einen unglückseligen Einfluß ausübte und vielleicht mit an ihrem Tode schuld gewesen ist. Sie starb am 12. November 1895 in Berlin an einer Meningitis infolge einer Mittelohrentzündung, wahrscheinlich weil die rettende Operation wegen des Widerstandes der Patientin zu spät ausgeführt wurde. In Würzburg hat sie mir zwei Söhne geschenkt.

Der älteste Hermann Otto Lorenz wurde geboren am 16. Dezember 1888 und war von Anfang an ein kräftiges gesundes Kind. Dem entsprach auch seine spätere Entwicklung. Er hat nur eine gefährliche Krankheit mit zwei Jahren durchgemacht, einen Magen- und Darmkatarrh, da der Hausarzt törichterweise von vornherein ein Stopfmittel gab und dadurch einen hartnäckigen, gefährlichen Darmverschluß herbeiführte. Das Kind wäre sicher gestorben und zwar an Verdurstung, wenn wir ihm nicht auf Rat von Leube zuguterletzt per anum eine große Menge Wasser hätten zuführen können.

Die Geburt des zweiten Knaben, die am 5. Juli 1891 stattfand, war verfrüht und das Kind infolgedessen schwach. Es hat sich zwar später ziemlich rasch erholt. Aber als wir nach Berlin übersiedelten, hatte der jetz ganz kräftige Knabe unter der schlechten Milch der Großstadt zu leiden. Er bekam infolgedessen im Juni 93 einen anhaltenden Darmkatarrh und im Herbst desselben Jahres das Scharlachfieber, gefolgt von einem sehr häßlichen Bronchialkatarrh. Diese Umstände haben vielleicht ungünstig auf das Nervensystem des Kindes gewirkt. Trotzdem entwickelte er sich zu einem großen, starken und geistig regsamen Jüngling, der mit Leichtigkeit die Schule absolvierte und mit 18 Jahren die Universität bezog. Ohne sich eine Erholungspause nach dem Gymnasium zu gönnen, stürzte er sich mit Feuereifer auf das Studium der Medizin. Alles sprach bei ihm für eine hoffnungsvolle Zukunft, als der Sommer 1910 eine jähe Schädigung der Gesundheit brachte. Er hatte hartnäckig auf der Absicht bestanden, in dem Sommer als Mediziner das vorgeschriebene halbe Jahr mit der Waffe zu dienen und geriet zu seinem Unglück in ein Infanterieregiment zu Jena. Durch den forcierten Militärdienst des Sommers ist der noch nicht genug entwickelte, ungewöhnlich große junge Mann überanstrengt worden. Er bekam Herzbeschwerden und wurde vom Militär entlassen. Die

Krankheit war an und für sich wohl nicht schlimm, aber sie übte auf das Gemüt des jungen Mannes einen verderblichen Einfluß aus, denn er sah, wie es bei jungen Medizinern nicht selten ist, die Krankheit in einem besonders düstern Bilde. Er setzte zwar seine Studien fort und bestand im Frühjahr 1912 in Heidelberg mit Auszeichnung die ärztliche Vorprüfung, studierte dann zwei Semester in Würzburg und kam wegen militärischer Dinge im Sommer 1913 wieder nach Berlin. Aber hier brach er nervös zusammen. Seine Arbeitskraft war erschöpft. Er glaubte immer kränker zu werden, ließ sich im Herbst nach Nauheim, später nach Meran schicken und verfiel in eine tiefe Melancholie. Weder ich noch die behandelnden Ärzte haben seinen Zustand richtig erkannt, sonst hätte man vielleicht der fortschreitenden geistigen Erkrankung vorbeugen können. So aber kam es im November in Meran zum offenkundigen Ausbruch der Geisteskrankheit. Durch einen mehrmonatlichen Aufenthalt bei Binswanger in Jena gelang eine zeitweise Heilung. Er blieb noch den Sommer als Studierender in Jena. Obschon aus dem Militärverhältnis entlassen, hatte er sich auf eine Anfrage der Militärmedizinalbehörde verpflichtet, im Kriegsfalle ärztliche Hilfsdienst zu leisten. Dementsprechend wurde er im September 1914 als Unterarzt an ein Lazarett zu Erfurt kommandiert. Diese Tätigkeit hat ihm anfangs zugesagt und gutgetan, aber nach 5 Monaten kam er in Streit mit dem Vorgesetzten und im Anschluß daran erfolgte ein neuer Ausbruch seiner Krankheit. Er ist dann wiederholt bei Binswanger gewesen, ohne aber seine Arbeitskraft wiederzugewinnen. Manchmal schien es, als sei die Krankheit ganz gehoben, so ruhig und vernünftig wußte er sich zu geben. Als ich im August 1916 mit ihm einen mehrwöchigen Aufenthalt in St. Blasien genommen hatte, war ich voller Hoffnung, daß die Heilung anhalte. Aber plötzlich kam ein Rückfall, schlimmer als die vorhergehenden. Wir kehrten nach Wannsee zurück, und er ging dann bald wieder zu seinem Freund Binswanger. Hier verschlimmerte sich der Zustand, er mußte in eine geschlossene Anstalt aufgenommen werden und das Bewußtsein, nun aller Wahrscheinlichkeit nach ein verlorener Mensch zu sein, hat ihn im Zustand tiefer Depression zu dem Entschluß geführt, freiwillig in den Tod zu gehen. Er ist am 4. November 1916 im Alter von 25 Jahren gestorben und ruht auf dem kleinen Friedhof zu Wannsee. Ein lieber, guter Sohn, ein talentvoller und strebsamer junger Mann ist mit ihm dahingegangen.

Mein dritter Sohn Alfred Leonhard Joseph wurde geboren am 3. Oktober 1894 zu Ambach am Starnberger See. Ich war damals bereits in Berlin und meine Frau wollte nur für einige Wochen ihre Eltern in Ambach besuchen. Da sie in hoffnungsvollem Zustand war und schon früher einmal Unglück gehabt hatte, so hielt ich die Reise für zu gewagt. Aber mit Rücksicht auf das hohe Alter ihres Vaters ließ sie sich nicht abhalten. Die Folge war, daß sie sich in Ambach gleich zu Bett legen und 5 Monate bis zur Geburt des Knaben liegen mußte. Auch dieser kam sehr zart auf die Welt. Er wurde später ebenfalls ein kräftiger Mann, hatte aber eine sehr empfindliche Haut und litt von Zeit zu Zeit an nervösen Kopfschmerzen. In frühester Jugend hatte er im Sommer in Wannsee zweimal hintereinander einen malariaartigen Zustand durchgemacht, bis wir unseren Wohnsitz dort auf die Höhe verlegten. Auch er war begabt, fleißig und ein guter Schüler, verließ mit 18 Jahren das Gymnasium und wollte wie der älteste Bruder Chemiker werden. Ich riet ihm aber davon ab wegen der großen Empfindlichkeit seiner Haut und seiner Kopfnerven. Er studierte deshalb zunächst Physik, verließ diese aber schon nach zwei Semestern und wurde Mediziner. Während er in Heidelberg studierte, brach der Krieg aus. Er meldete sich zunächst freiwillig, wurde aber zurückgestellt. Im Januar 15 trat er bei einem Artillerieregiment in Berlin ein. Nach erfolgter Ausbildung ging er im September desselben Jahres ins Feld und fand bei der Munitionskolonne, bei der sein Bruder Hermann Leutnant war, als Sanitätsgefreiter Verwendung. Hier wurde er bald zum Unteroffizier befördert. Im Juli 1916 hat er während seines letzten Urlaubs in Heidelberg die ärztliche Vorprüfung mit Auszeichnung bestanden. Im August desselben Jahres wurde er mit der Munitionskolonne nach Rumänien geschickt, und die beiden Brüder haben damals den Vormarsch in der Dobrudscha mitgemacht. Hier wurde er zum Feldunterarzt befördert. Während der ältere Bruder nach einer 6-wöchentlichen Ausbildung in Berlin Gasschutzoffizier wurde und dann als solcher den Vormarsch der deutschen Armee in Rumänien im Hauptquartier von Falkenhayn mitmachte, kam Alfred in verschiedene Spitäler und zuletzt zu seinem Unglück in ein Seuchenlazarett zu Bukarest, wo die sanitären Verhältnisse nach seiner eigenen Schilderung recht schlecht waren. Hier hat er sich an fleckfieberkranken Türken infiziert und ist nach 14tägiger Krankheit am 29. März 1917 gestorben. Er wurde auf dem Ehrenfriedhof zu Bukarest beerdigt, und sein Bruder Hermann, der damals in Focsani stand, konnte ihm mit kurzem Urlaub die letzte Ehre erweisen. Er war ebenfalls ein sehr lieber, verständiger und begabter Mensch von vornehmer Gesinnung und sehr geschickt im Verkehr mit dem Volke. Wahrscheinlich wäre er ein ausgezeichneter Arzt, vielleicht auch ein erfolgreicher Forscher geworden.

Im Sommer 1888 war der hochbetagte Robert Bunsen vom Lehramt zurückgetreten. Die Professur wurde zunächst Victor Meyer angeboten, der aber nach einigem Zögern ablehnte und in Göttingen bleiben wollte. Darauf erhielt ich den Ruf und der betreffende Referent des badischen Ministeriums kam zur Unterhandlung mit mir nach Würzburg. Die Bedingungen waren im allgemeinen recht günstig und obschon ich gerne in Würzburg war, hatte Heidelberg doch für mich und noch mehr für meine Frau eine gewisse Anziehungskraft. Wir sind deshalb zusammen im Frühjahr 89 nach Heidelberg gefahren, um uns über alle Einzelheiten zu unterrichten. Im Hotel trafen wir bereits Exzellenz Bunsen, der seine Dienstwohnung aufgegeben und einstweilen Quartier im Gasthaus genommen hatte. Der verehrungswürdige alte Herr empfing uns mit großer Höflichkeit und suchte uns die Vorzüge der Heidelberger Stelle möglichst klar zu machen.

Als ich nach der ersten Unterredung meine Frau frug, welchen Eindruck sie von dem großen Chemiker empfangen hatte, erwiderte sie lachend: „Erst möchte ich ihn waschen und dann küssen; denn er ist ein gar lieber Mann."

Am nächsten Morgen zeigte uns Bunsen das von ihm erbaute und so lange benutzte Laboratorium am Wredeplatze. Er war ganz verliebt in das alte Haus, das allerdings die Weihe einer großen Tradition und gewaltigen wissenschaftlichen Arbeit trug. Aber die Hilfsmittel waren doch im Vergleich zur Neuzeit recht bescheiden. Die Ventilation wurde noch wie in den alten Alchemistenküchen durch einen großen Rauchfang besorgt und auf meine Frage, ob das genüge, erklärte der alte Herr: „Wir haben hier die reine Gartenluft." Im Gegensatz dazu meinte dann der Assistent, an den ich mich noch vertraulich wandte, daß der Gestank meist unausstehlich sei.

Als wir zum Schluß auch die Dienstwohnung besichtigen wollten, führte uns zwar Bunsen bis zur Eingangstür, zog dann aber einen riesigen Schlüsselbund hervor und prüfte jeden einzelnen Schlüssel, ob er zum Schloß passe. Der Versuch fiel negativ aus und das Resultat wurde bei mehrmaliger Wiederholung nicht besser, so daß ich schließlich den alten Herrn bitten mußte, seine Bemühungen einzustellen. Hinterher habe ich erfahren, daß die Operation mit dem Schlüssel eine Komödie war. Bunsen wollte uns einfach die Wohnung nicht zeigen, weil er fürchtete, sie würde meiner Frau ebenso wie Frau Meyer mißfallen und einen Grund zur Ablehnung des Rufes bilden. Darin hatte er sich allerdings geirrt. Nicht die Wohnung, sondern die Maßregeln, die man zur Erweiterung des Instituts getroffen hatte, und die nach meiner Ansicht nur eine Flickerei bedeuteten, waren für mich bestimmend, in Würzburg zu bleiben, nachdem man mir dort einen Neubau des Instituts in Aussicht gestellt hatte. Inzwischen war bei Meyer ein gründlicher Stimmungswechsel eingetreten, er bedauerte außerordentlich, nicht nach Heidelberg gegangen zu sein und besuchte mich umgehend in Würzburg, um mich darüber aufzuklären. Wir wurden rasch einig;

denn mir war es lieb, absagen zu können, ohne daß der alte Bunsen und die Heidelberger Fakultät sich über mangelnde Schätzung der dortigen Professur zu beklagen hätten. Mit dem Ministerium zu Karlsruhe wurde die Angelegenheit telegraphisch erledigt und Meyer ist dann, wie bekannt, nach Heidelberg gegangen. Das alte Bunsen'sche Laboratorium blieb im Betrieb, wurde aber durch einen Neubau für die organische Abteilung ergänzt. Ich habe mir später das vergrößerte Laboratorium angesehen und die Überzeugung gewonnen, daß meine ursprüngliche Ansicht richtig war. Man hätte mit demselben Gelde einen sehr viel größeren und zweckmäßigeren Neubau 10 Minuten vor der Stadt errichten können.

Die Universitätsverwaltung zu Würzburg und das Kultusministerium zu München gaben ihren Dank für mein Bleiben sofort zu erkennen; denn sie beantragten den Neubau des chemischen Instituts, wofür die Stadt einen prächtigen Bauplatz am Pleichering für die Überlassung des alten Gebäudes in der Maxstraße zur Verfügung stellte. Außerdem wurde für mich eine Gehaltszulage von 1000 Mk. in Aussicht gestellt, ohne daß ich in diesem Punkt eine Forderung gestellt hatte. Aber beide Positionen bedurften der Genehmigung durch den bayerischen Landtag. Sie wurden von der ultramontanen Mehrheit in schroffer Weise abgelehnt und die Gehaltszulage sogar zum zweiten Male, als die Kammer der Reichsräte die kleine Summe in den Etat wieder eingestellt hatte. Der Grund dieser schlechten Behandlung ist mir erst einige Jahre später bekannt geworden. Zum Trost dafür wurde mir zunächst ein bayerischer Orden verliehen und bald nachher auch die Gehaltszulage gewährt, nachdem die Mittel dafür durch einen Sterbefall frei geworden waren. Die Forderung für den Neubau mit 650 000 M. mußte allerdings um 2 Jahre, d. h. bis zur nächsten Haushaltsperiode verschoben werden. Die Sache nahm dann einen sehr lustigen und für bayerische Verhältnisse so charakteristischen Verlauf, daß ich sie hier in den Einzelheiten mitteilen will.

Um die von der Regierung betonte Notwendigkeit eines Neubaues für das chemische Institut an Ort und Stelle zu prüfen, erschien eine Kommission des Landtages in Würzburg. Sie bestand aus dem Ministerialbeamten Dr. Bumm, dem Vertreter der ultramontanen Mehrheit Dr. Daller und dem Vertreter der liberalen Minderheit Dr. Schauß. Letzterer hatte die Freundlichkeit, mich vor der offiziellen Besichtigung im Institut aufzusuchen, da er mich von der Münchener Zeit her persönlich kannte. Er eröffnete die Unterredung mit der Frage: „Wie kommen Sie als gut katholischer Mann dazu, Ihre Kinder protestantisch taufen zu lassen?" Auf meine Erklärung, daß ich stets protestantisch gewesen sei, erwiderte er: „Herrgott, dann ist Ihnen ja vor zwei Jahren schweres Unrecht geschehen", wobei er die schlechte Behandlung von seiten des bayerischen Landtages im Auge hatte. Als ich aber dann zufügte, daß meine Frau katholisch sei, erklärte er lachend: „Nun, das ist noch viel schlimmer, dann haben Sie also die Strafe von damals redlich verdient."

Bald nachher hatten wir die Ehre, die Kommission zu empfangen. Damit das in würdiger Weise geschehe, waren nach Verabredung mit den Assistenten und Studenten Vorbereitungen getroffen, nicht mit Blumen oder weißgekleideten Jungfrauen, sondern auf viel wirksamere Weise mit den stärksten Riechstoffen der Chemie. Brom, Schwefelwasserstoff, Ammoniak, Mercaptan, Skatol, Isonitril, Kakodyl hatten dazu gedient, die verschiedenen Räume des Instituts mit einer infernalischen Atmosphäre zu erfüllen, um den Mitgliedern der Kommission die ungenügende Größe und schlechte Ventilation überzeugend ad nasum zu demonstrieren. Ich sehe noch die erstaunten Gesichter der Herren, die sich tapfer durch die Gerüche durcharbeiteten, und als wir schließlich im Keller angelangt waren, atmete die ganze Gesellschaft auf und erklärte, daß hier die Luft bei weitem am besten sei.

Meine Aufgabe war es selbstverständlich, Herrn Dr. Daller für den Neubau zu gewinnen, und ich bot dafür alle mir zu Gebote stehende Beredsamkeit auf. Als kluger Mann hielt er sich für verpflichtet, auch billigere Möglichkeiten, z. B. einen Umbau des alten Instituts, zu erwägen.

Als ich diesen Gedanken zurückwies, weil das Haus in der Grundlage verfehlt sei, stellte er die überraschende, aber gewiß nicht unberechtigte Frage: „Wer garantiert uns denn dafür, daß solche Eselei nicht wieder passiert?" Ich war selbstbewußt genug, für meine Person die Eselei entschieden abzulehnen und hatte den Eindruck, daß sowohl Dr. Daller, wie die beiden anderen Herren von meinen Darlegungen befriedigt seien. In der Tat wurde mehrere Wochen später die Summe für den Neubau von dem Landtag ohne jeden Abstrich bewilligt, und ich war auf diesen scheinbaren Erfolg meiner Bemühungen ziemlich stolz.

Als ich aber ein halb Jahr später Adolf von Baeyer in München besuchte und ihm den Verlauf der Verhandlungen über den Neubau des Würzburger Instituts erzählte, erfuhr ich mit einem gewissen Gefühl der Enttäuschung, daß der Gesinnungswechsel bei Dr. Daller nicht durch meine Beredsamkeit und sonstige Veranstaltungen herbeigeführt worden wäre, sondern durch den Eingriff eines meiner alten Schüler aus der Münchener Periode, Dr. Brandl, durch den ich die früher erwähnte kleine Arbeit über die Bestimmung des Fluors in Silicaten hatte ausführen lassen. Dieser Herr war mit Dr. Daller engbefreundet und hatte ihm ernste Vorwürfe wegen der schlechten Behandlung gemacht, die mir von Seiten des bayerischen Landtages zwei Jahre vorher zuteil geworden war. Das war der Grund, warum Dr. Daller nunmehr für den Neubau des Instituts eintrat und die ungekürzte Bewilligung der Bausumme befürwortete.

Für die Begründung der Forderung an den Landtag hatten der Universitätsbaumeister von Horstig und ich einen provisorischen Bauplan entworfen und erhielten nun von der Regierung in München plötzlich den überraschenden Auftrag, diesen mit der vom Landtag bewilligten Summe auszuführen. Ich erklärte darauf kurzweg der Universitätsverwaltung, daß der Plan durchaus unreif sei und erst durch ein neues, nach jeder Richtung hin durchdachtes Projekt ersetzt werden müsse. Aber dazu hatte der Universitätsbaumeister keine rechte Lust, weil ihm inzwischen auch die Aufgabe zugefallen war, ein neues Haus für die gesamte Universität zu bauen. Es mußte also für das Institut ein besonderer Baumeister gesucht werden. Diese Aufgabe fiel wieder mir zu, und durch fleißige Erkundigungen bei Sachverständigen gelang es mir auch, einen im bayerischen Staatsdienst stehenden, jungen tüchtigen Baumeister zu ermitteln. Mit meinem Vorschlag, diesem Herrn unter günstigen Bedingungen den Institutsbau in Würzburg anzuvertrauen, fand ich beim Kultusministerium in München größtes Entgegenkommen. Aber die Bauverwaltung glaubte sich durch mein Vorgehen in ihren Rechten beeinträchtigt und schlug meine Bitte rundweg ab. Infolgedessen wandte ich mich im Einverständnis mit der Universitätsverwaltung in Würzburg an den hervorragenden Architekten Professor Hase in Hannover, mit dem ich vorher in einem Briefwechsel über seinen Sohn, der in Würzburg Chemie studierte, gestanden hatte. Er empfahl uns einen tüchtigen jungen Baumeister aus dem Kreise seiner Schüler, und es kam mit dem Herrn ein Vertrag zustande. Als dieser aber zur Genehmigung den Behörden in München vorgelegt wurde, entstand dort eine große Aufregung über die Wahl eines preußischen Baumeisters und über das eigenmächtige Vorgehen von Professor Fischer. Der Vertrag wurde annulliert und die Münchener Baubehörde sandte, da sie selbst keinen geeigneten Mann zur Verfügung hatte, einen jungen schwäbischen Baumeister nach Würzburg, der sich aber hinterher als unfähig erwies. Der ganze Streit mit der Baubehörde zu München hatte aber das Gute, daß auf ihre ernsten Vorstellungen hin sich nun der tüchtige Universitätsbaumeister von Horstig noch bereit erklärte, auch den Neubau des chemischen Instituts zu übernehmen. Mit ihm zusammen habe ich dann einen neuen Plan ausgearbeitet, der nach meinem Gefühl gründlich durchdacht war und sowohl in den Raumverhältnissen, wie in den technischen Einrichtungen dem Bedürfnis der Universität Würzburg entsprach. Mit kleinen Abänderungen ist er auch wirklich ausgeführt worden, allerdings erst nach meinem Weggange von Würzburg, und ich glaube, daß auch heute noch das Würzburger Institut zu den besteingerichteten Laboratorien Deutschlands gehört.

Nach alter Gewohnheit habe ich in Würzburg die Ferien regelmäßig zu kürzeren oder längeren Reisen benutzt. Ostern ging es meistens nach dem Süden. Die Reise nach Corsica 1886, wo der Skatolgeruch mich begleitete, ist früher schon erwähnt. Nach 14-tägigem Aufenthalt in Ajaccio erhielt ich den Besuch von W. Königs und R. von Pechmann, die von Nizza kamen und nach einer stürmischen Nacht in ziemlich erschütterter Verfassung auf der Insel landeten. Einige Tage später sind wir zusammen auf einem Privatwagen von Ajaccio quer durch die Insel über Corte nach Bastia gefahren. Es ist eine Strecke von etwa 120 km, und die Fahrt dauerte zwei Tage. Es war für die beiden kleinen corsischen Pferde eine achtungswerte Leistung; denn die schöne Straße führte über eine beträchtliche Paßhöhe von mehr als 1000 m. Eisenbahnverbindungen gab es damals noch nicht auf der Insel.

Wegen der Armut des Landes kamen wir etwas verhungert in Bastia an, von wo es noch in der gleichen Nacht weiter nach Livorno ging. Hier sind wir für die kleinen Entbehrungen der sonst genußreichen Landfahrt entschädigt worden durch ein opulentes Mahl, das Pechmann zusammen mit dem Chef de cuisine des Hotels zusammengestellt hatte; denn er war nicht allein ein guter Chemiker, sondern auch in der Kochkunst wohl unterrichtet. Ich bin später noch einige Male mit diesem verdienten Fachgenossen auf Reisen zusammengetroffen. Königs hat ihm einen trefflichen Nekrolog gewidmet. Ich kann hier nur bestätigen, daß er ein sehr angenehmer Reisegesellschafter war, der gerne das Amt des Reisemarschalls übernahm und dessen Anordnungen man ruhig vertrauen konnte, wenn man nicht aufs Sparen angewiesen war.

Im nächsten Jahre verbrachte ich die Osterferien zu Bordighera an der Riviera di Ponente in Gesellschaft von Baeyer und später bin ich dort wiederholt mit Baeyer und einmal auch mit Victor Meyer zusammengetroffen. Zu unserm Kreise gesellte sich auch der Dichter Ludwig Fulda, der uns bei Tisch mit allerlei Späßen und Schnurren trefflich unterhielt. Der Tag diente dann regelmäßig für Spaziergänge und Ausflüge in die prächtige Umgebung, und die Abende verbrachten wir ebenso regelmäßig in der Bierstube des Hotels mit Glücksspiel, wobei der höchste Einsatz allerdings auf 2 Soldi normiert war. Baeyer galt als der Sachverständigste, da er in jungen Jahren einmal einen mehrwöchentlichen Aufenthalt in Monaco gehabt und dort mit den Croupiers der Spielbank verkehrt hatte. Er war deshalb der Bankhalter, und es machte ihm großen Spaß als solcher an manchen Abenden einen

Gewinn von 4 bis 5 Frs. einzustreichen. Ich bin selbst immer ein Feind des Spiels gewesen, weil es von meinem Vater als eines der schlimmsten Laster so oft gerügt worden war. Aber an die Spielabende in Bordighera, bei denen allerdings die Leidenschaft nicht allzu sehr erregt wurde, denke ich doch mit Vergnügen zurück.

Für manche Leute ist freilich ein solches Spiel mit dem Spiele ein gefährliches Ding. So hat unser Freund Pechmann schwer darunter gelitten; denn der größere Teil seines Vermögens war der Spielbank zu Monte Carlo zum Opfer gefallen, und auch in späteren Jahren, wenn er in die Gegend der Bank kam, bedurfte es sorgfältiger Überwachung durch seine Freunde, wenn die Reisekasse nicht gefährdet sein sollte. Auch unter meinen späteren Berliner Kollegen habe ich solche Spielfreunde kennen gelernt, die im übrigen sehr kluge und vernünftige Leute waren. Denn die Leidenschaften sind von dem Verstande ziemlich unabhängig.

Selbstverständlich fehlte es auch bei dem Zusammenleben an der Riviera, besonders auf den Spaziergängen, nicht an wissenschaftlichen Gesprächen, und es gab wohl kaum ein wichtiges Problem der Chemie, das wir nicht behandelt hätten.

In besonderer Erinnerung ist mir eine stereochemische Frage geblieben. Im voraufgegangenen Winter 1890/91 hatte ich mich mit der Aufgabe beschäftigt, die Konfiguration der Zucker aufzuklären, ohne ganz zum Ziele zu gelangen. Da kam mir in Bordighera der Gedanke, die Entscheidung über die Konfiguration der Pentosen durch ihre Beziehungen zu den Trioxyglutarsäuren zu treffen. Leider konnte ich wegen Mangel eines Modells nicht feststellen, wieviel solcher Säuren nach der Theorie möglich seien, und ich legte deshalb die Frage Baeyer vor. Er griff solche Dinge mit großer Wärme auf und konstruierte gleich aus Zahnstochern und Brotkügelchen Kohlenstoffatommodelle. Aber nach langem Probieren gab auch er die Sache auf, angeblich, weil es ihm zu schwer wurde. Es ist mir erst später in Würzburg durch lange Betrachtung von guten Modellen gelungen, die endgültige Lösung zu finden.

Später bin ich mit Baeyer in den Osterferien wiederholt an den Genfer See, nach Territet bei Montreux gegangen, wo wir einigemale auch mit Carl Graebe, der damals Professor in Genf war, zusammentrafen. Bei Sonnenschein war der Aufenthalt am Genfer See außerordentlich erfrischend und die herrliche Umgebung gab reichlich Gelegenheit zu schönen Spaziergängen und größeren Ausflügen. Graebe war ein belebendes Element in unserem Kreise, und wenn an der langen Tafel des Grand Hotels sein helles, lautes Lachen ertönte, so wurde die Aufmerksamkeit der ganzen Gesellschaft dadurch geweckt. Durch sein heiteres Wesen, seine gefälligen Umgangsformen, das kluge Urteil in wissenschaftlichen und auch rein menschlichen Dingen hat er mir so gut gefallen, daß ich mich gerne um seine Freundschaft bewarb und mit ihm wiederholt noch in späteren Jahren während der Herbstferien im Schwarzwald zusammengetroffen bin.

Mit 65 Jahren verließ er Genf, wo er mehr als ein Vierteljahrhundert als Professor der Chemie gewirkt hatte, und zog sich nach seiner Vaterstadt Frankfurt a. M. zurück, wo er jetzt noch hochbetagt, aber in körperlicher und geistiger Frische weilt. Seine großen wissenschaftlichen Arbeiten gehören der Geschichte an. Aber auch jetzt noch ist er schriftstellerisch tätig, und im vergangenen Herbst erfuhr ich zu meiner großen Freude in Baden-Baden von ihm, daß er eine Geschichte der organischen Chemie geschrieben habe, welche sicherlich das gleichartige vortreffliche Buch von Ed. v. Hjelt in glücklicher Weise ergänzen wird und von der ich hoffe, daß der Druck durch die Kriegszeit nicht verhindert wird.

Im Anschluß an einen Aufenthalt in Territet haben Baeyer und ich 1892 an dem internationalen Kongreß der Chemiker zur Reform der Nomenklatur der Kohlenstoffverbindungen in Genf teilgenommen. Er unterschied sich in vorteilhafter Weise von den lärmenden und verwirrenden internationalen Zusammenkünften wissenschaftlicher oder technischer Art, die in den letzten 20 Jahren stattfanden und die ich, wenn irgend möglich, vermieden habe. Der in Genf versammelte bestand aus etwa 60 Personen, die alle im gleichen Hotel untergebracht waren und die wie eine große Familie mehrere Tage zusammen verlebten. Auch mehrere Damen, u. a. auch meine eigene Frau nahmen an den bescheidenen, aber behaglichen, geselligen Veranstaltungen teil.

An der Spitze des Kongresses stand Charles Friedel aus Paris, ein geborener Elsässer und ein sehr sympathischer Mann; ich kannte ihn schon von einem früheren Besuch in Paris. Er begrüßte mich mit der ruhigen Freundlichkeit, die seinem Wesen eigen war. Wir haben uns lange unterhalten, weil ich ihm viel von seiner Vaterstadt Straßburg erzählen konnte, wobei er bis zu Tränen gerührt die Abtrennung seiner Heimat von Frankreich beklagte.

Auch die meisten anderen europäischen Länder waren vertreten. Eine Hauptrolle bei den Verhandlungen spielte Adolf Baeyer, der ebenfalls sich mit Nomenklaturfragen schon vielfach beschäftigt hatte und durchweg mit seinen Vorschlägen durchdrang. Über die Einzelheiten zu berichten, habe ich um so weniger Veranlassung, als die Verhandlungen ziemlich ausführlich in den chemischen Zeitschriften geschildert sind.

Von dem Ergebnis des chemischen Kongresses ist manches geblieben und wohl dauerndes Eigentum der chemischen Sprache geworden. Aber die konsequente Durchführung einer rationellen Nomenklatur nach der chemischen Konstitution hat sich doch als unmöglich erwiesen, da sie schließlich zu Namen führte, die wegen ihrer Länge unbrauchbar sind. Auch das, was man vorzugsweise ins Auge gefaßt hatte, die Registrierung der Kohlenstoffverbindungen mit Hilfe solcher Namen, ist, wie man weiß, inzwischen abgelöst worden durch die praktische einfachere Registrierung nach der empirischen Formel, wie sie von M. M. Richter zuerst angewandt wurde. Aber auch hier wächst die Zahl der unter gleicher Formel aufgeführten Isomeren mit erschreckender Schnelligkeit, und schon muß man daran denken, neben der empirischen Formel noch ein zweites Registrierungsmittel zu finden, um die Aufsuchung der einzelnen Stoffe zu erleichtern.

Die Tage des Genfer Kongresses werden trotzdem allen Teilnehmern in bester Erinnerung geblieben sein; denn er war in seinem harmonischen Verlauf und dem behaglichen Verkehr seiner Mitglieder ein würdiges Abbild der gemeinsamen Interessen, welche die Vertreter der Wissenschaft in allen Ländern miteinander verbinden sollte. Nach den traurigen Erfahrungen des Weltkrieges rufe ich mir diese besseren Zeiten gerne ins Gedächtnis zurück und hoffe, daß mit der Rückkehr des Friedens auch die Vernunft und das Gefühl der Solidarität bei den Gelehrten und ganz besonders bei den Naturforschern zurückkehren wird.

Während der Würzburger Periode bin ich in den Herbstferien meist nach einem Seebad in Belgien oder Holland und später nach Norderney gegangen. An einer solchen Badereise nach Scheveningen 1889 nahm auch meine Frau teil, und wir haben dort mit Arthur Dilthey und seiner Frau mehrere vergnügte Wochen zugebracht und hinterher die größeren holländischen Städte besucht. Bei dieser Gelegenheit habe ich auch meiner Frau Euskirchen und Umgebung gezeigt und sie unserer zahlreichen Familie am Niederrhein vorgestellt. Gewöhnlich aber ging sie in den Herbstferien mit den Kindern nach Ambach am Starnberger See zu ihren Eltern.

Zu der Jagd in Euskirchen, der ich früher so viel Erfrischung verdankte, bin ich seit meiner Erkrankung in Erlangen nicht mehr gekommen, weil ich mich nicht neuen Erkältungen aussetzen wollte. Statt dessen habe ich im September wiederholt die Naturforscherversammlung besucht, während der Würzburger Zeit diejenige zu Berlin 1886 und Heidelberg 1889. Letztere stand im Zeichen der Physik und Chemie; denn Heinrich Hertz und Victor Meyer hielten die beiden Hauptvorträge, der erste über seine große Entdeckung der elektrischen Wellen und der andere über allgemeine Probleme der Chemie.

Auch H. von Helmholtz und Werner von Siemens kamen in Gesellschaft von Edison, der gerade den Phonographen erfunden hatte und dieses merkwürdige Instrument durch einen Gehilfen vorführen ließ. Ferner habe ich hier den Physiker Bolzmann aus Wien zuerst sprechen hören. Er hatte die merkwürdige Gewohnheit, jeden Satz in der höchsten Stimmlage zu beginnen und im tiefsten Bariton zu beenden. Das war so anstrengend, daß ihm schon nach einigen Minuten der Schweiß über das Gesicht rann, und wir Zuhörer hatten trotz der großen Achtung vor dem Redner Mühe, unsere Heiterkeit zu verbergen.

Bunsen war nicht anwesend. Wahrscheinlich hatte er noch genug von den Strapazen der 500-jährigen Jubelfeier der Universität Heidelberg, die einige Jahre zuvor im August 1886 stattgefunden und an der ich auch teilgenommen hatte.

In Gesellschaft von Baeyer machte ich damals meinen ersten Besuch bei Bunsen, der uns sehr freundlich und mit Bergen von Zigarren empfing, aber in der Unterhaltung wohl wegen seiner Schwerhörigkeit zurückhaltend war. Wir trafen dort Sir Henry Roscoe, der von England herübergeeilt war, um seinem alten Lehrer und Freunde Bunsen die Repräsentation während dieser Festtage zu erleichtern. Roscoe war ein sehr liebenswürdiger Mann, der sich uns jungen Fachgenossen schnell anzupassen wußte und uns durch die Erzählung von Schnurren aus seiner Heidelberger Studienzeit oder von seinen Erlebnissen in England viel Spaß machte. Ich habe ihn noch mehrmals in England selbst wiedergesehen und er hat mir auch einige sehr freundlich gehaltene Briefe geschickt.

Natürlich fehlte es damals in Heidelberg nicht an den üblichen Festen auf dem alten Schloß, an einem gewaltigen Kommers, dem der Großherzog in eigener Person präsidierte, und ähnlichen akademischen Veranstaltungen.

Die Naturforscherversammlung in Berlin vom Jahre 1886, die erste in der neuen Reichshauptstadt, war außerordentlich stark besucht. Sie trug einen anderen Charakter wie in Heidelberg, verlief aber für uns Chemiker auch sehr interessant; denn die Verhandlungen in unserer Sektion waren reich an wissenschaftlichem Inhalt und die einheimischen Mitglieder der chemischen Gesellschaft gaben sich alle Mühe, durch behaglichen geselligen Verkehr und Veranstaltungen von lustigen Festen, z. B. einem Bierabend der durstigen chemischen Gesellschaft unter dem Vorsitz von C. Scheibler, den auswärtigen Fachgenossen den Aufenthalt in Berlin zu verschönern. Hier habe ich A. W. von Hofmann von neuem kennen gelernt und seine große Gewandtheit in geschäftlichen und repräsentativen Dingen bewundert.

Die chemische Gesellschaft hatte bei dieser Gelegenheit eine Ausstellung von wissenschaftlichen Präparaten veranstaltet, zu der ich neben anderen Dingen die frisch bereiteten synthetischen Indole beisteuerte. Darunter befand sich eine stattliche Menge von ganz reinem Skatol, und der Entdecker dieses Stoffes, Professor Brieger, war sehr befriedigt, daß auch mein Präparat den von ihm geschilderten üblen Geruch besaß; denn A. von Baeyer hatte einige Zeit vorher mitgeteilt, daß reines Skatol nicht unangenehm rieche. Er war offenbar der nicht seltenen Täuschung zum Opfer gefallen, die durch die ganz verschiedene Wirkung von Riechstoffen in konzentrierter oder verdünnter Form auf das Geruchsorgan entstehen kann.

Im Anfang des Jahres 1890 konnte ich in den Berichten der chemischen Gesellschaft die Synthese der Mannose und Lävulose mitteilen. Die Folge davon war eine Einladung des Vorstandes, einen zusammenfassenden Vortrag über die Kohlenhydrate in Berlin zu halten. Dieser fand statt am 23. Juli 1890, und ich konnte, unterstützt von meinem Mitarbeiter Dr. J. Tafel, die wichtigsten Phasen der Untersuchung durch Experimente illustrieren. Es war das erste Mal, daß ich in der chemischen Gesellschaft sprach, und als Dank dafür erntete ich von Seiten des Vorsitzenden, Herrn A. W. von Hofmann, einige sehr freundliche Worte der Anerkennung. Hinterher fand dann, wie es Sitte war und auch geblieben ist, ein in einfachen Formen gehaltenes Abendessen zu Ehren des Vortragenden statt.

Als ich nach Würzburg zurückkehrte und meine Frau, die neugierig auf den Verlauf des Vortrages war, mich danach fragte, habe ich mir eine kleine Neckerei erlaubt und ihr gesagt, die Leute hätten bei den Hauptstellen der Rede „Au" gerufen. Darauf gewaltige Entrüstung und Vorwürfe gegen die unhöflichen Preußen, wozu sie sich als Bayerin völlig berechtigt fühlte; denn auch mich selbst hat sie zuweilen im Zorn als Preußen tituliert. Ich habe sie allerdings hinterher über den Scherz aufgeklärt, aber die gute Laune war doch verdorben.

Die erste Synthese der natürlichen Zucker hat mir auch die erste öffentliche Anerkennung von Seiten des Auslandes eingetragen; denn ich erhielt bald nachher von der Chemical Society zu London die Davy-Medaille und wurde von der wissenschaftlichen Gesellschaft zu Upsala zum korrespondierenden Mitglied gewählt.

Obschon Würzburg nicht an gerade der großen Heerstraße lag, so ist mir doch mancher liebe Besuch von Fachgenossen dort zuteil geworden; so kam Eduard Hjelt eines Tages als früherer Studierender der Universität, später Victor Meyer, dann Otto N. Witt, H. W. Perkin jun. und mancher andere. Am meisten überraschte mich Ernst Haeckel aus Jena, der früh morgens mit der Reisetasche und in aller Eile erschien, um sich nach Ludwig Knorr zu erkundigen. Aus der kurzen, aber sehr lebhaften Unterhaltung ist mir ein Ausspruch Haeckels im Gedächtnis geblieben: „Wenn Ihr Chemiker synthetisch das richtige Eiweiß macht, dann krabbelt's." Knorr erhielt einige Monate später in der Tat einen Ruf nach Jena und ist im Herbst 1889 dahin übergesiedelt. Es sind jetzt nahezu 30 Jahre, daß er ein angesehenes Mitglied der Thüringer Hochschule bildet.

Die sonderbarsten Besuche erhielt ich aus Amerika. Eines Tages erschien ein Professor der Physiologie, der von einem reichen Mann Geld erhalten hatte, um eine Universität in Worcester U. S. A. zu gründen. Er hatte die romantische Idee, ein ganzes Schiff mit europäischen Professoren, Assistenten, Instrumenten, Präparaten und ähnlichen Dingen zu beladen und mit diesem Apparat dann seine Universität auszustatten. Die Unterhaltung mit mir eröffnete er mit der Frage: „Wollen Sie mit mir als Professor nach Amerika gehen?", worüber ich so überrascht war, daß ich das Ganze für einen Scherz hielt, bis er sein ausführliches Programm entwickelte. Er war übrigens ein gebildeter und weitgereister Mann, der viel Interessantes zu erzählen wußte.

Bald nachher erschien eine amerikanische Dame, die sich als Miss Helene Abott und Fachgenossin vorstellte. Zu ihrem besonderen Schutze hatte sie sich ein zweites weibliches Wesen mitgebracht, das sich bei näherer Besichtigung als eine Negerin entpuppte. Sie erklärte in Würzburg wissenschaftlich arbeiten zu wollen und war erstaunt, daß Frauen noch nicht zu den Vorlesungen zugelassen seien. Ich habe ihr dann das Laboratorium gezeigt und sie den jüngeren Herren, Knorr, Wislicenus, Tafel vorgestellt. Sie machte ganz verständige Bemerkungen und zeigte, daß sie keine schlechten theoretischen Kenntnisse besaß. Nach ihrem Weggang wurde Kriegsrat gehalten, ob wir ihr vom Senat der Universität den Zutritt in das Laboratorium erwirken sollten. Einzelne waren mit Begeisterung dafür, aber die bedächtigen Elemente konnten die Befürchtung nicht unterdrücken, daß sie in dem bis dahin so gut harmonierenden Kreise leicht Verwirrung anrichten könne. Entsprechend dem Majoritätsbeschluß habe ich ihr dann abgeschrieben und erhielt darauf von ihr eine zwar höfliche, aber ziemlich energisch gehaltene Antwort, worin sie die Rückständigkeit Deutschlands in bezug auf das Frauenstudium rügte. Sie ist später die Gattin von Arthur Michael geworden, aber sie sind, soweit ich unterrichtet bin, nach einiger Zeit wieder auseinander gegangen.

Im Frühjahr 1892 mußte ich wegen eines Anfalls von Influenza einige Tage zu Bett liegen und meine Frau las mir gerade aus den eben erschienenen Berichten der chemischen Gesellschaft den Nekrolog von Peter Gries vor, von dem ich selbst den wissenschaftlichen Teil geschrieben hatte. Aber viel interessanter war der persönliche Teil von A. W. von Hofmann verfaßt und mit köstlichen Humor gewürzt. Wir haben darüber gerade herzlich gelacht, als ein Telegramm von Tiemann einlief, das den plötzlichen Tod von Hofmann meldete. Ich konnte wegen meiner Krankheit nicht zur Beerdigung hingehen, was ich um so mehr bedauerte, da ich dem Vorstand der chemischen Gesellschaft angehörte und von Hofmann bei meinem letzten Besuch in Berlin so freundlich empfangen worden war.

Mein Gesundheitszustand war damals nicht befriedigend. Wie sich hinterher herausstellte, war ich das Opfer einer chronischen Vergiftung durch Phenylhydrazin geworden. Während manche meiner Mitarbeiter und auch einige Diener sehr empfindlich gegen die Base waren, und darauf mit nervösen Beschwerden oder mit starken Anschwellungen von Hand und Arm reagierten, schien ich sehr widerstandsfähig gegen das Gift zu sein; denn seine schädliche Wirkung hatte sich bis 1891 auf ein Ekzem der Finger und inneren Handflächen beschränkt. Um so schlimmer gestaltete sich die chronische Vergiftung, die im Herbst 1891 auftrat und in sehr lästigen Störungen der Darmtätigkeit, namentlich in nächtlichen Koliken und Durchfällen sich äußerte. Die Krankheit erreichte im Winter 1891/92 ihren Höhepunkt und spottete aller normalen ärztlichen Behandlung. Erst die Anwendung von Prießnitzumschlägen brachte mir Erleichterung und den lang entbehrten Schlaf zurück. Die Vergiftung ist zum Teil durch Dämpfe, aber wie ich später feststellen konnte, noch viel mehr durch die Haut, d. h. von den Händen aus, zustande gekommen. Ich habe darunter viele Jahre gelitten und schließlich hat sich eine Idiosynkrasie gegen Phenylhydrazin und ähnliche Stoffe herausgebildet. Es war die zweite Schädigung, die von meinem Beruf kam, und ich wäre ihr wahrscheinlich erlegen, wenn nicht die Ursache erkannt worden wäre und ich dann die Berührung mit der schädlichen Base möglichst vermieden hätte. Die Vergiftung hatte natürlich auch recht schlecht auf mein Nervensystem eingewirkt und der Aufenthalt in der Dienstwohnung des Würzburger Instituts, die fortwährend mit der Laboratoriumsatmosphäre erfüllt und außerdem ungewöhnlich heiß war, wurde mir im Sommer 1892 so unangenehm, daß ich meinen Haushalt in ein gemietetes Landhaus vor der Stadt mit großem Garten verlegte.

Hier erschien an einem schönen Junitag plötzlich Geheimrat Friedrich Althoff vom Kultusministerium in Berlin. In scheinbar ganz naiver Form erzählte er mir, er habe einen zufälligen Aufenthalt in Würzburg nur benutzen wollen, unsere Bekanntschaft von der Berliner Naturforscherversammlung zu erneuern. Er sprach sich sehr erfreut über die einfache süddeutsche Lebensweise aus, über die Bescheidenheit der Professoren hierzulande, kam dann auf die Berliner Verhältnisse, das dortige chemische Institut und die Absicht des Kultusministers, für die Pflege der Chemie in Preußen möglichst viel zu tun. Es würde ihn interessieren, auch meine Ansicht darüber zu hören, worauf ich ihm freimütig erklärte, daß das von Hofmann erbaute Institut den Bedürfnissen der Gegenwart keineswegs mehr genüge. Erst zum Schluß stellte er an mich die Frage, ob ich den nötigen Neubau als Nachfolger von Hofmann nicht selbst besorgen wolle. Die Fakultät habe mich neben Kékulé und Baeyer vorgeschlagen, aber den Wunsch ausgesprochen, daß mit Rücksicht auf das hohe Alter der beiden Erstgenannten ich tatsächlich berufen würde. Über das Angebot, das an und für sich ja recht ehrenvoll war und auch in so entgegenkommender Form gemacht wurde, war ich selbst keineswegs erfreut; denn nun stand ich vor der Notwendigkeit, zwischen Würzburg, wo ich mich so glücklich fühlte, und Berlin, wovor mir graute, zu entscheiden.

Mein Entschluß wäre rasch gefaßt gewesen und zugunsten von Würzburg ausgefallen, wenn ich allein gestanden hätte und nur meinem Gefühl gefolgt wäre. Aber meine Frau war ehrgeiziger und ich mußte Althoff wenigstens das Versprechen geben, nach Berlin zu kommen, um die Verhältnisse an Ort und Stelle kennen zu lernen. Das geschah auch 8 Tage später. Im Ministerium zu Berlin war man in jeder Beziehung entgegenkommend. Der Minister Exzellenz Bosse empfing mich wegen Zeitmangel Sonntags morgens um 8 Uhr, um mir zu versichern, daß er alles tun würde, meine Bedingungen, insbesondere auch den Neubau des Instituts zu erfüllen. Auch die Berliner Fachgenossen haben mir stark zugeredet, den Ruf anzunehmen. Dazu noch keineswegs entschlossen, fuhr ich nach München, wohin mich der dortige Minister eingeladen hatte. Ich war erstaunt über die wenig geschickte Art, in der er mich zur Ablehnung des Berliner Rufes bereden wollte. Zunächst mußte ich $1^1/_2$ Tage warten, bevor er mich überhaupt empfing und dann behauptete er, ich wäre durch die Bewilligung des Neubaues in Würzburg verpflichtet, dort zu bleiben. Ich antwortete ihm, daß der Bau doch nicht mir persönlich bewilligt sei, wenn das aber zuträfe, so könne man ihn ja aufgeben, da er noch garnicht begonnen sei. Kurzum ich kam von München etwas verstimmt nach Würzburg zurück. Inzwischen war dort mein alter Vater eingetroffen, der von dem Berliner Ruf gehört und sich sofort aufgemacht hatte, um mir zuzureden, ein so gutes Geschäft nicht leichtfertigerweise auszuschlagen. Zum zweiten Mal würde mir die Stelle in Berlin nicht mehr angeboten. Andererseits könne ich ja, wenn es mir dort nicht gefiele, jederzeit wieder wechseln.

Die Vorzüge Berlins konnte auch ich mir nicht verhehlen. Das rege wissenschaftliche Leben der Reichshauptstadt und die in Aussicht gestellten großen Mittel, die Möglichkeit, einen größeren Kreis von Schülern um mich zu versammeln, hatte in der Tat für einen Mann in meinem Lebensalter (ich war noch nicht 40 Jahre alt) viel Verlockendes.

So kam ich denn nach 8-tägigem Schwanken zu dem Entschluß, meine persönliche Neigung beiseite zu setzen und den Ruf anzunehmen. Ich bin dann zum zweiten Mal und zwar in Begleitung meiner Frau nach Berlin gefahren, um eine Wohnung zu mieten, da die Dienstwohnung von Frau von Hofmann noch bis zum Mai 1893 besetzt war, und um im Institut eine Reihe von kleinen baulichen Änderungen zu vereinbaren, die in den Herbstferien getroffen werden sollten.

Nachdem ich nunmehr fest gebunden war, redeten schon einige Kollegen offener über die Berliner Zustände, und mein alter Lehrer Kundt überraschte mich mit der Bemerkung: „Na Fischer, Sie werden sich wundern über den Pack Arbeit, den man hier einem Professor aufladet." Als ich darauf etwas erschrocken an ihn die Frage richtete, warum er mir das nicht vor 14 Tagen gesagt hätte, als ich ihn im Vertrauen auf die alte Freundschaft um Aufklärung über die Berliner Zustände gebeten hatte, erwiderte er lachend: „Ja, dann wären Sie nicht gekommen."

Bei diesem letzten Aufenthalt genoß ich auch die erste Probe von dem geselligen Verkehr im Berliner Gelehrtenkreise; denn Frau von Helmholtz hatte Kunde von unserem beabsichtigten Besuch in Berlin erhalten und telegraphisch meine Frau und mich zu einer Abendgesellschaft eingeladen. Diese fand statt in der prächtigen Dienstwohnung der physikalisch-technischen Reichsanstalt zu Charlottenburg, und wir trafen dort einen interessanten Kreis von Leuten, darunter auch den alten Werner von Siemens, mit dem ich mich lange über technisch-elektrochemische Probleme unterhielt, und der hinterher meine Frau durch besondere Liebenswürdigkeit auszeichnete. Hier mußte ich meine erste Berliner Tischrede halten als Antwort auf einige Worte der Begrüßung, die Helmholtz an meine Frau und mich richtete. Ausgehend von dem Keppler'schen Gesetz der Planetenbewegung und ihrem Einfluß auf die Entwicklung der Physik konnte ich ein astronomisches Abbild der Gesellschaft entwickeln, in dessen Mittelpunkt Frau von Helmholtz als Sonne kam. Scheinbar habe ich dadurch ihre Gunst gewonnen; denn sie ist mir später immer in sehr freundlicher Weise entgegengekommen.

Bei der Abreise von Berlin war ich in sehr gedrückter Stimmung, hervorgerufen durch den schlechten Zustand des chemischen Instituts und durch manche überraschende Auskunft über die Verpflichtungen, die den Nachfolger von Hofmann erwarteten. Wenn ich mich nicht geschämt hätte, mein Wort zu brechen, so würde ich auf dieser Rückfahrt telegraphisch bei dem Kultusministerium in Berlin meine Zusage widerrufen haben. Aber dazu war es jetzt zu spät, besonders auch, weil ich das Gefühl hatte, daß man im Ministerium zu München über meine Annahme des Berliner Rufes verschnupft war.

Mein Vater war inzwischen von Würzburg wieder abgereist, ganz befriedigt von dem Erfolge seiner Überredungskunst. Für ihn selbst war eine Veränderung des Wohnsitzes eine Kleinigkeit; denn er hatte sich ja gerade zu der Zeit entschlossen, Euskirchen nach 56-jährigem Aufenthalt zu verlassen und nach Straßburg i. Els. überzusiedeln.

Für mich begann nun eine unbehagliche Zeit, die Vorbereitungen des Umzuges nach Berlin; denn ein Professor der Chemie wandert nicht nur mit seiner Gelehrsamkeit und den Büchern, sondern auch mit Präparaten, Apparaten und Assistenten. Als letztere folgten mir Dr. Oscar Piloty, der früher erwähnte Däne Dr. Fogh und Dr. Lorenz Ach. Wislicenus war inzwischen außerordentlicher Professor geworden und als solcher in Würzburg gebunden. Auch für Julius Tafel, den ich gerne mitgenommen hätte, konnte in Berlin keine passende Stellung geschaffen werden. Dagegen war es mir lieb, den Diener J. Wetzel, der sich später durch einige zweckmäßige Glasapparate bekannt gemacht hat, als Präparator nach Berlin verpflanzen zu können. Mit dem Umzug des Haushalts hatte ich nichts zu tun, weil meine Frau ihn als ihr Recht und ihre Pflicht in Anspruch nahm.

Der Abschied von der lieben Stadt Würzburg, den Kollegen und Studenten war herzlich aber kurz. Die beiden Kinder kamen zu den Großeltern an den Starnberger See und ich selbst zog mit meiner Frau und einer Köchin in ein kleines, allerliebstes Holzhaus, 10 Minuten von Berchtesgaden am Untersberg in prächtiger Umgebung gelegen. Hier haben wir wie auf der Hochzeitsreise ganz für uns gelebt und bei herrlichem Wetter 6 Wochen zugebracht. Es war die richtige Vorbereitung für die kommende Berliner Periode. Zwar hatte mir die deutsche chemische Gesellschaft für die Feier ihres 25-jährigen Bestehens im November d. J. die Gedächtnisrede auf A. W. von Hofmann übertragen. Aber es kam mir bald zum Bewußtsein, daß F. Tiemann als Freund und Schüler des Verstorbenen viel mehr für diese Aufgabe berufen und auch gerne bereit sei, sie zu übernehmen. Nach Vereinbarung mit dem Vorstande der Gesellschaft haben wir deshalb getauscht, und die Entbindung von der Rede war, offen gestanden, für mich eine angenehme Erleichterung.

Berliner Zeit

Mitte September ging meine Frau nach Würzburg zurück, zur Anordnung des Umzugs nach der Berliner Wohnung, die wir im Hause eines Dr. von Dechend, Mitarbeiter von Tiemann bei den Berichten, in der Königin Augustastraße nahe beim Tiergarten für ein Jahr gemietet hatten. Ende September verließ auch ich Berchtesgaden und traf in Berlin meine Frau ganz erschöpft und in Tränen, hervorgerufen durch die Widerwärtigkeiten im Verkehr mit den Berliner Umzugsleuten. Als Süddeutsche war sie mit diesen rohen Menschen nicht fertig geworden, und erst der Hilfe von Dr. Piloty, der durch seine gewaltige Körpergröße und seine Energie auch den Berliner Packern imponierte, hatte sie es verdankt, daß sie mit ihren Wünschen und Anordnungen durchdrang. Dazu kam die gedrückte Stimmung, die damals auf der Reichshauptstadt lastete wegen der großen Choleraepidemie in Hamburg, die eine strenge Beaufsichtigung des Personenverkehrs nötig machte. Die geschickte Bewältigung dieser sanitären Aufgabe, welche Berlin trotz einigen 100 Fällen eingeschleppter Cholera vor der Epidemie bewahrte, hat mir hinterher sehr imponiert. Aber für alle Leute, die damals namentlich aus den einfachen süddeutschen Verhältnissen heraus Berlin zuwanderten, war es eine unbehagliche Zeit. Meine Frau wurde zum zweiten Male davon betroffen, als sie gleich hinterher die Kinder von den Großeltern am Starnberger See abholte; denn die Kinderfrau, die sie mitbrachte, konnte das Eisenbahnfahren nicht vertragen, erkrankte unterwegs und wurde von Leipzig an dauernder Aufsicht unterstellt.

Die Eingewöhnung des Haushaltes und der Familie in die Berliner Verhältnisse ging Dank der Hilfe unserer Freunde rascher und besser von statten, als wir gedacht. Auch die kleinen Umbauten im Laboratorium, die besonders auf eine bessere Ventilation gerichtet waren, wurden rechtzeitig fertig. Und in der letzten Woche des Oktobers konnte das Institut für den regelmäßigen Betrieb wieder eröffnet werden.

Tiemann, der bis dahin die Leitung des Praktikums in Händen gehabt hatte, zog sich davon ganz zurück, blieb aber im Institut als

Privatgelehrter zur Fortsetzung seiner erfolgreichen wissenschaftlichen Studien und hielt auch noch eine kleine Spezialvorlesung.

Mit S. Gabriel und den übrigen von Hofmann ererbten Assistenten Dr. Pulvermacher und Dr. Richter konnte ich mich leicht über die Verteilung der Arbeiten einigen. Die von mir mitgebrachten beiden Assistenten Piloty und Fogh übernahmen den Unterricht in dem Hauptsaal, der der analytischen Chemie eingeräumt war, Gabriel blieb in der organischen Abteilung und ich selbst übernahm außer den beiden großen Experimentalvorlesungen die Aufsicht über das Ganze. Als Privatassistent bei meinen eigenen Untersuchungen stand mir Dr. Lorenz Ach, der schon in Würzburg dieselbe Stellung bekleidet hatte, zur Seite. Von den Würzburger Studierenden war nur der Engländer Crossley, der später in London Professor an dem pharmazeutischen Institut wurde, mit übergesiedelt.

Die Wintervorlesung über anorganische Chemie habe ich mit einem kurzen Nachruf auf meinen großen Vorgänger begonnen, aber nach einem ganz anderen, schon in Würzburg benutzten Schema gehalten, weil diese Form für eine vorzugsweise aus Medizinern, Apothekern und Oberlehrern zusammengesetzte Zuhörerschaft nach meiner Erfahrung leichter verständlich und auch dem Entwicklungsgang unserer Wissenschaft mehr angepaßt war. Es machte mir aber besondere Freude, dabei die Hilfe von Dr. C. Harries, der schon bei Hofmann Vorlesungsassistent gewesen war, zu haben; denn ich konnte nun viele der von Hofmann ausgebildeten Experimente zusammen mit den schon von mir gebrauchten dem Vortrag angliedern. So ist das reichhaltige Buch von Vorlesungsversuchen entstanden, das mit zweckmäßigen Ergänzungen seitdem in Berlin gebraucht wird und das auch manchem jüngeren Fachgenossen als Muster gedient hat.

Kaum war der Betrieb des Instituts voll im Gange, so brach eine kleine Revolte bei den Dienern aus. Sie erklärten plötzlich, die Arbeit nicht leisten zu können und zu wollen. Offenbar hatten sie die etwas zu freundliche Umgangsform, die ich von Süddeutschland her gewöhnt war, mißverstanden und hielten nun eine Machtprobe mir gegenüber für angebracht. Zudem waren sie durch das vorangegangene Regiment in bezug auf Arbeitsleistung etwas verwöhnt. Ich mußte nun den Ton wechseln, habe einen Diener, der kündbar war, sofort entlassen, einen anderen, der leider schon festangestellt war, durch den Minister pensionieren lassen, und mir im Handumdrehen andere Aushilfen verschafft. Das war für Berlin die richtige Methode, und ich habe später mit den Dienern, für die ich übrigens auch nach bestem Können sorgte, niemals mehr ernste Zerwürfnisse gehabt. Im Gegenteil, ich muß sagen, daß sie bei richtiger Behandlung brauchbarer und leistungsfähiger als in Süddeutschland waren.

Das nach den Anordnungen von A. W. Hofmann erbaute chemische Laboratorium in der Georgenstraße, das den Beinamen „I. Chemisches Institut der Universität" führte, zum Unterschied von dem durch Rammelsberg geplanten und später von Landolt benutzten „II. Chemischen Institut" in der Bunsenstraße, galt in bezug auf Architektur und Fassade als Sehenswürdigkeit, war aber für chemische Zwecke recht unpraktisch gebaut. Überall fehlte es an Luft und Licht und ein großer Teil des bebauten Raumes bestand aus dunklen und unbenutzbaren Korridoren. Nur die beiden Hauptarbeitssäle im ersten Stock nach der Georgenstraße und das geräumige Privatlaboratorium konnten als normale Arbeitsräume angesehen werden. Dagegen war der große Vorlesungssaal so dunkel, daß selbst mittags von 11 bis 12 meist künstliche Beleuchtung angewandt werden mußte. Ganz ungenügend war auch die Ventilation, und meine erste Sorge war deshalb, gerade so wie im Würzburger Institut, die Anlage einer ganzen Reihe von Kapellen, in der einfachen Form, wie sie früher geschildert wurde. Um ihnen genügenden Zug zu sichern, wurde eine besondere Luftzufuhr durch ein in die Wand geschlagenes Loch und einen im Innern der Säle über Manneshöhe nach aufwärts geführten Holzkanal angelegt.

Selbst die Heizung befand sich in einem traurigen Zustand; denn die dafür vorhandenen Torföfen funktionierten so schlecht, daß ein Teil der Studenten sich Privatgasöfen angeschafft hatte, die natürlich dem Institut durch den Gasverbrauch teuer zu stehen kamen. Sie mußten durch neue Kohlenöfen ersetzt werden.

Der laufende Etat, der vorher etwa 15 000 M. betrug, war infolge meiner Forderung auf 20 000 M. vom Minister erhöht worden. Auch zwei neue Assistenten wurden bewilligt, aber wie ich später zu meinem Bedauern erfuhr, dem chemischen Institut zu Göttingen abgeknappst.

Für die Ergänzung des Inventars hatte ich ebenfalls eine bescheidene Summe — wenn ich nicht irre, waren es 15 000 M. — bei der Berufung ausbedungen. Als ich diese aber in Anspruch nehmen wollte, gab es einen heftigen Zusammenstoß mit Geheimrat Althoff im Kultusministerium. Die Summe schien ihm nachträglich zu hoch, und er verstieg sich sogar zu der Forderung, daß ich diese Anschaffungen aus eigener Tasche machen sollte. Es kam dann zu einer Aussprache, die Herrn Althoff darüber belehrte, daß ich nicht im geringsten gesonnen sei, mich schlecht behandeln zu lassen und auf irgend eine Bedingung der Berufung zu verzichten. Wir haben später noch manche Meinungsdifferenz miteinander gehabt, aber unsere Unterhaltung spielte sich von da ab immer in gemäßigter Form ab. Allmählich gewöhnte er sich auch daran, bei mir in chemischen Dingen Rat zu holen. Er hatte leider die unbequeme Gewohnheit, die Besucher in einem recht unbehaglichen

Vorzimmer stundenlang warten zu lassen. Als mir das zum zweiten Mal passierte, bin ich weggegangen und auf seine spätere Anfrage, warum ich das getan, erklärte ich ihm ganz offen, er könne mich wohl warten lassen, wenn ich von ihm etwas erreichen wolle; wenn er aber meine Hilfe beanspruche, so müsse er mich gleich empfangen, denn meine Zeit sei ebenso kostbar und ebenso knapp wie die seine. Das hat er eingesehen, und je länger ich mit ihm in Verkehr stand, umsomehr habe ich ihn schätzen gelernt. Er war ein sehr kluger, ideenreicher Mann, der für jeden noch so verfahrenen Karren ein Mittel der Fortbewegung zu finden wußte. Dazu kam eine außerordentliche Arbeitskraft und Ausdauer in der Verfolgung seiner Pläne. Zudem ließ er sich bei allen wichtigen Entscheidungen nur von sachlichen Rücksichten bestimmen. Mit den äußeren Formen nahm er es nicht genau, und er hat manche Angehörige der preußischen Hochschulen durch sein suburbanes Wesen stark vor den Kopf gestoßen. Trotzdem bin ich der Ansicht, daß seine Tätigkeit für die Blüte der preußischen Hochschulen, insbesondere auch für ihre Ausstattung mit Instituten, Bibliotheken, Seminaren von größter Bedeutung gewesen ist.

Trotz der äußeren rauhen Form war er im Grunde genommen ein gütiger Mann, der überall half, wo er nur konnte. Das bewies seine Fürsorge für Witwen und Waisen, für altersschwache Professoren und Diener und die wenig bekannte Tatsache, daß er einen großen Teil seines Vermögens für solche Zwecke verschenkt hat.

Als meine Frau im jugendlichen Alter starb, hatte er solches Mitleid mit den drei kleinen hinterlassenen Kindern, daß er wohl ein Jahrzehnt hindurch diese alljährlich besuchte und ihnen auch vom Kultusministerium öfter durch Übersendung von hübschen Büchern eine Überraschung bereitete. Die Jungens waren darüber immer hocherfreut, weil sie glaubten, daß der Kultusminister in eigener Person ihnen auf diese Weise seine Anerkennung für gute Leistungen in der Schule aussprechen wollte. In Übereinstimmung mit vielen Kollegen habe ich Althoff damals für die bedeutendste Persönlichkeit im preußischen Kultusministerium gehalten.

Auch mit seiner Frau, einer geb. Ingenohl aus Neuwied, deren Mutter, eine geb. von der Leyen, aus der Gegend von Flamersheim stammte und eine Jugendfreundin meines Vaters war, bin ich genau bekannt geworden, besonders während eines Aufenthaltes in Meran, von dem später noch die Rede sein wird. Sie ist eine liebe, kluge Frau und lebt trotz eines schweren Herzleidens noch jetzt hochbetagt in Steglitz. Das Andenken ihres Mannes hat sie durch eine sehr geschickte Lebensbeschreibung mit vielen interessanten Originalbriefen des Verstorbenen geehrt.

Die Professur der Chemie war schon von meinem Vorgänger her mit vielen Nebenämtern verbunden, die auch mir übertragen wurden. Dahin gehörte zunächst ein Lehrstuhl an der sogen. Pepinière, der jetzigen Kaiser Wilhelm-Akademie für das militärärztliche Bildungswesen. Sie wird mit einem kleinen Gehalt honoriert und die Studierenden der Akademie besuchen, wie die übrigen Mediziner die regelmäßigen Vorlesungen über Experimentalchemie. Außerdem besteht an der Akademie ein Wissenschaftlicher Senat, in den ich ebenfalls gewählt wurde, und dem ich jetzt noch angehöre. Hier werden unter dem Vorsitz von Exzellenz von Schjerning militärärztliche Fragen verschiedenster Art in akademischer Form besprochen und daran schließt sich ein heiteres, auch von mir stets gerne besuchtes Abendmahl in den prächtigen Kasinoräumen der Akademie.

Das zweite medizinische Nebenamt erhielt ich in der wissenschaftlichen Deputation für das Medizinalwesen, die damals dem Kultusministerium angegliedert war. Die Aufgabe der Deputation war recht mannigfaltig. Sie besorgte die Prüfung der preußischen Kreisärzte und galt für die preußische Zivil- und Gerichtsverwaltung als höchste sachverständige Instanz in allen medizinischen Fragen. Als Mitglieder gehörten ihr an die meisten Kliniker, dann der Hygieniker und der Chemiker der Universität. Für alle medizinischen Fragen war also ein Fachmann vorhanden. Was außerhalb der engeren Medizin lag, fiel dem Chemiker zu, und so habe ich Gutachten der verschiedensten Art erstatten müssen, manchmal über Dinge, die mir recht fern lagen, und für die ich mir erst die richtigen Sachverständigen suchen mußte. Im Gedächtnis geblieben ist mir ein Gutachten über die Königin Louisen-Quelle, den sogen. Gesundbrunnen, der in früherer Zeit weit außerhalb der Stadt lag, und wohin damals die Berliner Bevölkerung bei schönem Wetter in Scharen pilgerte, um das berühmte Wasser zu trinken.

Das Wasser dieser Quelle war um die Mitte der 90er Jahre von einem Apotheker als Tafelwasser unter der Bezeichnung „Natürliches Mineralwasser" auf den Markt gebracht worden. Da er aber dem Wasser Kohlensäure zugeführt hatte, so erfolgte die Anzeige bei der Behörde wegen Betrugs. Der Mann wurde vom Gericht zu einer kleinen Geldstrafe verurteilt, aber was viel schlimmer war, der Polizeipräsident erließ eine öffentliche Warnung vor dem Tafelwasser, das nichts weiter als gewöhnliches Brunnenwasser sei. Dadurch wurde das Geschäft in kürzester Zeit ruiniert. In seiner Not wandte sich der Besitzer an das Ministerium des Innern. So kam die Sache an die wissenschaftliche Deputation, und ich wurde zum Berichterstatter bestellt. Die Darlegungen des Klägers waren so ungeschickt, daß ich anfänglich glaubte, er sei im Unrecht. Ich entschloß mich aber zu einer lokalen Besichtigung der Quelle und sicherte mir dafür die Hilfe eines Beamten von der geologischen Landesanstalt. Nach langem Suchen fanden wir die ehemals so berühmte Quelle, die einem ganzen Stadtteil Berlins den Namen gegeben hat, die aber sehr wenig Leute noch kannten, in dem Kellergeschoß eines großen Mietshauses. Wir erkannten alsbald, daß es sich hier um eine wirkliche, aus ziemlich großer Tiefe kommende starke Quelle handelte, deren Wasser sehr wohlschmeckend und sicherlich hygienisch ganz unbedenklich war. Auch die Sättigung mit Kohlensäure wurde in sachverständiger und sauberer Weise ausgeführt. Dementsprechend fiel mein Gutachten über die Quelle günstig aus. Der Polizeipräsident mußte sein abfälliges Urteil widerrufen und das Tafelwassergeschäft war gerettet.

Aus diesem Beispiel habe ich ersehen, wie zweckmäßig es ist, sich bei solchen Gutachten nicht auf die Akten zu verlassen, sondern die tatsächlichen Verhältnisse, womöglich durch Augenschein gründlich kennen zu lernen.

In der Regel aber waren meine Gutachten für die Deputation recht langweilig und auf Kleinigkeiten, z. B. Zänkereien zwischen Apothekern und Drogisten beschränkt. Ich zog es deshalb vor, nach Ablauf meiner 5-jährigen Amtszeit aus der Deputation auszutreten. Mein Nachfolger wurde Landolt. Später, als Althoff der Vorsitzende der Deputation geworden war, hat er mich nochmals halb zwangsweise als Mitglied in diese berufen, aber nach einigen Jahren bin ich wieder ausgetreten und die Stelle ist dann dem Pharmakologen aus der medizinischen Fakultät übertragen worden.

Die Einführung in die philosophische Fakultät, die anfangs November 1892 stattfand, vollzog sich in einfachster Form, hatte aber gleich einen Haufen von Geschäften zur Folge; denn die Chemiker werden durch die zahlreichen Promotionsprüfungen besonders stark in Anspruch genommen. Bei der mündlichen Prüfung der Chemiker waren Landolt und ich immer gleichmäßig beteiligt, aber in der schriftlichen Beurteilung der Dissertationen fiel mir meistens das Hauptreferat zu, weil die große Mehrzahl Themata aus der organischen Chemie behandelten.

Von der Schwerfälligkeit der Geschäftsführung war ich in hohem Grade überrascht. Da die Fakultät ungeteilt ist und damals etwa 50, jetzt aber weit mehr als 60 ordentliche Mitglieder hat, so kann man sich denken, welche ausführlichen Debatten entstehen, wenn sogen. prinzipielle Fragen behandelt werden. Dazu kommt noch die Gewohnheit, nicht allein alle kleinen Geschäfte, sondern sogar die Abstimmung über das Resultat jeder einzelnen Doktorprüfung durch die gesamte Korporation vorzunehmen.

Das hängt zusammen mit der Einrichtung der sogen. Sedecim, d. h. mit dem Rechte der 16 ältesten anwesenden Mitglieder, die Promotionsgelder unter sich zu teilen. Für den Eintritt in diesen Kreis der Auserwählten ist aber nicht das Lebensalter, sondern die Zahl der Jahre maßgebend, die der Betreffende als Ordinarius an irgend einer deutschen Universität zugebracht hat. Daß dieser patriarchalische Verteilungsmodus der Billigkeit entspricht, kann niemand behaupten. Ich habe die ganze Einrichtung immer für ein Haupthindernis bei allen Reformvorschlägen in dem Geschäftsgang der Fakultät angesehen. Das Festkleben an Satzungen und Traditionen tritt überhaupt in dieser Körperschaft so stark hervor, daß es häufig an den Zopf erinnert und zuweilen geradezu lächerlich wirkt. Z. B. galten noch bei meinem Eintritt in die Fakultät die deutschen Bundesstaaten als Ausland und dementsprechend konnte ein Angehöriger dieser Staaten ohne Abiturium in Berlin promovieren. Erst ein Jahr später wurde nach einem Antrage des Physikers Kundt dieser Unsinn abgeschafft.

Die Größe der Körperschaft bringt natürlich eine Unmenge von Geschäften mit sich, deren Erledigung viel kostbare Zeit in Anspruch nimmt. Der philosophische Dekan ist deshalb ein viel geplagter Mann, und die Fakultät hält während des Semesters in der Regel jede Woche eine Sitzung ab, die für die Prüfungen 2 Stunden und für die geschäftlichen Dinge ungefähr die gleiche Zeit dauert. Wer vorher an der Sitzung der Akademie der Wissenschaften teilnimmt, und bei den Prüfungen beschäftigt ist, hat das Vergnügen, jeden Donnerstag von 4 bis 10 Uhr sitzen zu müssen. Dazu kamen für mich noch die zahlreichen Prüfungen von Medizinern, Apothekern und Lehramtskandidaten. Ich habe die ersten 12 Jahre meines Berliner Aufenthaltes über nichts so sehr geseufzt, wie über den Verlust an Zeit und Arbeitskraft, der auf diese Weise entstand. Später ist es mir gelungen, von der Mehrzahl dieser lästigen Geschäfte entbunden zu sein.

Wegen der Größe der Fakultät müssen alle wichtigen Angelegenheiten und namentlich die Berufungsgeschäfte besonderen Kommissionen überwiesen werden. Ich habe im Laufe der Zeit vielen angehört und mich immer gefreut über die rein sachliche, von jedem Klickenwesen weit entfernte Art der Verhandlung. Dagegen konnte ich mich manchmal des Eindrucks nicht erwehren, daß die Berufungsgeschäfte nicht mit der sorgfältigen Liebe behandelt werden, wie das an kleinen und mittleren Universitäten geschieht, wo jeder Einzelne an der Person des neuen Kollegen ein direktes Interesse hat. So erklärt sich auch die, nach meiner Auffassung falsche Gewohnheit, nach Berlin Männer zu berufen hauptsächlich nach dem wissenschaftlichen Ansehen, aber in einem Alter, wo man von ihnen keine großen Dienste weder für die Wissenschaft, noch für den Unterricht zu erwarten hat. Das gilt besonders für die Naturforscher, die früher verbraucht sind, als die Vertreter der Geisteswissenschaften, und ich habe mich immer wieder verpflichtet gefühlt, Bedenken gegen die Berufung alter Personen zu erheben.

Im allgemeinen spielen die Naturforscher in der Berliner Fakultät nicht die Rolle, die sie beanspruchen könnten. Die Vertreter der Geisteswissenschaften sind zahlreicher und sicherlich zum Reden mehr geneigt, vielleicht auch in der Form gewandter. Da sie ferner mehr Zeit haben und die Sitzungen regelmäßiger und andauernder besuchen, so führen sie hier das große Wort, und ich habe wiederholt gegen die Verletzung der Interessen der Naturwissenschaften Einspruch erheben müssen. Das hat zuweilen zu ziemlich erregten Debatten geführt und mir wahrscheinlich bei der Gegenpartei einige Antipathie eingetragen.

Die Akademie der Wissenschaften

Die meisten Mitglieder der Akademie sind auch Professoren an der Universität. Zur technischen Hochschule hat die Körperschaft erst persönliche Beziehungen mit der Einrichtung der technischen Abteilung in der mathematisch-naturwissenschaftlichen Klasse im Jahre 1900 erhalten.

Dazu kommen in der Regel noch einige Männer, die keine Lehrtätigkeit ausüben, da die Akademie in der Wahl von Mitgliedern ganz frei ist. Zu meiner Zeit war die letzte Art von Mitgliedern vertreten durch den großen elektrischen Erfinder Werner von Siemens, den Botaniker Pringsheim, die Astronomen Auwers und Vogel und später durch den Elektriker von Hefner-Alteneck, sowie den Eisenbahnbaumeister Zimmermann.

In der Akademie herrscht noch heute der vornehme Geist ihres Gründers Leibniz, und ich habe mich während nunmehr der 25 Jahre, die ich ihr angehöre, stets über den unparteiischen und unabhängigen Sinn dieser Korporation gefreut. Jeder Versuch tatkräftiger Männer aus der Staatsregierung, die Maßregeln ihrer inneren Verwaltung zu beeinflussen, wurde einmütig abgewiesen. Auf der anderen Seite war man natürlich, sobald materielle Wünsche in Frage kamen, auf die Unterstützung der Staatsorgane angewiesen, und ich kann sagen, daß die Akademie in der mir bekannten Periode von Seiten des vorgeordneten Kultusministeriums dauernd großes Wohlwollen und tatkräftige Unterstützung gefunden hat. Früher war es Althoff im Kultusministerium, der ihre Geschäfte behandelte, und in den letzten 10 Jahren Fr. Schmidt, der heute selbst Kultusminister ist.

Die inneren Geschäfte der Akademie wurden fast ausnahmslos in rein sachlicher Weise und mit vornehmer Ruhe erledigt, die nicht selten einen Anflug von Langeweile zeigt. Zu erregten Debatten kam es nur zuweilen bei der Wahl neuer Mitglieder. Erst mit dem Krieg ist es anders geworden. Er hat in der ersten Zeit auch verwirrend auf die Denkweise mancher Akademiker zurückgewirkt; davon wird später noch die Rede sein.

Obschon von Leibniz gegründet und in den Einzelheiten von Friedrich dem Großen nach dem Muster der Pariser Akademie reorganisiert, verleugnet die Berliner Akademie doch nicht ihren preußischen Charakter. Das zeigt sich in der außerordentlich peinlichen, manchmal pedantischen Form der Geschäfte und noch mehr in der Bestimmung, daß jedes Mitglied alljährlich an einem bestimmten Tage nach der sogen. Lesekarte einen wissenschaftlichen Vortrag zu halten hat. Es liegt auf der Hand, daß ein solcher Zwang mit dem Wesen der wissenschaftlichen Forschung in Widerspruch steht. Er ist auch von manchem Mitglied der Akademie, wie mir bekannt, recht unbequem empfunden worden. Aber wiederholte Anläufe, ihn abzuschaffen, sind gescheitert. Ebenso überflüssig ist meines Erachtens die Bestimmung, daß Abhandlungen, die in den Schriften der Akademie veröffentlicht sind, von dem Urheber innerhalb der gesetzlichen Frist nur mit der Bewilligung der Akademie anderweitig publiziert werden dürfen. Allerdings wird sie von den Naturforschern wenig mehr beachtet, seit das Sekretariat derartige Verstöße gegen das Statut sehr milde handhabt. Das ist sehr vernünftig, denn der Naturforscher kann heutzutage nach einer Veröffentlichung in der Akademie nicht 1 bis 2 Jahre warten, bevor er seine Resultate auch in einer Fachzeitschrift mitteilt, da die Berichte der Akademie einen zu kleinen Leserkreis besitzen. Diese könnte sich wohl damit begnügen, die Priorität in der Veröffentlichung zu besitzen. Die erschwerten Bedingungen der Publikation haben zur Folge gehabt, daß ich ebenso wie die meisten anderen Berliner Naturforscher in den akademischen Schriften nur einen kleinen Teil meiner Untersuchungen niedergelegt habe, und das ist wohl der Grund, weshalb die Schriften der Akademie als Publikationsorgan nicht die Bedeutung erlangt haben, wie es bei ähnlichen ausländischen Akademieberichten der Fall ist. Im engeren Kreise haben wir öfter die Frage erwogen, ob es zweckmäßig sei, eine Änderung in der Redaktion der Sitzungsberichte eintreten zu lassen in der Weise, daß über alle wichtigen Untersuchungen, die in den von den Mitgliedern der Akademie geleiteten Instituten ausgeführt werden, kurze Mitteilungen in die Sitzungsberichte zu machen wären. Aber die geringe Neigung der Akademie, kurze, manchmal vorläufige Publikationen aufzunehmen, die besonders in der philosophisch-historischen Klasse zutage trat, ist derartigen Abänderungsplänen zu hinderlich gewesen. Dazu kommt die bei solchen Korporationen nicht seltene Macht der Gewohnheit, die alles Neue bekämpft, und die manchmal recht absonderliche Formen annehmen kann. Ein kleines Beispiel mag das illustrieren.

In dem alten Hause unter den Linden, das jetzt durch den Neubau ersetzt ist, waren die Versammlungsräume der Akademie auf viel kleinere Verhältnisse zugeschnitten. Beleuchtung und Heizung befanden sich ungefähr im gleichen Zustand wie vor 100 Jahren, und die Folge war, daß besonders in den öffentlichen Sitzungen in dem überfüllten Saale eine entsetzliche Luft herrschte und manchem Akademiker, ebenso wie der anwesenden Zuhörerschaft die meist zweistündige Dauer der Sitzung zu einer kleinen Plage werden ließ. Man beschloß also eine Änderung eintreten zu lassen, und da ich damals durch die Studien für den Neubau des chemischen Instituts in Fragen der Heizung und Ventilation Kenntnisse erworben hatte, so wurde ich beauftragt, zusammen mit dem Physiker F. Kohlrausch eine solche Einrichtung auch für den akademischen Sitzungssaal zu planen. Mit Hilfe eines geschickten Ingenieurs war diese Aufgabe rasch gelöst. Aber als der Plan zur Begutachtung der gesamten Akademie vorgelegt wurde und sich dabei herausstellte, daß die Luft des Saales während der Sitzung 2 bis 3 Mal in der Stunde erneuert werden sollte, trat eine peinliche Sorge vor Zugluft ein, und als gar ein angesehenes Mitglied der Korporation erklärte, er würde dadurch kalte Füße bekommen, war trotz der Versicherung, daß die eingeführte Luft angewärmt sei und durch das Ausströmen aus zahlreichen kleinen Öffnungen den Charakter des Zugwindes ganz verliere, unser Vorschlag nicht mehr zu retten.

Einen zweiten Mißerfolg ähnlicher Art habe ich als Mitglied einer Kommission für den Neubau des Akademischen Hauses Unter den Linden gehabt. Zwar sind Heizung und Ventilation hier nach unseren Vorschlägen in vernünftiger Weise ausgeführt worden, aber bezüglich der Akustik konnte ich nicht durchdringen. Nach meinen Erfahrungen beim Neubau des chemischen Instituts hatte ich verlangt, daß die Sitzungssäle mit kassetierten Holzdecken und auch mit möglichst viel Holzbekleidung an den Wänden ausgestattet würden, da diese wie Resonanzböden wirken. Obschon die Mitglieder der Akademie diesmal auf meiner Seite waren, ist es mir nicht gelungen, den Eigensinn der Architekten zu überwinden. Sie behaupteten von Akustik sehr viel zu verstehen, haben unseren Rat in den Wind geschlagen und es auch richtig fertig gebracht, daß die Akustik in dem Neubau recht viel zu wünschen übrig läßt.

Im Sommer 1893, bald nach meiner Aufnahme in die Akademie, habe ich in der mathematisch-naturwissenschaftlichen Klasse meinen ersten Vortrag über die Synthese der Alkylglukoside gehalten, die der Ausgangspunkt für viele andere Arbeiten geworden ist.

In der Leibniz-Sitzung, die im Juli desselben Jahres stattfand, habe ich auch meine Antrittsrede gehalten, die in den Sitzungsberichten veröffentlicht ist und bei der ich nicht allein die Richtung meiner eigenen Untersuchungen kennzeichnete, sondern auch in ganz kurzer Form einen Überblick über die Entwicklung der Chemie während der letzten 50 Jahre gab. Die Antwort darauf erteilte Dubois-Reymond und begrüßte mich als den zweiten Zuckerchemiker, gleichsam als den Erben von S. Marggraff in der Akademie. Ich habe daraus von neuem gesehen, welch populärer Stoff der Zucker ist, und als ich einige Zeit später einen Vortrag über die Stereochemie der Zucker hielt, kam Helmholtz zu mir, um seiner Freude Ausdruck zu geben, daß die Chemie derartige komplizierte Fragen des molekularen Baues behandeln könne. Selbstverständlich war das Urteil eines solchen Mannes für mich und meine Wissenschaft besonders ehrenvoll. Ich habe zwar gleich gemerkt, daß er die Sache nur halb verstanden hatte, weil ihm die Tatsachen, auf denen die Spekulation beruhte, zu fremd waren, aber mit dem Feingefühl des Genies hatte er doch den großen Fortschritt erkannt, den die Lehre von van't Hoff und Le Bel und ihre Spezialanwendung auf so komplizierte Gebilde wie den Zucker der Chemie gebracht haben. Seitdem ist mir, wie allen Mitgliedern der Akademie, alljährlich die Aufgabe zugefallen, an einem bestimmten Tage einen Vortrag abwechselnd vor der mathematisch-physikalischen Klasse und der gesamten Akademie zu halten. Ich habe mich dabei stets bemüht, allgemeine, meinem Arbeitsgebiet naheliegende Probleme in möglichst volkstümlicher Form zu behandeln und in der Regel auch die Aufmerksamkeit der Zuhörerschaft gefunden. Die in den Sitzungsberichten gleichzeitig von mir veröffentlichten Experimentalarbeiten geben davon kein richtiges Bild, da sie nur ein kleiner Ausschnitt aus dem Vortrag waren.

In jüngerer Zeit, wo auch das Programm der öffentlichen Sitzungen durch wissenschaftliche Vorträge erweitert wurde, habe ich auf Einladung der Akademie als Erster der mathematisch-physikalischen Klasse in der Friedrichsitzung Januar 1907 über die Chemie der Proteine und ihre Bedeutung für die Biologie gesprochen.

Die Geschäfte der Korporation werden nach den Beschlüssen der Gesamtheit größtenteils von vier Sekretären geführt, aber wo fachmännischer Rat nötig ist, treten auch die einzelnen Mitglieder in Aktion, entweder für ihre Person allein, oder als Teile besonderer Kommissionen. Wo chemischer Rat nötig war, hat die Körperschaft nie versäumt, mich zu fragen. Einer meiner ersten Ratschläge hat leider eine peinliche Folge gehabt.

Bei der Verleihung des Ordens pour le merite Friedensklasse an ausländische Gelehrte wird nämlich die Akademie regelmäßig ersucht, Vorschläge zu machen, und als es sich hierbei um die Wahl eines Chemikers handelte, hielt ich mich für verpflichtet, in erster Linie L. Pasteur und außerdem noch Frankland und van't Hoff vorzuschlagen. Akademie und Regierung schlossen sich dem an, und der Orden wurde durch den deutschen Botschafter in Paris zuerst Pasteur angeboten. Dieser lehnte ab, was sein gutes Recht war. Aber er beging gleichzeitig die Taktlosigkeit, seinen Entschluß nicht geheim zu halten. Die Sache kam in die Öffentlichkeit und wurde von der französischen Presse zu einer großen chauvinistischen Kundgebung ausgenutzt. Als ich einige Jahre später Paris besuchte, um die dortigen Laboratorien zu besichtigen und H. Moissan aufsuchte, richtete er an mich sofort die Frage, weshalb die Berliner Akademie keinen Franzosen mehr zum Mitglied wähle. Als ich ihm darauf antwortete, daran sei das Benehmen von Pasteur schuld und die Sorge der Berliner Gelehrten vor weiteren Skandalen, die der Würde der Wissenschaft nur schaden könnten, gab er mir recht und betonte, daß auch die große Mehrheit der französischen Forscher es für falsch halte, nationalistische Tendenzen in der Wissenschaft gelten zu lassen. Seitdem haben eine ganze Reihe von Franzosen, u. a. Marcellin Berthelot, nach Vorschlag unserer Akademie den Orden erhalten und angenommen.

Viel erfreulicher verlief ein anderer Antrag auf Ehrung eines Ausländers, den ich viele Jahre später gemeinsam mit Walter Nernst stellte. Er betraf die Verleihung der goldenen Leibniz-Medaille an Herrn Ernest Solvay in Brüssel und gleichzeitig an Herrn H. von Böttinger als Anerkennung ihrer Verdienste um die Förderung der Wissenschaften durch Zuwendung von reichen materiellen Mitteln. Beide Herren erschienen in der Leibniz-Sitzung an 1. Juli 1909, um diese Medaillen in Empfang zu nehmen, und abends hatten Nernst und ich das Vergnügen, in den Räumen des Automobilklubs zu Ehren der beiden Herren eine gesellige Zusammenkunft mit vielen Mitgliedern der Akademie, mit hervorragenden Chemikern und anderen Personen des wissenschaftlichen Berlins zu veranstalten. Mir fiel die Aufgabe zu, die beiden Herren in einer Tischrede zu begrüßen. Das Fest verlief in harmonischer und heiterer Stimmung, und die dabei gehaltenen Reden sind in einer kleinen Schrift „Festmahl zu Ehren der Herren E. Solvay und H. v. Böttinger am 1. Juli 1909 im Kaiserlichen Automobilklub zu Berlin" zusammengefaßt worden. Als Dank für die Ehrung machte Herr von Böttinger eine Stiftung von 30 000 M., die er auf meinen Rat zur Erwerbung von Mesothorium bestimmte. Ich habe den Ankauf des Präparates besorgt und dasselbe verwaltet, bis die Akademie den Neubau Unter den Linden wieder bezog und dadurch die Möglichkeit erhielt, ihren gesamten materiellen Besitz im eigenen Hause aufzuheben. Das Mesothorium ist inzwischen auf meinen Rat von der Akademie mit großem Gewinn wieder verkauft und dafür ein entsprechender Vorrat des viel haltbareren Radiums angeschafft worden.

Außer dem Mesothorium habe ich auch länger als 20 Jahre die akademische Instrumentensammlung, die in meiner Dienstwohnung zuerst in der Dorotheenstraße und später in der Hessischenstraße untergebracht war, verwalten müssen. Die Inventarisierung der einzelnen Objekte war verhältnismäßig einfach, aber der Verkehr mit den Gelehrten, denen die Instrumente überlassen wurden, brachte zuweilen unbequemen Briefwechsel mit sich. Auch diese Verpflichtung bin ich inzwischen los geworden, weil die Instrumentensammlung ebenfalls in das neue Haus der Akademie übergeführt werden konnte, und jetzt von dem Archivar der Korporation verwaltet wird.

Als Inhaber des akademisch-chemischen Laboratoriums und der damit verbundenen Dienstwohnung in der Dorotheenstraße, einem der ältesten wissenschaftlichen Laboratorien Berlins, das von S. Marggraff erbaut und in dem Achard zum erstenmal Rohrzucker in größerer Menge (1700 Pf.) aus Zuckerrüben herstellte, stand ich zu der Akademie in einem besonderen Verhältnis. Das trat besonders scharf hervor, als der Plan eines Neubaues des chemischen Instituts gereift war, und die Verlegung des Instituts an eine andere Stelle nötig wurde. Für die Bewilligung der Bausumme stellte damals die Finanzverwaltung die Bedingung, daß die Akademie auf den Besitz in der Dorotheenstraße verzichten, und dafür einen Anteil an dem neuen Gebäudekomplex in der Hessischenstraße erhalten soll. Es ist begreiflich, daß die Aufgabe des alten Besitzes, der 150 Jahre der Akademie gehörte, und wo nicht allein der akademische Chemiker, sondern zeitweise auch der Astronom Unterkunft gefunden hatte, von manchem Mitglied der Korporation schmerzlich empfunden wurde, und daß deshalb die Zustimmung der Akademie zu dem neuen Plane von einer besonderen Verhandlung abhängig gemacht wurde. Als die Stimmung zu schwanken schien, traten zwei Männer auf das Entschiedenste für den Plan ein, der Mediziner Rudolph Virchow und der Theologe Adolf Harnack mit der Bemerkung, daß die Akademie verpflichtet sei, einer Wissenschaft wie der Chemie, wenn sie in eine Notlage geraten sei, zu Hilfe zu kommen. Diese großzügige Auffassung drang durch. Der Plan wurde einstimmig genehmigt, und ich habe mich dauernd der Akademie für diese vornehme Handlung verpflichtet gefühlt. Die Sympathie, die ich von Anfang an für sie gehabt habe, ist dadurch noch um ein Erhebliches gesteigert worden. Infolgedessen habe ich auch mit besonderer Freude eine Ehrung entgegengenommen, die mir von der Akademie durch Verleihung der Helmholtz-Medaille im Januar 1909 erwiesen wurde. Die Medaille wird nicht nach dem Statut, aber gewohnheitsgemäß alle zwei Jahre abwechselnd einem Physiker oder Biologen gegeben. Demnach habe ich sie in letzter Eigenschaft erhalten, weil meine chemischen Arbeiten die Biologie vielfach berühren. Durch den Besitz der Medaille erhält man gleichzeitig das Recht, für neue Verleihungen Vorschläge zu machen, und ich habe schon das nächste Mal davon Gebrauch gemacht, um van't Hoff diese Ehre zuteil werden zu lassen. Er empfing die Medaille einige Wochen vor seinem Tode. Es war die letzte Ehrung, die ihm erwiesen wurde, und die ihm nach dem Zeugnis der Gattin noch viele Freude bereitet hat. Selbstverständlich habe ich auch immer mitgewirkt, wenn es sich um die Wahl von Chemikern zum ordentlichen oder korrespondierenden Mitglied handelte, und mit großer Befriedigung kann ich bezeugen, daß in allen diesen Geschäften bei den Chemikern der Akademie volles Einverständnis leicht zu erzielen war. Speziell erwähnen will ich die Wahl zu auswärtigen Mitgliedern, die auf meinen Vorschlag 1900 für M. Berthelot und 1905 für A. von Baeyer stattfand. Zusammen mit van't Hoff hatte ich das Vergnügen, die letzte Wahl bei der Feier des 70-jährigen Geburtstages von Baeyer im Oktober 1905 in München verkünden zu können.

An der Feier des 200-jährigen Stiftungsfestes der Akademie konnte ich leider nicht teilnehmen, weil ich wegen einer hartnäckigen Schwäche des Stimmbandes einen längeren Aufenthalt an der Riviera nehmen mußte. Die Feier brachte der Akademie eine Vermehrung der Mitglieder, speziell für die mathematisch-naturwissenschaftliche Klasse die Abteilung für technische Wissenschaften, in die als erste Mitglieder die Herren Müller-Breslau, Professor an der technischen Hochschule zu Charlottenburg, Martens, Direktor des Materialprüfungsamtes zu Lichterfelde und von Hefner-Alteneck, elektrotechnischer Erfinder, früher Beamter der Firma Siemens & Halske, gewählt wurden. Dadurch erhielt die Akademie Gelegenheit, hervorragende auswärtige Vertreter der Technik durch Ernennung zu korrespondierenden Mitgliedern zu ehren, und ich habe mir wiederholt erlaubt, solche Wahlen für technische Chemiker, z. B. für Ludwig Mond, Ernest Solvay und Auer von Welsbach anzuregen.

Die Bedeutung einer wissenschaftlichen Akademie ist selbstverständlich in erster Linie bedingt durch das wissenschaftliche Ansehen der einzelnen Mitglieder. Zur Zeit meines Eintritts gehörten zur Berliner Akademie eine verhältnismäßig große Anzahl von Männern ersten Ranges. An der Spitze der Naturforscher stand Hermann von Helmholtz, ein wissenschaftliches Universalgenie; denn er hat nicht allein in der Physik, Physiologie und Mathematik grundlegend gewirkt, sondern auch in der Erkenntnistheorie Achtenswertes geleistet. Wenn man von Alexander von Humboldt absieht, so war er der vielseitigste Naturforscher des 19. Jahrhunderts, nicht allein in Deutschland, sondern wahrscheinlich in der Welt, und diese Vielseitigkeit hat der Gründlichkeit seiner Forschung nicht den geringsten Abbruch getan. Dazu besaß er in hohem Maße die Gabe, naturwissenschaftliche Erkenntnis in leicht verständlicher Form und vornehmer Sprache weiteren Kreisen zugänglich zu machen, und ich kenne wenig naturwissenschaftliche Schriften, die auf mich in jüngeren Jahren so anregend gewirkt haben, wie die von Helmholtz publizierten Vorträge über verschiedene Zweige der Physik, Physiologie und Mathematik. Vielleicht hat Justus Liebig durch seine populären Schriften über die Bedeutung der Chemie für den Ackerbau, die Künste und Gewerbe größeren Einfluß auf die Entwicklung des wissenschaftlichen und wirtschaftlichen Lebens in Deutschland gehabt, aber an Feinheit der Darstellung und Schönheit der Form können sie den Vorträgen von Helmholtz nach meinem Empfinden nicht gleichgestellt werden. Zu der Zeit, als ich Helmholtz kennen lernte, war er schon 72 Jahre alt und eine in jeder Beziehung abgeklärte Persönlichkeit. Es war für uns Jüngere stets ein besonderes Vergnügen, wenn er in der Fakultät oder Akademie das Wort ergriff und in ruhiger besonnener Art seine Meinung äußerte. Ich habe wiederholt Gelegenheit gehabt, mit ihm Privatgespräche zu führen, bei denen er stets ein wohlwollendes Interesse an der Chemie kund gab. Diese persönliche und innige Berührung zwischen einzelnen Mitgliedern der Akademie vollzog sich in ungezwungener Form bei den Nachsitzungen, die in einem Kaffee stattfanden, zuerst im Hotel de Rome und später in verschiedenen anderen Kaffees Unter den Linden oder in der Potsdamer Straße. Diese Nachsitzungen waren nicht selten belehrender und vor allen Dingen unterhaltender als die amtliche Hauptsitzung. Als Helmholtz im August 1894 starb, war es mir als Vorsitzender der chemischen Gesellschaft, zu deren Ehrenmitgliedern er gehörte, eine angenehme Pflicht, ihm einen Nachruf zu widmen, der in den Berichten der Gesellschaft gedruckt wurde, und in dem ich meiner Bewunderung für den großen Mann vollen Ausdruck gegeben habe.

Eine zweite sehr interessante Persönlichkeit in dem akademischen Kreise war Rudolph Virchow, pathologischer Anatom, Hygieniker, Anthropologe und Politiker, ein Mann von einer ungewöhnlichen Arbeitskraft, der trotz der Zersplitterung seiner Tätigkeit alles, was er anfaßte, mit großer Gründlichkeit und Überlegung besorgte. Noch im Alter von 80 Jahren pflegte er nicht allein den ganzen Tag, sondern auch die halbe Nacht der Arbeit zu widmen, besaß dafür allerdings auch das Talent, jede freie Minute sogar in den Sitzungen oder bei Gesellschaften zum Schlaf benutzen zu können. Er war scharf in seinem Urteil und konnte gegen Auswüchse der Medizin und Hygiene oder gegen Mißgriffe der Staatsverwaltung in schärfster Weise auftreten. Aber ich habe stets den Eindruck bekommen, daß er sich nur durch sachliche Gründe und durch vornehme politische, soziale oder wirtschaftliche Grundsätze führen ließ. Zu mir persönlich ist er immer sehr freundlich gewesen, und seiner wirkungsvollen Unterstützung bei dem Neubau des chemischen Instituts habe ich schon gedacht. Aus der Akademie erinnere ich mich eines scharfen und erfolgreichen Protestes, den er gegen die Aufnahme einer von S. Schwendener präsentierten Abhandlung eines gewissen Dr. Pinkus über den vermeintlichen Zusammenhang zwischen dem Wachstum der Kopfhaare und dem Schicksal ihres Trägers in die Sitzungsberichte erhob. Seine Kritik war in biologischen Kreisen gefürchtet, aber von der großen Mehrzahl als sachlich anerkannt. Bei mir hat er einmal an einem Abendessen teilgenommen, das ich zu Ehren von William Ramsay veranstaltete. Als wir aus dem warmen Eßzimmer in die etwas unterkühlten Gesellschaftszimmer übersiedelten, sagte er scherzhaft zu mir: „Sie scheinen wie der Schah von Persien über alle Klimatas zu verfügen", worauf ich ihm erwiderte, daß in der Chemie die Periode der extremen Temperaturen begonnen habe.

Ein anderer hervorragender Mediziner in der Akademie war der Physiologe E. Dubois-Reymond, besonders bekannt geworden durch die zahlreichen akademischen Reden, die er als Sekretär der Korporation gehalten hat. Er war zweifellos auch ein geistreicher Mann und von ungewöhnlicher formaler Gewandtheit. Von Zeit zu Zeit lese ich noch heutzutage einzelne seiner Reden mit Genuß, wenn sich auch nicht leugnen läßt, daß die Effekthascherei dabei eine gewisse Rolle spielt, und die übertriebene mechanistische Tendenz einer überwundenen Periode der Wissenschaft angehört.

Von der Chemie verstand Dubois-Reymond wenig, aber er war doch einsichtsvoll genug gewesen, ihr in seinem neuerbauten physiologischen Institut eine ziemlich große Abteilung mit selbständiger Leitung zu gewähren, und wenn man sich daran erinnert, daß E. Baumann und A. Kossel als Vorsteher dieser Abteilung einen großen Teil ihrer wissenschaftlichen Arbeiten ausgeführt haben, so muß man Dubois-Reymond zugestehen, daß er in der Wahl seiner Mitarbeiter eine glückliche Hand gehabt hat.

Die allgemeine Vorlesung über die Fortschritte der Naturwissenschaften, die er jahrzehntelang als Publikum gelesen hat, war berühmt nach Inhalt und Form und wurde von den Mitgliedern aller Fakultäten gerne besucht. Daß sie seit seinem Tode in Berlin fehlt, ist zweifellos eine Lücke in dem naturwissenschaftlichen Unterricht der Universität, aber ich wüßte augenblicklich niemand in unserem Kreise zu nennen, der imstande wäre, wie Dubois diese Aufgabe zu lösen.

Unter den jüngeren Naturforschern der Akademie traf ich zwei alte Bekannte, den Botaniker Engler, der gleichzeitig mit mir Privatdozent in München gewesen war und den Physiker August Kundt, meinen lieben Lehrer aus Straßburg, dessen ich früher schon gedacht habe. Er hatte sich mit der ihm eigenen Energie in Berlin nicht allein in die

Aufgabe des Lehrers und Forschers, sondern auch in den Strudel des gesellschaftlichen Lebens gestürzt. Dem war seine Gesundheit nicht gewachsen. Ein schweres Herzleiden machte sich schon damals bemerkbar und im Mai 1894 ist er daran gestorben. Es war mir eine angenehme Pflicht, ihm in den Berichten der chemischen Gesellschaft einen kurzen Nachruf zu widmen. Die Chemiker werden immer dankbar anerkennen, daß er zusammen mit Warburg die Einatomigkeit des Quecksilberdampfes nach seiner allgemeinen Methode der Schallgeschwindigkeit eindeutig bewies und damit das gleiche für die später entdeckten Edelgase möglich machte. Kundt war ein ausgezeichneter Lehrer und lieber Mensch, der auf jüngere Naturforscher eine große Anziehungskraft ausübte und dessen Tod eine große Lücke in unseren naturwissenschaftlichen Kreis riß. Das zeigte sich sofort bei der Berufung eines Nachfolgers. Auf Anregung von Helmholtz präsentierte die Fakultät dem Ministerium nur einen Kandidaten, Friedrich Kohlrausch. Aber ehe er zusagte, starb Helmholtz und Kohlrausch wurde nun sein Nachfolger an der physikalisch-technischen Reichsanstalt zu Charlottenburg.

Die Wahl des Physikers an der Universität mußte dann einstweilen verschoben werden, weil keine Persönlichkeit in Deutschland vorhanden war, die alle Beteiligten befriedigte. Infolgedessen tauchte im Laufe des Winters der Gedanke auf, J. H. van't Hoff für die Stelle zu gewinnen, nachdem bekannt geworden war, daß er nicht ungern Amsterdam verlassen würde. Die Unterrichtsverwaltung ging rasch auf diesen Plan ein, und zunächst wurde Professor M. Planck nach Amsterdam geschickt, um mit van't Hoff die Möglichkeit einer solchen Berufung zu überlegen. Den weiteren Verlauf der Verhandlung habe ich in einer Gedächtnisrede auf van't Hoff geschildert, die ich bald nach seinem Tode in der Leibnizsitzung der Akademie hielt. Er lehnte die Professur an der Universität ab, wurde aber dafür ein Jahr später als Akademiker nach Berlin gerufen und erhielt nun eine außerordentliche Professur an der Universität mit dem Charakter als Honorarprofessor, um sein Gehalt auf eine angemessene Höhe bringen zu können. An der Berufung van't Hoffs habe ich selbstverständlich regen Anteil genommen und bin dafür durch ein schönes Freundschaftsverhältnis belohnt worden, das bis zu seinem Tode dauerte, und das mich mit diesem bedeutenden Naturforscher in nahe Berührung gebracht hat. Abgesehen von einigen kleinen Schwächen war er ein prächtiger Mann von origineller Denkart und kaum minder originell in seinen Lebensgewohnheiten. Eine öffentliche Vorlesung über chemische Theorien hat er regelmäßig in unserem Institut gehalten, während er sich für seine Experimentalstudien zusammen mit Meyerhofer ein bescheidenes Privatlaboratorium eingerichtet hatte.

Nachdem van't Hoff die physikalische Professur an der Universität abgelehnt hatte, wurde Emil Warburg berufen. Als er 10 Jahre später als Nachfolger von Kohlrausch an die physikalisch-technische Reichsanstalt ging, wurde Drude Professor der Physik an der Universität, fand aber schon nach kurzer Zeit ein tragisches Ende. Seitdem ist Rubens, der Entdecker der Reststrahlen bei uns der Vertreter der Experimentalphysik, während die theoretische Physik in der ganzen Zeit meiner Berliner Tätigkeit in würdigster Weise von M. Planck geleitet und gepflegt wurde. Mit allen diesen Physikern, welche durchweg verständige, wohlwollende und ruhig denkende Menschen waren, habe ich im freundschaftlichen Verhältnis gestanden. Besonders gilt das von Friedrich Kohlrausch, mit dem ich schon in Würzburg befreundet war. In der großen Stadt ist zwar der ungezwungene leichte Verkehr, wie Würzburg ihn bot, kaum möglich, aber mit Kohlrausch bin ich doch wieder in sehr nahe persönliche Berührung gekommen durch einige Versuche, die wir im Jahre 1898/99 miteinander ausführten, und die nichts geringeres bezweckten, als eine Verwandlung chemischer Elemente ineinander. Lange bevor man solche Erscheinungen bei den radioaktiven Substanzen beobachtete, hatte ich mir die Meinung gebildet, daß auf der Sonne und anderen Fixsternen, nach den dort herrschenden extremen Temperaturen und den besonderen Druckverhältnissen sich elementare Verwandlungen vollziehen und daß der wirkliche Zustand der Sonne in bezug auf die chemischen Elemente ein Gleichgewichtszustand sei. Als Hypothese habe ich diese Ansicht auch häufig in meinen Vorlesungen geäußert, allerdings niemals darüber publiziert, weil ich keine tatsächlichen Gründe dafür anführen konnte. Ich war nun weiter auf den Gedanken gekommen, daß man vielleicht elementare Verwandlungen erreichen könne, wenn man auf eine kleine Menge von Stoff eine ungeheure Menge von Energie konzentriere und schlug Kohlrausch vor, solche Versuche mit Wasserstoff in sehr verdünntem Zustand bei langer Einwirkung von Kathodenstrahlen auszuführen. Der Gedanke schien nicht unsinnig, und wir hatten uns einen hübschen Apparat zusammengebaut, der es gestattete, den mit Kathodenstrahlen behandelten Wasserstoff auf den Druck von 6 bis 8 mm zu konzentrieren und dann spektralanalytisch zu prüfen. Speziell hatten wir die Hoffnung, den Wasserstoff ganz oder teilweise in Edelgas überzuführen. Die Herstellung des Apparates, der Kathodenstrahlen, der Vakuumpumpe usw. fiel natürlich dem Physiker zu. Ich übernahm dafür die sorgfältige Reinigung der Materialien, der Glasgefäße, des Quecksilbers und des Wasserstoffes, der zuletzt immer aus Paladiumwasserstoff bereitet wurde. Wir haben so manchen Nachmittag bis in die späten Abendstunden hinein zusammen gearbeitet, leider ohne ein sicheres Resultat zu erzielen. Besondere Schwierigkeiten machten uns die Kathodenstrahlen, weil mit der Reinheit der Kathoden der Widerstand außerordentlich wuchs, und schließlich solche Spannungen angewandt werden mußten, daß die Gefäße es nicht mehr aushielten. Wir haben nichts darüber publiziert und das ist der Grund, warum ich hier so ausführlich darüber gesprochen habe. Etwa 10 Jahre später sind ähnliche Versuche von englischen Physikern veröffentlicht worden, welche der Meinung waren, daß sie dabei die Entstehung von Edelgasen sicher beobachtet hätten. Sie scheinen sich aber getäuscht zu haben; denn später hat man von dieser Entdeckung nichts weiter als abfällige Kritik gehört, und daß ein solcher Irrtum leicht eintreten kann, wird jeder, der sich mit ähnlichen Versuchen beschäftigt hat, zugeben.

Eine andere kleine Arbeit habe ich viele Jahre später mit Heinrich Rubens angestellt. Sie betraf die Prüfung der von dem Physiologen Rosenthal aufgestellten Behauptung, daß Wechselströme eine Hydrolyse von Stärke und ähnlichen Kohlenhydraten bewirken können, wenn die Schwingungszahl eine angemessene, aber für jeden Fall verschiedene Höhe erreicht hat. Unsere Versuche fielen ganz negativ aus, wurden aber nicht publiziert, weil uns die Sache nicht mehr interessierte, sobald der Irrtum von Rosenthal festgestellt war. Einige Jahre später hat ein Physiko-Chemiker die Rosenthal'sche Behauptung auch öffentlich widerlegt. Die Ursache des Irrtums ist ihm aber verborgen geblieben. Ich glaube dieselbe gefunden zu haben durch die Beobachtung, daß der verwandte Stärkekleister in der Tat reduzierende Eigenschaften annehmen kann, wenn man ihn im offenen Gefäß einer langen Behandlung mit Wechselstrom unterwirft; denn dann bildet sich in der umgebenden Luft Ozon, das die Stärke in der Wärme rasch angreift und in reduzierende Substanzen verwandelt. Man vermeidet diesen Fehler vollständig, sobald man die den Stärkekleister enthaltenden Gefäße verschließt. Die Wirkung des Ozons auf Stärkekleister, die meines Wissens schon in rohem Umriß bekannt ist, verdient übrigens eine nähere Untersuchung bezüglich der dabei entstehenden Produkte, und ich hoffe selbst noch Gelegenheit zu haben, darüber neue Erfahrungen zu sammeln.

Mein Spezialkollege sowohl in der Akademie wie Fakultät war der Chemiker Hans Landolt, ungefähr 20 Jahre älter wie ich. Wie früher erwähnt, hätte ich 1880 sein Nachfolger in Aachen werden können, als er an die Landwirtschaftliche Hochschule zu Berlin übersiedelte. Von dort wurde er als Nachfolger von Rammelsberg an die Universität berufen und mit der Direktion des II. chemischen Instituts der Universität in der Bunsenstraße betraut. Als Sachverständiger war er bei meiner Berufung nach Berlin und auch bei der Aufnahme in die Akademie besonders stark beteiligt.

Landolt, der Sprosse einer alten Züricher Familie, verleugnete trotz seines langen Aufenthaltes in Deutschland nicht den Schweizer. Er war ein kluger, kritisch veranlagter Kopf und ein Mann von aufrichtigem angenehmen Charakter. Als Forscher nahm er keine hervorragende Stellung ein, aber seine Bücher, besonders die großen, mit Börnstein zusammen publizierten Tabellen erfreuen sich in den Fachkreisen großer Wertschätzung. Wir sind sehr gut miteinander ausgekommen und haben nur äußerst selten in geschäftlichen Dingen verschiedene Ansichten vertreten. Er war kein Führer, weder in der Wissenschaft noch in der Berliner Gelehrtenkorporation, aber er erfreute sich wegen seines besonnenen Urteils und seiner allgemeinen Charaktereigenschaften in unserem Kreise allgemeiner Beliebtheit. Das zeigte sich bei der Feier seines 70. Geburtstages, den er kurz nach einer schweren Gallensteinoperation erlebte. Mir war es bei dieser Gelegenheit ein Vergnügen, beim Festessen die Rede auf den Jubilar zu halten und darin auch des schönen kollegialen Verhältnisses zu gedenken, das zwischen uns dauernd bestanden hat. Auch zu seiner Gattin und der Tochter, die an Professor O. Liebreich verheiratet war, bin ich in ein freundschaftliches Verhältnis getreten. Frau Landolt hat sich ein Verdienst um den geselligen Verkehr der chemischen Privatdozenten und Assistenten erworben, die sie häufig zu sich einlud, und ihnen damit den gesellschaftlichen Anschluß ermöglichte, den ich leider bei meiner zurückgezogenen Lebensweise nicht bieten konnte.

Im Alter von 74 Jahren gab Landolt die Professur an der Universität auf und zog sich zurück in die physikalisch-technische Reichsanstalt zu Charlottenburg, um die subtilen Versuche über etwaige Gewichtsveränderung der Gesamtmaße bei chemischen Reaktionen fortzusetzen, die er während mehr als 10 Jahre im II. chemischen Insitut angestellt hat. Für diese Versuche war er besonders veranlagt. Es machte ihm große Freude, die Wägungen immer mehr zu verfeinern und etwaige Fehlerquellen aufzuspüren. Nach manchen Schwankungen kam er, wie bekannt, zu dem Resultat, daß bei chemischen Vorgängen keine durch unsere jetzigen Hilfsmittel nachweisbare Gewichtsveränderung eintritt. Das war nichts Neues; denn derselbe Satz gehört zu den Grundlehren der Chemie, aber es ist doch ein zweifelloses Verdienst, wenn gerade solche fundamentalen Sätze von Zeit zu Zeit mit den fortgeschrittenen technischen Hilfsmitteln geprüft werden, und der Name Landolt wird sich wahrscheinlich in der Geschichte der Wissenschaft mehr an diese Versuche knüpfen als an seine sonstigen wissenschaftlichen Arbeiten.

Landolts Nachfolger an der Universität wurde Walter Nernst, der ausgezeichnete Physiko-Chemiker zu Göttingen. Bei dieser Gelegenheit wurde das II. chemische Institut umgetauft in „Physikalisch-chemisches Institut der Universität" und damit verlor auch das I. chemische Institut seine I und erhielt wieder den alten einfachen Namen „Chemisches Institut der Universität".

Wie begreiflich wurde Nernst auch bald in die Akademie der Wissenschaften gewählt und die Chemie war dann durch eine Gruppe von 4 Männern, Landolt, van 't Hoff, Nernst und mich vertreten, die ihr in allen fachlichen Fragen den gebührenden Einfluß sichern konnten. Nernst, den ich schon von Würzburg her kannte, weil er dort beim Doktorexamen in Chemie von mir geprüft wurde, ist mir später ein lieber und hilfsbereiter Kollege gewesen. Sowohl in der Fakultät wie in der Akademie haben wir fast immer einmütig zusammen gehandelt und nur bei der Besetzung von chemischen Lehrstühlen an anderen preußischen Universitäten standen zuweilen die Ratschläge, die wir dem Kultusminister gaben, miteinander in Widerspruch. Am schärfsten trat das zutage bei der Besetzung der Professur in Breslau nach dem Abgang von Buchner. Die Fakultät hatte Abegg vorgeschlagen und dieser Kandidat fand die lebhafteste Unterstützung von Nernst. Ich war aber der Meinung, daß die Professur der Experimentalchemie und die Direktion des chemischen Instituts einem Experimentalchemiker übertragen werden sollte, während man für Abegg eine besondere ordentliche Professur für physikalische Chemie schaffen sollte. Nach langem Schwanken hat der Minister Professor Biltz aus Kiel nach Breslau berufen und damit den von mir ausgesprochenen Grundsatz als gerechtfertigt anerkannt. Nernst soll diese Wendung der Dinge nicht allein für die Person des Herrn Abegg, sondern auch für die physikalische Chemie als Zurücksetzung empfunden haben, wie mir allerdings nur auf Umwegen bekannt geworden ist. Ich will hier offen bekennen, daß ich niemals der Entwicklung der physikalischen Chemie hinderlich entgegengetreten bin, sondern sie stets als notwendigen Bestandteil des Unterrichts und der Forschung anerkannt und auch bei den Behörden empfohlen habe. Ich hätte sogar in dem neuen Institut eine besondere Abteilung für physikalische Chemie eingerichtet, wenn nicht die Herren van't Hoff und Landolt sich dagegen ausgesprochen hätten mit der Begründung, daß man ihnen die Vertretung der physikalischen Chemie allein überlassen sollte. Die Berechtigung dieses Wunsches lag auf der Hand, und so habe ich meine Absicht fallen lassen, obschon es damals ziemlich leicht gewesen wäre, Professor Bredig aus Leipzig für unser Institut zu gewinnen. Meine Hoffnung, daß mit der Berufung von Nernst alle studierenden Chemiker in Berlin seine Vorlesungen und Übungen besuchen und sich genügend in physikalischer Chemie unterrichten würden, hat sich allerdings nicht erfüllt.

Die große Stadt bringt es mit sich, daß der Student sich so weit wie möglich auf den Besuch eines Fachlaboratoriums beschränkt und dazu kommt noch, daß im Berliner Doktorexamen die chemische Technologie als Nebenfach gewählt werden darf, so daß für physikalische Chemie kein Raum mehr bleibt. Ich halte das, offen gestanden, für eine Lücke unseres Unterrichtes, und ich habe mich selbst immer bemüht, durch die Handhabung des Doktorexamens den Studenten die Wichtigkeit der physikalischen Chemie klar zu machen, aber ich kann mir doch nicht verhehlen, daß es wirksamer wäre, sie obligatorisch zu machen und hoffe, daß diese Maßregel nach dem Kriege auch getroffen wird.

Nernst ist kein einseitiger Fachgelehrter, sondern interessiert sich auch für industrielle und wirtschaftliche Dinge, wozu er durch die Erfindung seiner elektrischen Lampe genügende praktische Gelegenheit fand. Er ist ferner Automobilist und Landwirt, Jäger und war während des Krieges sogar Soldat. Das entspricht ganz seinem lebhaften Geiste, seiner großen Energie und seinem leidenschaftlichen Empfinden. Zusammen mit seiner liebenswürdigen Gattin und seinen Kindern hat er auch ein gastfreies Haus in Berlin eingerichtet, wo die jungen Chemiker und Physiker leicht Aufnahme finden. Er war anfangs der Meinung, daß ich auf seine gesellschaftlichen Bemühungen eifersüchtig werden könnte und sprach sich darüber freimütig aus. Meine Antwort, daß ich nicht den geringsten Wert darauf lege, gesellschaftlich eine Rolle zu spielen und ihm nur dankbar sein könnte, wenn er sich dieser Verpflichtung unterziehe, war für ihn ebenso überraschend wie beruhigend.

Auf unsere gemeinsame Arbeit bei dem Plane zur Errichtung einer chemischen Reichsanstalt und bei der Gründung des Kaiser Wilhelm Instituts für Chemie werde ich später zurückkommen. Dasselbe gilt für die Herren Haber, Beckmann und Willstätter, die erst mit der Gründung der Kaiser Wilhelm Institute nach Berlin kamen, und dann in die Akademie aufgenommen wurden.

Zu den Mineralogen und Geologen bin ich in kein näheres Verhältnis getreten. Zwar habe ich mich ernstlich bemüht, das Interesse der jungen Chemiker für Kristallographie und Mineralogie zu wecken und auch eine kleine Anzahl von Mineralien, die ich der Güte des Herrn Kollegen Klein verdankte, in einem Verbindungssaal des chemischen Instituts aufgestellt, aber es ist mir nicht gelungen, eine rege Beteiligung der Chemiker an kristallographischen oder mineralogischen Vorlesungen oder Übungen zu erreichen. Das hängt zusammen mit den Verhältnissen der Großstadt und der für Chemiker nicht zweckmäßigen Promotionsordnung der philosophischen Fakultät.

Unter den Biologen haben mich am meisten die Physiologen S. Schwendener, Haberlandt und der Mediziner M. Rubner interessiert. Sie waren alle drei gut unterrichtete Gelehrte und verdiente Forscher. Ich bin mit den beiden letzteren, namentlich während des Krieges bei mannigfachen Beratungen über Ernährung häufig zusammengekommen.

In der philosophisch-historischen Klasse der Akademie war bei meinem Eintritt Theodor Mommsen die hervorragendste Persönlichkeit. Trotz der demokratischen Verfassung der Akademie war sein Einfluß in der philosophisch-historischen Klasse so groß, daß er seine Meinung in fast allen wichtigen Fragen durchzusetzen wußte. Am deutlichsten trat für uns Naturforscher seine Macht in die Erscheinung, wenn er die Wahl von neuen Mitgliedern für seine Klasse zu vereiteln suchte. Seine Machtstellung verdankte er übrigens den ungewöhnlichen wissenschaftlichen Leistungen und der außerordentlichen Schlagfertigkeit seiner Kritik, in der er den formgewandten und leidenschaftlichen Verfasser der römischen Geschichte nicht verleugnete. Wer dieses Werk gelesen hat, kann sich dem Eindruck seines Genies und seiner gewaltigen Arbeitskraft nicht entziehen. Ich selbst habe nur hier und da kurze Gespräche mit ihm geführt und sein Erstaunen hervorgerufen durch die Mitteilung, daß ich als Gymnasiast auf Anregung eines Oberlehrers selbst Abklatsche von römischen Inschriften meiner Heimat angefertigt hatte. Noch ein zweites Mal ist es mir gelungen, ihn in eine gewisse Aufregung zu versetzen, als ich in der philosophischen Fakultät einen harten Kampf gegen die Promotion von russischen Chemikern ohne ausreichende Vorbildung führte und dabei auf das frühere Vorgehen von Mommsen gegen den allerdings viel schlimmeren Unfug der Promotion in absentia an einigen deutschen Universitäten hinwies. Er trat infolgedessen auch wirklich bei der Abstimmung auf meine Seite, und ich gewann dadurch eine kleine Majorität in dem von mir um die Ehre meiner Wissenschaft geführten Kampf.

Auch den Historiker Sybel, den Verfasser der Geschichte der französischen Revolution und der Gründung des Deutschen Reiches habe ich in der Akademie kennen gelernt, nachdem ich seine Werke teilweise vorher gelesen. Die letzten Vorträge, die er uns hielt, waren durch Feinheit der Form ausgezeichnet.

Treitschke kam erst später in die Akademie, nachdem Mommsen gestorben und sonstiger Widerstand durch den Erfolg seiner deutschen Geschichte beseitigt war. Er ist selten in unserem Kreise erschienen, weil er schon krank war und bald nachher starb. Ich kannte ihn aber längst von der philosophischen Fakultät her. Er war damals schon so schwerhörig, daß er von den Verhandlungen kaum ein Wort verstand, hatte aber doch den Ehrgeiz, wenn irgend möglich, in die Diskussion einzugreifen und ließ sich für den Zweck von einem seiner Nachbarn Stichworte aufschreiben, die den Lauf der Debatte kennzeichnen sollten. Ich selbst habe an einem Abend diese Aufgabe übernommen und muß gestehen, daß sie recht schwer zu erfüllen war. Wenn Treitschke aus dem Stichwort den Stand der Verhandlung richtig entnahm, so war ich erstaunt zu sehen, mit welcher Schnelligkeit er den ganzen Komplex der Fragen erfasste und mit seiner großen Beredsamkeit sich darüber verbreitete. Es kam aber auch vor, daß er durch Mißverständnis des Stichwortes ganz falsch ging und dann über Dinge redete, die mit der verhandelten Frage in gar keinem Zusammenhang standen. Charakteristisch für ihn war die große Leidenschaft, mit der er alle ihn interessierenden Dinge ergriff und mit dichterischem Schwung besprach. Dieser Eigenschaft war zum großen Teil der starke Einfluß zuzuschreiben, den seine Vorträge auf die akademische Jugend ausübten.

Eine ganz andere Natur war der Germanist Erich Schmidt, ausgezeichnet durch körperliche Schönheit, klangvolle Stimme und liebenswürdiges Wesen. Er galt mit Recht als ausgezeichneter Redner und vorzüglicher Kenner der deutschen Literatur. Von der Berliner Gesellschaft wurde er reichlich verwöhnt, und die fast täglichen Abendessen, die er jahrelang in dem Kreise der Hautefinance oder des hohen Beamtentums mitmachte, haben zweifelsohne mit zu seinem verhältnismäßig frühen Tode beigetragen.

Von der so häufigen Pedanterie des Philologen war nicht das geringste bei ihm zu bemerken, obschon er eine gründliche fachwissenschaftliche Bildung besaß und auch dauernd wissenschaftlich tätig blieb. In geschäftlichen Dingen, falls sie ihn interessierten, bewies er große Gewandtheit und Sicherheit, und ich habe ihn für den besten Dekan gehalten, den die philosophische Fakultät während meiner 25-jährigen Angehörigkeit besessen hat.

Da seine Frau eine geborene Strecker und Halbschwester der Frau Leube war, so bin ich mit ihm hier und da auch in gesellschaftlichen Verkehr gekommen und kann nur sagen, daß ich ebenso wie viele andere Menschen von seiner strahlenden Persönlichkeit eingenommen war.

In gewissem Gegensatz zu Erich Schmidt stand der klassische Philologe Diels, ein prächtiger Mann, klug, wohlwollend, durchaus nicht einseitig in Wissen und Anschauungen und sehr brauchbar als Sekretär der Akademie. Manche Reden, die er in dieser Eigenschaft hielt, darf man wegen der feinen Form und des gedanklichen Inhalts als Perlen bezeichnen. Mit seinen Söhnen hat er besonderes Glück gehabt; denn sie sind alle anerkannte Gelehrte geworden und einer von ihnen, Otto, war mir jahrelang ein lieber Schüler und Mitarbeiter.

Eine interessante Persönlichkeit war auch der Nationalökonom C. Schmoller, gleichzeitig bekannt als guter Historiker. Wenn er in unverfälschter schwäbischer Mundart in der Akademie oder Fakultät über wissenschaftliche oder geschäftliche Dinge sich äußerte, so konnte man immer sicher sein, kluge und ausgereifte Dinge zu hören. Viel stürmischer war sein Spezialkollege Ad. Wagner, der aber nicht der Akademie angehörte, und der in hohem Alter trotz körperlicher Hilflosigkeit durch seine leidenschaftliche Sprache uns immer noch imponierte. Allerdings habe ich sein Urteil in manchen Dingen als befangen oder als verschroben ansehen müssen.

Unter den Vertretern der Geisteswissenschaften hat mich von Anfang an ein Mann besonders angezogen, ehe ich wissen konnte, daß ich später mit ihm in enge, gemeinschaftlicher Arbeit gewidmete Beziehung treten würde. Das ist der Theologe und Historiker Adolf von Harnack. In der wissenschaftlichen Welt gilt er mit Recht als Polyhistor, denn außer seinem Spezialfach hat er sich ausgezeichnete Kenntnisse auf philologischem, literarischem und ethischem Gebiet angeeignet und daneben ist es ihm noch gelungen, einen Überblick über die modernen Naturwissenschaften zu gewinnen. Wie er mir öfter erzählte, verdankt er das nicht so sehr der Ehe mit einer Enkelin von Liebig, als vielmehr der Kinderstube; denn seine drei Brüder haben fast gleichzeitig mit ihm Naturwissenschaft oder Medizin in Dorpat studiert. In dieser jungen Gesellschaft, die bei der beschränkten Wohnung der Eltern auch räumlich immer eng vereint war, fand, wie leicht begreiflich, dauernd ein reger Ideenaustausch statt. Infolge seines vorzüglichen Gedächtnisses hat Adolf von Harnack diese Eindrücke als dauernden Besitz bewahren können. Mir ist er menschlich zuerst näher getreten bei dem Tode meiner Frau, deren Familie er von seinem mehrjährigen Aufenthalt in Erlangen her wohl kannte, insbesondere durch seine Bemühungen, eine Erzieherin für meine Kinder zu gewinnen. Später verpflichtete er mich zu Dank durch sein mannhaftes Eintreten für den Neubau des chemischen Instituts, das wesentlich dazu beitrug, die Zustimmung der Akademie zu diesem Plane zu erlangen. Für das 200-jährige Jubelfest schrieb er die Geschichte der Akademie, die ich ziemlich fleißig gelesen habe, weil sie mir als vortreffliche Quelle für die Geschichte des akademisch-chemischen Laboratoriums diente. Mehr als irgend ein anderer Vertreter der philosophisch-historischen Klasse hat Harnack sich dauernd bemüht, auch die Interessen der Naturwissenschaften zu fördern, und er war deshalb der berufene Mann, an die Spitze der Kaiser Wilhelm Gesellschaft zur Förderung der Wissenschaften zu treten, als diese im Frühjahr 1911 gegründet wurde. Davon werde ich später ausführlich berichten. Auf dem Gebiete der Geisteswissenschaften halte ich Harnack augenblicklich für den hervorragendsten deutschen Gelehrten, und als Organisator wissenschaftlicher Arbeiten nimmt er sicherlich den ersten Platz ein.

Mit der Gründung der Kaiser Wilhelm Institute für Chemie und physikalische Chemie erweiterte sich natürlich auch der Kreis der Chemiker, und von ihnen sind drei Männer, Beckmann, Haber und Willstätter in die Akademie aufgenommen worden, über die ich später im Zusammenhang mit der Kaiser Wilhelm Gesellschaft mich äußern werde.

Chemische Gesellschaft

Zur Zeit meiner Übersiedlung nach Berlin war die wissenschaftliche Chemie hauptsächlich vertreten durch die deutsche chemische Gesellschaft, die Hofmann gegründet und 25 Jahre in musterhafter Weise und mit großem Erfolg geleitet hatte.

Die Feier ihres 25-jährigen Bestehens, die früher schon erwähnt wurde, war das erste wissenschaftliche Fest, das ich in meiner neuen Stellung miterlebte. Bald nachher begann die Alltagsarbeit, an der ich namentlich während der ersten 10 Jahre regelmäßig teilnahm, nicht allein, weil ich wiederholt zum Präsidenten und Vizepräsidenten gewählt wurde, sondern weil die chemische Gesellschaft damals noch in unserem Institut in der Georgenstraße zu Gast wohnte. Ihre Büroräume waren nämlich die Dienstwohnung des ersten Assistenten des Instituts, auf deren Benutzung früher F. Tiemann und später S. Gabriel zugunsten der Gesellschaft verzichtet hatten. Für die regelmäßigen Sitzungen diente der Hörsaal des Instituts, und die Gesellschaft bezahlte dafür eine ganz geringe Miete und eine kleine Entschädigung für Beleuchtung und Heizung. Da außerdem die meiste Arbeit für die Geschäfte des Präsidiums, des Sekretariats und der Redaktion der Berichte ehrenamtlich geleistet wurde, so waren die Ausgaben der Gesellschaft abgesehen von dem Druck der Berichte sehr klein, und dadurch ist es ihr möglich gewesen, im Laufe von etwa 25 Jahren durch Ersparnisse ein Barvermögen von 200 000 M. zu erwerben.

Durch den Tod von Hofmann war zunächst keine Störung der Geschäfte eingetreten, da F. Tiemann in gewohnter Weise die beiden Ämter als Sekretär und Redakteur fortführte. Ich selbst hielt mich anfangs zurück und beteiligte mich nur an den Vorstandssitzungen, um die Art der Geschäftsführung und die daran beteiligten Personen als möglichst neutraler Beobachter kennen zu lernen. Aus demselben Grunde lehnte ich es auch Ende des Jahres 1892 ab, für die Wahl zum Präsidenten als Kandidat aufgestellt zu werden. Infolgedessen fiel die Leitung der Präsidialgeschäfte an Landolt. Das war allerdings ein Fehler, wie ich hinterher eingesehen habe; denn trotz aller seiner guten

Eigenschaften war Landolt zu sorglos, und auch wegen seiner Schwerhörigkeit in bewegter Debatte zu ungeschickt, um einem Sturm vorzubeugen, der für die Gesellschaft zu großem Schaden hätte werden können. Dieser wurde heraufbeschworen durch die Meinung einiger zu den Gründern der Gesellschaft gehörigen Vorstandsmitglieder, daß Tiemann seine Ämter zu autokratisch verwalte. Den äußeren Anlaß dazu bot der Vorschlag, M. Berthelot und C. Friedel in Paris zu Ehrenmitgliedern zu ernennen. Ich hatte ihn zusammen mit Tiemann überlegt und er war in einer Vorstandssitzung besprochen worden, bei der leider die meisten Mitglieder fehlten, angeblich, weil die Ernennung von Ehrenmitgliedern nicht auf der Tagesordnung gestanden habe. Als es dann bei der Generalversammlung zur Wahl kam, wurde der Vorschlag von der Opposition, die sich ganz im stillen organisiert hatte, bekämpft, und die Wahl mußte vertagt werden. Dabei kam es zu stürmischen Szenen, die der Vorsitzende Landolt nicht meistern konnte, und die Versammlung schloß in recht unerfreulicher Weise mit einem Mißton, der die Gefahr einer dauernden Spaltung im Vorstand anzuzeigen schien. In derselben Sitzung wurde ich zum Präsidenten gewählt, nahm auch die Wahl an, aber mit dem Gefühl höchsten Unbehagens und dem Entschluß, unter allen Umständen ähnliche Auftritte unmöglich zu machen.

Ich habe dann mehrere Monate im stillen ziemlich schwere Arbeit tun müssen, um die Gegensätze, die im Vorstand so stark aufeinander geplatzt waren, wieder auszugleichen und es zu ermöglichen, daß Tiemann vorläufig Redaktion und Sekretariat behielt. Dazu war es nötig, daß ich mich als Präsident mit dem ganzen Material der Verwaltung, namentlich auch mit der Tagesordnung und den Protokollen der wissenschaftlichen Sitzungen und der Vorstandssitzungen genau vertraut machte. Das geschah im besten Einvernehmen mit Tiemann, dessen große Arbeitskraft und treue Fürsorge für die Gesellschaft ich stets anerkannt habe. So hat sich auch ein persönliches Verhältnis freundschaftlicher Art zwischen uns entwickelt, und als er im Herbst 1899 plötzlich starb, hielt ich mich für verpflichtet, die Gedenkrede auf ihn zu halten[1]), die später durch einen ausführlichen Nekrolog von Witt in einigen Punkten ergänzt wurde. Es ist deshalb nicht nötig, daß ich ihm hier noch weitere Worte freundschaftlichen Gedenkens widme.

Eine unserer ersten gemeinschaftlichen Aufgaben war die Propaganda für die Wahl von Berthelot und Friedel, die bis zur nächsten ordentlichen Generalversammlung vertagt war. Wir hielten uns dazu um so mehr verpflichtet, als an beide Herren schon die private Anfrage ergangen war, ob sie eine solche Wahl annehmen würden. Um die

[1]) Berichte der Deutschen chemischen Gesellschaft Bd. 32 S. 3239.

Opposition des kleinen Kreises Berliner Fachgenossen zu beseitigen, wandten wir uns an die große Zahl der auswärtigen Mitglieder, die ausnahmslos dem Vorschlag zustimmten. Ende 1894 wurden dann die beiden französischen Gelehrten zusammen mit Mendelejeff und Beilstein mit einer ungewöhnlich großen Majorität gewählt.

Bald nachher wurde die Gesellschaft vor eine ganz neue und weitreichende Aufgabe gestellt durch das Anerbieten des Herrn F. Beilstein, ihr alle seine Rechte an dem eben in 3. Auflage erschienenen Handbuch der organischen Chemie zu übertragen, falls sie bereit sei, Ergänzungsbände dazu erscheinen zu lassen und später eine neue Auflage zu veranstalten. Beilstein fühlte sich damals schon zu alt, diese große Arbeit selbst noch zu leisten. Er hatte sich zuerst an Herrn Paul Jacobson gewandt mit der Anfrage, ob er die Redaktion des Buches übernehmen wollte. Obschon Jacobson durch die Herausgabe des bekannten vortrefflichen Lehrbuches der organischen Chemie von Victor Meyer und Paul Jacobson besonders gut vorbereitet war, so schien ihm doch die neue Aufgabe für eine einzelne Person zu schwierig, und er machte deshalb den Vorschlag, daß die chemische Gesellschaft sich der Sache annehmen und durch Schaffung eines Zentralbüros für chemische Berichterstattung noch erweitern sollte. Zugleich erklärte er sich bereit, unter passenden Bedingungen an die Spitze eines solchen Büros zu treten. Damit gewann ein Gedanke praktische Form, der mir schon lange vorgeschwebt und den ich auch wiederholt mit Tiemann und anderen Fachgenossen im kleinen Kreise besprochen hatte, Zentralisierung der bis dahin stark verzettelten Berichterstattung für die wissenschaftliche Chemie, wodurch eine Verbilligung und eine weitere Verbreitung der referierenden Organe möglich werde. Bis dahin gab es den altbewährten Jahresbericht für Chemie, von Liebig und Wöhler begründet, der seit dem Jahre 1848 regelmäßig erschien und unmittelbar an den Jahresbericht von Berzelius anknüpfte. Daneben existierte das von Arendt gegründete Zentralblatt, das allwöchentlich erschien, jede einzelne Publikation besonders behandelte und dadurch die Literatur den Chemikern zwar nicht systematisch, aber nach kurzer Frist übermittelte. Etwas Ähnliches, aber in viel unvollständigerer Form boten die Berichte der chemischen Gesellschaft in ihrem Referatenteil. Endlich fanden sich noch solche Referate in der Zeitschrift des Vereins zur Wahrung der Interessen der chemischen Industrie, der Chemikerzeitung, der Apothekerzeitung und ähnlichen Organen. Es lag auf der Hand, daß diese Zersplitterung der Berichterstattung unzweckmäßig war, und daß mit denselben Mitteln viel Vollkommeneres geschaffen werden könnte.

Tiemann kam zu der gleichen Ansicht, und so haben wir beide 1895 den Entschluß gefaßt, den Vorschlag von Jacobson zu verwirklichen und dafür die Zustimmung des Vorstandes und später der Generalversammlung der Gesellschaft zu erwirken. Das ist nicht ohne erhebliche Mühe und Sorge gelungen. Schon im Vorstand erklärten sich einige vorsichtige Mitglieder, z. B. der sonst so wohlwollende C. Liebermann, gegen den Plan, weil sie in dem neuen buchhändlerischen Unternehmen eine Gefahr für das Wohlergehen und besonders für die Finanzen der Gesellschaft erblickten. Andererseits konnten wir durch Kalkulation feststellen, daß die chemische Gesellschaft in der Lage wäre, für das chemische Zentralblatt und für das Beilstein-Handbuch eine viel größere Anzahl von Abnehmern zu gewinnen und dadurch eine erhebliche Herabsetzung des Preises zu ermöglichen. Niemand konnte leugnen, daß das ein Gewinn für unsere Wissenschaft sei und auch das Ansehen der Gesellschaft vergrößern müsse. So mehrte sich denn rasch die Zahl der Anhänger des neuen Planes, und in der Generalversammlung vom 13. Dezember 1895 konnte ich als Vorsitzender die erste öffentliche Ankündigung davon machen. Die Entscheidung fiel in einer außerordentlichen Generalversammlung vom 19. Juni 1896. Unter Leitung von Dr. Jaffé trat hier die Opposition nochmals scharf in die Erscheinung. Aber ich konnte als Vorsitzender alle Klagen und Befürchtungen widerlegen, und schließlich wurde die für den Plan nötige Änderung der Statuten mit großer Majorität genehmigt. Zuvor war es notwendig gewesen, für beide Werke von der Verlagsbuchhandlung der Firma Leopold Voß in Hamburg das Verlagsrecht zu erwerben. Den entscheidenden Schritt dazu hatte ich zusammen mit Schatzmeister Dr. Holtz, der von Anfang an dem Plane wohlgeneigt war und ihn jederzeit energisch gefördert hat, in den Herbstferien 1895 getan. Im Anschluß an eine Versammlung des Vereins zur Wahrung der Interessen der chemischen Industrie zu Kiel, an der ich teilnahm, fuhr ich mit Dr. Holtz nach Hamburg und vereinbarte dort mit dem Leiter der Firma L. Voß für das chemische Zentralblatt einen festen Kaufpreis von 15 000 M. und für die Ergänzungsbände des Beilstein eine recht günstige Gewinnbeteiligung der Firma. Als diese dann später zögerte, die mündliche Vereinbarung schriftlich anzuerkennen und wir in Sorge kamen, den fertigen Vertrag der entscheidenden Generalversammlung im Juni 1896 nicht rechtzeitig vorlegen zu können, bat ich Tiemann, die Verhandlung mit gröberem Geschütz wieder aufzunehmen und zu Ende zu führen. Das gelang ihm auch, indem er die widerstrebende Firma durch höchst energische Telegramme zur Vernunft brachte.

Inzwischen waren mit Paul Jacobson und dem bisherigen Redakteur des chemischen Zentralblattes Professor R. Arendt in Leipzig Verhandlungen gepflogen worden. Beide traten anfangs 1897 in den Dienst der chemischen Gesellschaft. Arendt blieb in Leipzig und besorgte von dort aus bis zu seinem Tode die Redaktion des chemischen Zentralblattes. Jacobson siedelte nach Berlin über und übernahm nicht allein die Redaktion des Beilstein-Handbuches, sondern löste auch Tiemann ab als Generalsekretär der Gesellschaft in der Redaktion der Berichte und der Erledigung der Sekretariatsgeschäfte. Damit schied Tiemann aus den beiden Ämtern, die er fast 20 Jahre lang unbesoldet verwaltet hatte, und der Vorstand fühlte sich verpflichtet, ihm seinen wärmsten Dank durch die Überreichung einer stattlichen Adresse und durch Veranstaltung eines Festessens auszudrücken.

Diese Neuordnung der Geschäftsführung mit den erweiterten literarischen Aufgaben bezeichnet den Anfang einer neuen Periode in der Entwicklung der chemischen Gesellschaft. Für die Richtigkeit der Veränderung spricht der Erfolg. Zu den 4 Bänden der dritten Auflage des Beilstein-Handbuches hat die Gesellschaft inzwischen die gleiche Anzahl Ergänzungsbände herausgegeben und eine neue Auflage vorbereitet, die ein Wertobjekt von mindestens 1,2 Millionen sein und an Umfang die großen Konversationslexika erreichen wird.

Auch das chemische Zentralblatt hat sich in erfreulicher Weise entwickelt, erst unter der Leitung von Arendt und dann nach dessen Tode unter Führung von Professor A. Hesse. Die Zahl der Abnehmer ist auf mehr als das dreifache gestiegen, der Umfang fortwährend gewachsen und auch die Qualität der einzelnen Referate verbessert. Erst durch den Krieg ist in allen Punkten wieder eine Verminderung eingetreten. Dazu hat die chemische Gesellschaft noch zwei neue Werke übernommen, das Literaturregister der organischen Chemie, gegründet als Lexikon der Kohlenstoffverbindungen von M. M. Richter und jetzt weitergeführt von Professor Stelzner, ferner ein ähnliches Literaturregister der anorganischen Chemie, gegründet und geführt von Dr. M. K. Hoffmann.

Die Bedeutung und Nützlichkeit dieser verschiedenen Unternehmungen wird von keinem deutschen Chemiker geleugnet werden, und die chemische Industrie hat dieser Überzeugung später Ausdruck gegeben durch reiche materielle Unterstützungen. Sie begannen 1909 mit einer Spende der Firma Cassella & Co. in Frankfurt a. M. von 60 000 M. Bald nachher folgte eine durch den neuen Schatzmeister Dr. F. Oppenheim eingeleitete Sammlung von 200 000 M. als Beilstein-Fond, dazu kamen 2 Sammlungen für das Lexikon der anorganischen Verbindungen und für das Literatur-Register der organischen Chemie im Betrage von 75 000 M. und 120 000 M. Endlich ist während des Krieges der chemischen Gesellschaft bei ihrem 50jährigen Jubiläum zur Sicherstellung ihrer literarischen Bestrebungen die Riesensumme von $2^1/_2$ Millionen Mark von der Industrie und einigen Privaten gestiftet worden. Gleichzeitig ist ein ganzer Stab von wissenschaftlichen Beamten hauptsächlich für diese literarischen Dinge gebildet worden. Dadurch hat auch das nach dem Gründer Hofmann benannte Haus der Gesellschaft einen würdigen und ernsten Zweck erhalten. Der Plan seiner Gründung wurde sofort nach Hofmanns Tode gefaßt und fand auch im Kreise der auswärtigen Mitglieder vielfache Unterstützung. Ich selbst war für diesen Plan anfangs recht eingenommen, und habe das auch mit einem für meine damaligen Verhältnisse ziemlich hohen Beitrag (2000 M.) bekundet. Als ich aber nach Berlin kam, und in die Kommission zur Gründung des Hauses gewählt wurde, bin ich erschrocken über die nach meiner Ansicht leichtfertige Weise, in der das Unternehmen finanziert werden sollte. Die Sammlung hatte nur ungefähr 200 000 M. ergeben und nun sollte ein Haus errichtet werden, das einschließlich des Bauplatzes auf 800 000 bis 1 Million Mark geschätzt wurde. Damit wäre nicht allein das ganze Vermögen der chemischen Gesellschaft verbraucht worden, sondern auch noch eine erhebliche Schuldenlast entstanden. Ich hielt mich für verpflichtet, dagegen energisch Einspruch zu erheben, und habe es auch fertig gebracht, daß im Vorstand die Mehrzahl der Mitglieder zu der Ansicht kamen, das Hofmann-Haus müsse aus anderen Mitteln erbaut und der chemischen Gesellschaft kostenlos überwiesen werden. Dadurch ist der Bau sicherlich verzögert worden, aber das war kein Schaden, da die Gesellschaft in unserem Institut ein zwar bescheidenes, aber doch auskömmliches Heim besaß. Gleichzeitig wurden die Anhänger des luxuriösen Baues genötigt, eine andere Form der Finanzierung zu finden. Das ist auch den Bemühungen der Herren Holtz, Krämer, Martius und Tiemann gelungen, indem sie eine Gesellschaft m. b. H. zur Erbauung des Hauses ins Leben riefen, dessen Mitglieder Anteile von mindestens 5000 M. zu übernehmen hatten. Ursprünglich war dafür eine Verzinsung in Aussicht genommen, aber später haben die Herren es doch fertig gebracht, daß die meisten Besitzer der Anteile auf jede Entschädigung und Rückzahlung der Summe verzichteten. Ich selbst habe an diesem Geschäft keinen Anteil genommen. Auch der Bau des Hofmann-Hauses ist ohne meine spezielle Mitwirkung entstanden, nur bin ich, um einem Wunsch des Herrn Jacobson und der anderen wissenschaftlichen Beamten zu entsprechen, dafür eingetreten, daß in dem Hause ein kleines Laboratorium eingerichtet wurde, dessen Hilfsmittel auch den Experimentalvorträgen im großen Sitzungssaale zustatten kommen. Das Haus wurde 1900 mit einer Festsitzung eröffnet unter dem Vorsitz des Präsidenten J. Volhard, und A. v. Baeyer hielt damals die Hauptrede über die Geschichte des Indigos.

Die Schöpfung des Hofmann-Hauses ist in erster Linie das Verdienst von J. F. Holtz, der mit Tatkraft und Geduld die Sammlung des Geldes betrieb und auch später im Hofmannhaus-Verein die Hauptrolle spielte. Allerdings trug er auch die Schuld dafür, daß der Bau recht kostspielig ausfiel und daß die ursprüngliche Bausumme um etwa 80 000 M. überschritten wurde. Die Verzinsung einer Hypothek von 100 000 M. und die nicht unerheblichen Unterhaltungskosten fielen natürlich der chemischen Gesellschaft zur Last, so daß ihr das Haus doch jährlich 12 bis 15 000 M. Kosten macht. Ich habe keine Bedenken getragen, gegen die im wesentlichen durch das Hofmann-Haus stark gewachsenen Ausgaben der chemischen Gesellschaft Einspruch zu erheben, und es ist dadurch zeitweise zu einer Verstimmung zwischen Holtz und mir gekommen. Da aber die Finanzen der Gesellschaft trotz der vergrößerten literarischen Unternehmungen sich im neuen Jahrhundert günstig gestalteten, und durch die wachsende Zahl der Beamten der Bau immer besser ausgenützt werden konnte, so habe ich mich zufrieden gegeben und bin jetzt über das finanzielle Schicksal der Gesellschaft beruhigt. Aus Sparsamkeit wurde in den ersten Jahren der obere Teil des Hauses an die chemische Berufsgenossenschaft vermietet. Seitdem diese aber ihr direkt neben dem Hofmann-Haus gelegenes eigenes Heim bezogen hat, werden auch die oberen Räume nur von der chemischen Gesellschaft benutzt. Allerdings bildet einen Teil davon die Dienstwohnung von Professor Jacobson, die ihm in Anbetracht seiner Verdienste um die chemische Gesellschaft auch gelassen wurde, nachdem er vor einigen Monaten den größeren Teil seiner Ämter zugunsten seines Lehrbuches aufgegeben hat. Diese Dienstwohnung bildet aber eine Reserve an Raum, die später sicherlich zu den geschäftlichen Aufgaben der Gesellschaft herangezogen werden wird.

Der Hörsaal des Hofmann-Hauses ist durch zahlreiche Bilder verstorbener Chemiker geschmückt. Dasjenige von Georg Ernst Stahl, dem Begründer der Phlogistontheorie, habe ich gestiftet. Es ist eine von Walter Miehe ausgeführte Copie des im Besitz der Kaiser Wilhelm-Akademie für kriegsärztliche Wissenschaft befindlichen Originals. Das Bild zeigt den hervorragenden Mediziner und Chemiker, der in Berlin als Leibarzt des Königs Friedrich Wilhelm I. eine große Rolle spielte, in der kleidsamen Tracht des 18. Jahrhunderts, mit Sammetrock, Spitzenkragen und einer großen Allongeperrücke. Eine zweite Copie dieses Bildes, die ich von Fräulein Chales de Beaulieu vor 2 Jahren ausführen ließ und für das deutsche Museum zu München bestimmt hatte, ist weniger günstig ausgefallen, deshalb von Herrn Oscar von Miller als nicht geeignet für das Museum erklärt worden und befindet sich noch in meinem Besitz.

An den wissenschaftlichen Sitzungen der chemischen Gesellschaft habe ich in den ersten 8 Jahren meines Berliner Aufenthaltes, wo wir unter demselben Dache wohnten, fast ausnahmslos teilgenommen und auch einen großen Teil meiner wissenschaftlichen Versuche dort vorgetragen. Die erste Mitteilung im Januar 1893 betraf die Entdeckung des Amidoacetaldehyds. Mein Vorschlag, die physiologische Wirkung des salzsauren Salzes zu prüfen, gab Anlaß zu einer kleinen Debatte in der Sitzung. Es war mir aber leicht, die Opposition ad absurdum zu führen durch die Bemerkung, daß die Pflanzen meines Wissens Lebewesen ohne Nerven seien, und daß man alle physiologischen Fragen ohne vorgefaßte Meinung behandeln müsse. Von da an sind die jungen Chemiker in Berlin etwas vorsichtiger geworden, wenn sie in der chemischen Gesellschaft meine Mitteilungen kritisierten, obschon ich immer bereit war, jede vernünftige und in richtiger Form vorgebrachte Opposition anzuerkennen und sachgemäß zu behandeln. Die Sitzungen waren damals fast ausschließlich von Originalvorträgen ausgefüllt, die deshalb vielfach in übertriebener Weise ausgesponnen und langweilig wurden, während die von auswärts eingelaufenen zahlreichen Mitteilungen, deren Inhalt in der Regel viel interessanter war, von dem Schriftführer mit wenig Worten abgemacht wurden. Auf meinen Vorschlag wurde im Jahre 1896 die Änderung getroffen, daß die auswärtigen Mitteilungen von einer größeren Anzahl junger Chemiker in viel ausführlicherer Weise vorgetragen wurden, wodurch die Sitzungen zweifellos an Inhalt und Interesse gewannen. Andererseits bemühte sich der Vorstand mit vollem Rechte, die schon einige Jahre vor Hofmanns Tode eingeführten zusammenfassenden Vorträge über größere Arbeitsgebiete von den besten Fachleuten halten zu lassen. An der Auswahl der Redner habe ich in den ersten 26 Jahren ziemlich regelmäßig mitgewirkt. In besonderer Erinnerung sind mir geblieben die Vorträge von van't Hoff über die neue Theorie der Lösungen und von W. Ramsay über die Edelgase. van't Hoff kam bei der Gelegenheit zum ersten Mal nach Berlin, und ich habe bei dem kleinen Festmahl, das nach der Sitzung ihm zu Ehren veranstaltet wurde, anknüpfend an seinen Namen „vom Hofe" ihn als einen König der Wissenschaft gefeiert. Der Eindruck, den er damals bei uns hinterließ, war nicht ganz ohne Einfluß auf seine spätere Berufung nach Berlin.

Noch mehr wurde Ramsay gefeiert; denn an seiner Entdeckung nahm auch das große Publikum teil, und er konnte seinen Vortrag in modifizierter populärer Form einerseits vor dem Kaiserpaar im Hörsaal des chemischen Instituts und andererseits in der Urania wiederholen. An dem Vortrag für den Kaiser nahm nur das Präsidium der chemischen Gesellschaft teil, und es war das erste Mal, daß ich mit Wilhelm II. und der Kaiserin, allerdings nur ganz kurz, in Berührung kam. Der Wunsch des Kaisers, einen Vortrag von Ramsay zu hören, war uns von Professor Slaby übermittelt worden, und ich übernahm es, bei Ramsay telegraphisch anzufragen, ob er dazu bereit sei. Darauf kam die lakonische und für den Engländer charakteristische Antwort „Yes". Ramsay war während seines Berliner Aufenthaltes mein Gast, und ich bin ihm während dieser Tage, wo wir stundenlang zusammen plaudern konnten, persönlich nahe getreten.

Er war nicht allein ein vortrefflicher Naturforscher, sondern auch ein allgemein gebildeter, sehr sprachgewandter, kluger und besonnener Mann. Ein zweites Mal hat er während des internationalen Chemikerkongresses 1903 bei mir in der neuen Dienstwohnung, Hessischestraße 2 gewohnt, und der günstige Eindruck von früher wurde dadurch nur noch verstärkt. Ich habe ihn später in London wieder besucht und in ziemlich regelmäßigem Briefwechsel mit ihm gestanden. Insbesondere berichtete er mir mehrmals über seine radio-aktive Forschung. Der letzte Brief, den ich von ihm 6 Wochen vor Ausbruch des Weltkrieges empfing, war in einem ungewöhnlichen Ton gehalten und gab Kunde von der Leidenschaftlichkeit, mit der er politische Fragen ergriff. Damals war er aufs höchste erregt über die Homerulepolitik der englischen Regierung in Irland und stand auf Seiten der Ulster-Partei, die nach seiner Meinung dem Homerule bewaffneten Widerstand leisten müsse, und für die er entschlossen war, persönlich alle Hilfe zu leisten. Das war der Vorläufer der maßlosen Angriffe, die er bald nachher, veranlaßt durch den Krieg, gegen Deutschland richtete, und auf die ich einstweilen nicht eingehen werde.

Bei seinem ersten Besuch im Dezember 1908 veranstaltete ich für ihn eine Abend-Herrengesellschaft, zu der einige Berliner Chemiker und eine größere Anzahl von Akademikern eingeladen wurden. Ramsay hatte die Zeit vergessen und kam deshalb erst nachhause, nachdem die ganze Gesellschaft längst versammelt war. Ich habe dann die Geschwindigkeit bewundern müssen, mit der er den Straßenanzug ablegte und sich in den Evening-dress stürzte. Ich glaube, daß die ganze Verwandlung keine 3 Minuten in Anspruch genommen hat.

Ein größeres Festmahl war ihm Tags zuvor von der chemischen Gesellschaft gegeben worden, und ich mußte dabei die Tischrede auf den Gast halten, die ich in eine launige Form kleiden konnte. Sie ist gedruckt worden, und ich habe mir erlaubt, einige Exemplare an hervorragende auswärtige Fachgenossen zu schicken. Die merkwürdigste Wirkung hat sie in München ausgeübt; denn Pettenkofer, dem ich ein Blatt zugeschickt hatte, sandte dieses an W. von Miller, der damals schon schwer krank war und dem er einen kleinen Spaß bereiten wollte. Miller, der an Darmkrebs litt, hatte den Ehrgeiz, die Vorlesung über Chemie an der technischen Hochschule mit Aufgebot aller Energie so lange zu halten, als seine erschöpften Kräfte es irgendwie gestatteten. Er ließ sich zu dem Zweck zuletzt auf einer Bahre in den Hörsaal tragen.

Eine seiner letzten Vorlesungen hat er hauptsächlich mit dem Vorlesen meiner Tischreden ausgefüllt, an die er allerdings belehrende Bemerkungen über die Edelgase und ihre Entdeckung knüpfte. Das hat mir verschiedene Zuschriften seiner Zuhörer, die davon sehr belustigt waren, eingebracht.

Weitere bemerkenswerte Vorträge wurden gehalten von Nernst, Wallach, Buchner, Willstätter, Richards, Sabatier, Brunck, Knietsch usw., die alle mit lebhaftem Interesse gehört und für die ich zuweilen als Vorsitzender den Rednern den Dank der Gesellschaft abstatten mußte. Einen Vortrag sonderlicher Art hielt R. Fittig über ungesättigte Säuren, Laktone usw., sehr inhaltreich, aber in der Form so merkwürdig, daß die Jugend sich belustigte und wir Älteren alle Mühe hatten, den Ernst zu bewahren.

Mein erster Vortrag über Kohlenhydrate, der auf diejenigen von Baeyer und Victor Meyer folgte, ist früher schon erwähnt. Einen zweiten habe ich im Januar 1906 über Aminosäuren, Polypeptide und Proteine gehalten. Er erregte einiges Aufsehen unter den Fachgenossen. Etwa 8 Tage nach dem Vortrag wurde durch eine Notiz aus Wien die Aufmerksamkeit der Presse und des großen Publikums darauf gelenkt, und nun erlebte ich etwas sehr Merkwürdiges. Meine vorsichtig gehaltene Rede wurde von der Presse in der phantastischsten Weise ausgeschmückt, die Synthese des Eiweiß als eine fertige Sache dargestellt und mit der Lösung der Nahrungsfrage das goldene Zeitalter proklamiert. Ich habe versucht, dagegen Einspruch zu erheben, aber ohne jeden Erfolg. Wie eine Sturzwelle ging die Flut der Zeitungsartikel über mich und alle Vernunft hinüber, und schließlich erhielt ich aus den Vereinigten Staaten Amerikas Zeitungsartikel, die ganze Seiten bedeckten und durch Bilder illustriert waren, auf denen man die Verwandlung der Kohle in die schönsten Erzeugnisse eines eleganten Speisehauses sehen konnte. Selbstverständlich waren alle diese Bilder von mir selbst, meinem Laboratorium, den Versuchstieren usw. frei erfunden. Ich habe damals vor der fürchterlichen Macht und Zügellosigkeit der Presse einen Schrecken bekommen.

Den dritten zusammenfassenden Vortrag hielt ich im September 1913 in einer Sitzung, die von der chemischen Gesellschaft auf der Naturforscherversammlung zu Wien veranstaltet wurde. Das Thema war die Synthese von Depsiden, Flechtenstoffen und Gerbstoffen. Assistiert wurde ich bei dem Vortrag von meinem Sohn Hermann O. L. Fischer, der an der Synthese der Flechtenstoffe experimentell beteiligt gewesen war. Die Sitzung fand statt in dem Hörsaal des chemischen Instituts der Universität Wien, das damals von Guido Goldschmidt geleitet wurde. Der Saal war derartig überfüllt, daß mir hinter dem Experimentiertisch kaum Raum blieb, um die zahlreichen Tabellen zu erläutern und die Präparate zu demonstrieren. Zudem herrschte eine derartige

Hitze, daß ich nach $1^1/_2$stündigem Vortrag so naß war, als hätte ich die Donau durchschwommen. Es war die erste besondere Sitzung, welche die chemische Gesellschaft auf einer deutschen Naturforscherversammlung abhielt und ist bisher auch die einzige geblieben, weil infolge des Krieges keine Naturforscherversammlung mehr zustande kam.

Als im Jahre 1900 das chemische Institut in den Neubau an der Hessischenstraße verlegt war, wurde die bis dahin bestehende fast tägliche Berührung der chemischen Gesellschaft und mir unterbrochen. Auf meinen Antrag erhielt zwar die Gesellschaft vom Kultusministerium die Erlaubnis, noch ein Jahr in den alten Räumen der Georgenstraße zu bleiben, bis das Hofmann-Haus ganz fertiggestellt war, aber mit der räumlichen Trennung hatte ich das Gefühl, daß ich nicht mehr im gleichen Maße wie bisher für die Schicksale der Gesellschaft verantwortlich sei, sondern daß die Sorge für sie sich jetzt mehr gleichmäßig auf alle Mitglieder des Vorstandes verteilen müsse. Ich habe dementsprechend nur selten mehr die mir angetragene Wahl zum Vorsitzenden angenommen und mich auch an den Sitzungen der Gesellschaft in loserer Form beteiligt, als es früher der Fall war. Nur wenn wichtige Interessen auf dem Spiele standen, habe ich mich nicht gescheut, meinen ganzen Einfluß aufzubieten, um das, was ich für richtig hielt, durchzusetzen.

An dem 50-jährigen Jubiläum der Gesellschaft, das im Mai d. Js. gefeiert wurde, konnte ich mich nicht beteiligen, weil ich damals zur Erholung von einer Lungenentzündung in der Südschweiz weilte. Ich hatte aber vorher beim Kultusminister einen Antrag gestellt, den wissenschaftlichen Beamten der Gesellschaft, an der Spitze Herrn P. Jacobson, durch Verleihung von Titeln eine öffentliche Anerkennung zu erwirken. Das ist auch geschehen, und zu meiner großen Überraschung wurde auch mir bei der Gelegenheit ein hoher preußischer Orden verliehen. Offen gestanden hätte ich es lieber gesehen, daß das nicht geschehen wäre.

Die von mir angeregte und auch ursprünglich übernommene Gedächtnisrede auf A. von Baeyer hatte ich leider wegen Krankheit wieder abgeben müssen. Sie ist dann aber von R. Willstätter in sehr würdiger Weise gehalten worden.

Auch beim 40jährigen Bestehen der Gesellschaft war ein kleines Fest veranstaltet worden, und ich hatte das Vergnügen, bei der Gelegenheit Herrn Professor Henry Armstrong, den Abgesandten der Chemical Society zu London, bei mir zu Gast zu haben. Freundschaftliche Beziehungen zu ihm hatte ich längst, teils durch seinen Sohn, der mehrere Jahre mein Mitarbeiter war, und auch durch meinen eigenen Sohn Hermann, dem er und seine Familie während eines halbjährigen Aufenthaltes in England viele Freundlichkeiten erwiesen hatten.

Das chemische Institut in der Georgenstraße

Mein Einzug in das Institut, die von mir angeregten kleinen baulichen Veränderungen, die Vermehrung der Assistenten und die Verteilung der Arbeit unter die vorhandenen Lehrkräfte sind schon früher geschildert. Alles war von mir nur als Provisorium gedacht, weil ich sicher damit rechnete, daß in wenigen Jahren ein Neubau entstehen werde; denn seine Errichtung war die erste und wichtigste Bedingung, die ich bei der Berufung stellte, und sie war nicht allein von den Räten des Kultusministeriums, sondern auch von dem Minister Dr. Bosse selbst ausdrücklich bewilligt worden. Das Provisorium hat aber schließlich $7^1/_2$ Jahre gedauert und bildet so einen ziemlich langen Abschnitt in meiner Berliner Tätigkeit.

Mit Ausnahme der beiden Hörsäle und der beiden von Tiemann als Privatlaboratorium benutzten Zimmer lagen alle benutzbaren Räume im ersten Stock. Zwei gut erleuchtete Unterrichtssäle, getrennt durch einen kleineren Raum, gingen nach der Georgenstraße. Die übrigen Unterrichtszimmer mit Nebenräumen lagen in den korridorähnlichen Seitenflügeln und dienten gleichzeitig als Durchgang. Recht hübsch und auch ziemlich zweckmäßig eingerichtet war das aus zwei Räumen bestehende Privatlaboratorium, das durch den Korridor und zwei Bibliothekszimmer direkt mit dem Wohnhaus in der Dorotheenstraße in Verbindung stand. Hier habe ich mich, unterstützt von dem aus Würzburg mitgekommenen Privatassistenten Dr. Lorenz Ach schnell eingerichtet. Drei junge Chemiker, die ihre Doktorarbeit ausführen wollten — es waren die Herren Haenisch, Kopisch und der von Würzburg mitgenommene Engländer Crossley — wurden als Gäste aufgenommen und schon im ersten Jahr hatte ich das Glück, zwei hübsche Reaktionen, die Bereitung des Amido-Acetaldehyds und ähnlicher Stoffe, ferner die Synthese der Alkoholglucoside aus Zucker und Alkoholen in Gegenwart von Salzsäure zu finden. Gleichzeitig leitete ich mehrere Doktorarbeiten in dem zweiten organischen Unterrichtssaal und überwachte zusammen mit dem hier tätigen Assistenten die präparativen Übungen der organischen Anfänger nach der von mir geschriebenen „Anleitung zur Darstellung organischer Präparate",

IM LABORATORIUM

die sich schon in Erlangen und Würzburg bewährt hatte. Auch um die analytische Abteilung, in der Dr. Piloty und Dr. Fogh tätig waren, habe ich mich damals, so weit es meine Zeit zuließ, gekümmert.

Die großen Experimental-Vorlesungen hielt ich wie Hofmann im Winter über anorganische und im Sommer über organische Chemie. Aber anstelle des dreimal zweistündigen Vortrages zog ich es vor, 5 Mal je eine Stunde in der Woche zu sprechen, weil dadurch die Vorbereitungen der Experimente viel sorgfältiger werden und deshalb auch recht viel Zeit erspart werden kann. In der Tat zeigte sich denn auch, daß ich in den 5 Wochenstunden ein erheblich größeres Pensum mit einer größeren Anzahl von Experimenten ausführen konnte, als früher in drei Doppelstunden geleistet wurde.

Die Zahl der Zuhörer stieg schon im zweiten und dritten Jahre sehr erheblich und dann wurde der Zugang so groß, daß aller verfügbare Raum von stehenden Personen erfüllt war und meist die Zugangstüren nicht mehr geschlossen werden konnten. Ganz ungenügend war die Kleiderablage, in der häufig am Schluß der Vorlesung große Unordnung herrschte, und leider auch mancher Diebstahl vorkam.

Unter den Zuhörern befanden sich in den letzten Jahren nicht allein Chemiker, Mediziner, Apotheker und Lehramtskandidaten, sondern auch Angehörige der juristischen und sogar der theologischen Fakultät. Das hing wohl damit zusammen, daß ich mich bemühte, gelegentlich die Bedeutung der Chemie für die gesamte menschliche Kultur in philosophischer und wirtschaftlicher Beziehung darzulegen.

Auch das Laboratorium war bald überfüllt, was bei der geringen Anzahl von Arbeitsplätzen (etwa 80) nicht überraschen konnte. Dieser Zudrang war das beste Agitationsmittel für den Neubau des Instituts, und als ich damit drängte, erhielt ich schon im Frühjahr 1893 den Auftrag, ein Programm dafür auszuarbeiten. In freudiger Stimmung habe ich mich dieser Mühe sofort unterzogen, mußte aber zu meiner Enttäuschung bald erkennen, daß der Auftrag nur zum Schein erfolgt und daß es eines viel gröberen Geschützes bedurfte, um das Kultusministerium in dieser allerdings ziemlich schwierigen Frage zu energischer Handlung zu bringen; denn die Entscheidung über alle derartigen Forderungen der Wissenschaft in Preußen, die Geld kosten, hat in letzter Linie der Finanzminister, und wie schwer dessen Zustimmung zu erlangen war, werde ich erst später schildern.

Zuvor war mir vom Kultusministerium ein recht unbequemes Geschäft aufgehalst worden, das mit der Weltausstellung zu Chicago von 1893 zusammenhing. Das Ministerium hatte es übernommen, hier eine Ausstellung des preußischen Unterrichtswesens und der wissenschaftlichen Forschung in Deutschland zu veranstalten. Dabei sollte die chemische

Gesellschaft beteiligt sein, und im Auftrage des Kultusministeriums bemühte ich mich, dafür die Zustimmung des Vorstandes zu gewinnen. Gewünscht wurde vor allen Dingen eine Ausstellung von interessanten, in Deutschland entdeckten Präparaten. Im Vorstand herrschte anfangs keine Neigung, auf die Wünsche des Ministeriums einzugehen. Erst als ich den Herren klar machte, daß die langjährige Benutzung von Institutsräumen für Zwecke der Gesellschaft nur im Einverständnis mit dem Minister möglich gewesen sei und auch in Zukunft davon abhänge, entschloß man sich zur positiven Mitwirkung in Chicago durch Veranstaltung einer Sammlung von interessanten Präparaten und Apparaten und die Vorlage eines Albums mit den Bildern deutscher Chemiker. Ich mußte mich aber verpflichten, die Rundschreiben mit der Aufforderung zur Beteiligung an die Mitglieder der Gesellschaft zu entwerfen und zu versenden, sowie die Präparate in Empfang zu nehmen. Diese Arbeit geschah mit Hilfe des Assistenten Dr. Richter in den Räumen des Instituts. Sie hat viele Wochen ärgerlicher Mühe und manche Sorge gebracht. Ein hübscher Ausstellungsschrank wurde von der Firma Chemische Fabrik auf Actien vormals E. Schering in Berlin kostenlos zur Verfügung gestellt, und das Ganze hat später in Chicago offenbar gefallen; denn die beteiligten deutschen Laboratorien wurden mit Ehrendiplomen belohnt, und außerdem trat an uns der Wunsch der Amerikaner heran, die Sammlung behalten zu dürfen. Darauf konnte sich die Gesellschaft aber nicht einlassen, da die Präparate Privateigentum der einzelnen Mitglieder geblieben waren und diese fast alle den Wunsch äußerten, sie zurückzuerhalten. Sie sind dann auch nach Schluß der Chicagoer Ausstellung unversehrt nach Berlin zurückgekehrt und von hier aus an die einzelnen Besitzer verteilt worden. Ich hatte aber von dem ganzen Geschäft so viel Unbequemlichkeiten, daß ich spätere Zumutungen ähnlicher Art von Seiten des Kultusministeriums energisch ablehnte. Infolgedessen ist die Organisation und Beteiligung der wissenschaftlichen Chemie an der Ausstellung zu Philadelphia vom Kultusministerium Herrn C. Harries übertragen worden.

Der erste Winter, den ich in Berlin zubrachte, war ungewöhnlich streng, und da ich damals die Dienstwohnung noch nicht hatte, sondern in der Königin Augustastraße wohnte, so mußte ich den Weg dorthin täglich 4—6 Mal machen, natürlich mit einem Wagen, der aber wegen des hohen Schnees fast $^1/_2$ Stunde fuhr. Bei der starken Kälte war ich dann am Ende der Fahrt ganz erfroren, bis meine Frau auf den glücklichen Gedanken kam, mir einen großen Pelzsack anzuschaffen, in den ich während der Wagenfahrt bis unter die Arme schlüpfte.

Der Frost hatte auch im Laboratorium sehr unangenehme Folgen; denn die Wasserleitung platzte an verschiedenen Stellen, und es entstand eine große Überschwemmung. Da unglücklicherweise alle Röhren in die Wände gelegt waren und von der Leitung kein Plan mehr existierte, so mußten oft die Bauleute tagelang suchen, bis die verletzten Stellen gefunden waren. Dadurch wurde ich in dem schon früher gefaßten Vorsatz bestärkt, beim Neubau alle Rohrleitungen frei zu legen und leicht zugänglich zu machen. Trotz solcher kleinen und unangenehmen Überraschungen gingen übrigens der Unterricht und die wissenschaftliche Arbeit im Institut ungestört vorwärts.

Nachdem Frau von Hofmann im Mai 1893 die Dienstwohnung in der Dorotheenstraße verlassen und diese dann einer sehr notwendigen Reparatur unterzogen war, konnte ich sie im September desselben Jahres in Benutzung nehmen und war von nun an auch imstande, nicht allein die Experimentalarbeit intensiver zu betreiben, sondern auch die Gesamtaufsicht über das Institut sorgfältiger einzurichten. Das führte alsbald zur Entdeckung von Diebstählen, die von dem verkommenen Pförtner des Instituts begangen oder versucht wurden. Er hatte aus der Instrumentensammlung der Akademie, die sich in einem Dachraum des Wohnhauses befand, eine Reihe von Kupferplatten weggenommen und verkauft. Ich wurde darauf aufmerksam, als ich die Aufsicht über die Instrumentensammlung mit dem Einzug in die Dienstwohnung übernahm. Außerdem hatte er während der herrenlosen Zeit den vermauerten Kupferkessel aus der Waschküche der Dienstwohnung entfernt, um ihn zu verkaufen, war aber dabei von dem Baubeamten überrascht worden. Infolgedessen kündigte ich ihm die Stelle und dann hatte der Mensch die Unverschämtheit, sich in einer Eingabe an das Kultusministerium über die Behandlung zu beklagen, die ihm von dem neuen Direktor des Institutes zuteil werde. Diese wurde mir vom Minister zur Berichterstattung übersandt, und als ich darauf erwiderte, daß der Grund der Entlassung in dem dringenden Verdacht der Diebstähle gelegen habe, kam die prompte Anfrage, weshalb keine Strafanzeige erfolgt sei. Ich hatte auf diese Maßregel verzichtet, weil mir der Missetäter leid tat; denn er war zu der ungeordneten Lebensweise und sittlichen Verkommenheit durch den Mangel an Aufsicht und auch durch die schlechte, feuchte und dunkle Kellerwohnung gekommen. Sein Nachfolger, ein Kupferschmied Prisemuth, der durch das Tragen einer Brille ein ganz gelehrtes Aussehen angenommen hatte, war nicht viel besser. Zuerst hatte ich mit ihm einen unangenehmen Auftritt wegen seiner Frau, mit der er häufig in Zwist lebte. An einem Sonnabend Nachmittag wurde ich nämlich aus der ärztlichen Vorprüfung, die im physiologischen Institut der Universität stattfand, nachhause gerufen, weil mit der Frau des Pförtners ein Unglück passiert sei. Als ich im Institut in der Georgenstraße erschien, war der Pförtner ganz ruhig damit beschäftigt, die Korridore zu säubern, und auf meine erregte Frage, wie es mit seiner Frau stehe, erhielt ich die Antwort: „Dumme Eifersucht, Herr Professor, sie hat mir schon öfter blamiert". In Wirklichkeit hatte die Frau einen Versuch des Selbstmords durch Trinken starker Salzsäure gemacht und dadurch eine schlimme Ätzung von Mund, Kehlkopf und Speiseröhre davongetragen. Der Herr Gemahl zeigte dafür nicht das geringste Mitgefühl, und als die Frau einige Monate später aus dem Krankenhause zu ihm zurückkehrte, war er scheinbar ganz befriedigt davon, daß sie die Stimme verloren hatte und nur noch wispern konnte. Das war eine interne Familienangelegenheit, die mich zu keiner amtlichen Maßregel berechtigte. Als ich aber $1^1/_2$ Jahre später erfuhr, daß der Pförtner trotz strengen Verbots Nachschlüssel vom Haupttor des Instituts an Studenten vermietete, blieb mir nichts anderes übrig, als ihn ebenfalls an die Luft zu setzen, obschon er als Handwerker ganz gewandt war. Von Tiemann wurde er als Diener für sein Privatlaboratorium noch geschätzt und gehalten, bis verschiedene Missetaten auch hier seine Entlassung nötig machten.

Ein anderer Diener, der älteste im Hause, erwies sich ebenfalls als wenig zuverlässig. Er log in der unverschämtesten Weise, selbst wenn man ihn direkt überführen konnte und stand auch als Verwalter der Glasapparate und Chemikalien in dem Ruf der Unzuverlässigkeit. Ich habe deshalb, als er 65 Jahre alt war, seine Pensionierung beantragt, aber es war recht schwer, ihn los zu werden, obschon die gesetzliche Handhabe durch das Alter gegeben war. Schließlich ist er durch die Verleihung des allgemeinen Ehrenzeichens beruhigt worden, und ich habe an diesem Beispiel gesehen, welche große Macht den Staatsbehörden ganz besonders auch bei den Subaltern-Beamten in der Verleihung von Ehrenzeichen und Orden zur Verfügung steht. Die Minderwertigkeit der alten Diener glaubte ich wenigstens zum Teil in der schlechten Beschaffenheit ihrer im Institut befindlichen Dienstwohnung suchen zu müssen. Ich habe deshalb beim Neubau dafür gesorgt, daß die Wohnungen aller Unterbeamten hygienisch ganz einwandfrei und auch in bezug auf Bequemlichkeit, Beleuchtung, Heizung und dergl. behaglich eingerichtet wurden. Seitdem habe ich so traurige Dinge, wie die eben geschilderten, nicht mehr erlebt, und erst während des Krieges mit seiner demoralisierenden Wirkung sind im Verhalten der Diener wieder einige unbequeme Züge hervorgetreten. Ich stimme deshalb dem Urteil von einsichtigen Volkswirten, daß eine gute Wohnung die Menschen bessere und eine schlechte sie ethisch niederdrücke, gerne zu.

Mit den Assistenten des Instituts, deren Zahl mit den Hilfsassistenten und Privatassistenten allmählich wuchs, bin ich fast ausnahmslos recht gut ausgekommen. Sie waren vom Minister immer nur auf zwei

Jahre angestellt, aber diese Anstellung ist bei den Herren, die sich der wissenschaftlichen Laufbahn widmen wollten, auf meinen Antrag beliebig oft erneuert worden. Den Wunsch auf Wiederholung der Anstellung habe ich meines Wissens nur ein einziges Mal direkt abgeschlagen, weil der Betreffende sich ungebührlich gegen den Minister selbst benommen hatte.

An der Spitze der Assistenten stand von Anfang an S. Gabriel, der schon unter Hofmann dem Institut 20 Jahre lang erst als Studierender und dann als Assistent angehört hatte. Er ist mir immer ein lieber Kollege und Freund gewesen. Ich benutze deshalb gerne diese Gelegenheit, ihm herzlichen Dank zu sagen für die stets in freundlicher Weise gewährte Hilfe, die er mir so oft bei Erkrankungen oder anderen Verhinderungen sowohl in den Vorlesungen, wie auch in der Verwaltung des Instituts gewährt hat. Da ich auch von seiner wissenschaftlichen Tüchtigkeit überzeugt war, so hätte ich ihm gerne zu einer selbständigen Tätigkeit verholfen, aber allen Empfehlungen zum Trotz ist es nicht gelungen, ihm einen Ruf an eine andere Hochschule zu verschaffen. Erst mit der Entstehung des neuen Instituts war es möglich, ihm eine dauernde Anstellung als Abteilungsvorsteher zu geben, und ich habe später auch dafür gesorgt, daß seine Verdienste um das Institut und den Unterricht durch Verleihung von Orden, des Titels Geheimer Regierungsrat und die Ernennung zum ordentlichen Honorarprofessor in der philosophischen Fakultät anerkannt wurden. Er ist zwar ein Jahr älter, aber mit Rücksicht auf seine gute Gesundheit gebe ich gerne der Hoffnung Ausdruck, daß er mindestens ebenso lange wie ich seine Tätigkeit im Institut beibehalten wird.

Die zwei anderen Assistenten, die ich aus der Hofmann'schen Zeit übernahm, Dr. Richter und Dr. Pulvermacher, sind nach Ablauf von 1—3 Jahren freiwillig aus dieser Stellung geschieden. Vom ersteren habe ich nie mehr etwas gehört. Der zweite war etwa 10 Jahre später Generalsekretär des internationalen Chemikerkongresses zu Berlin. Viel länger ist Dr. C. Harries geblieben. Er war schon bei Hofmann provisorisch Vorlesungsassistent. In der gleichen Eigenschaft erhielt er bei meinem Amtsantritt eine ordentliche Assistentenstelle, wurde später Unterrichtsassistent und im Jahre 1900 bei Eröffnung des neuen Instituts Abteilungsvorsteher für organische Chemie. Nach dem Rücktritt von Claisen folgte er einem Ruf nach Kiel als ordentlicher Professor und Direktor des dortigen Instituts. Vor $1^1/_2$ Jahren hat er diese Stellung aufgegeben und sich als Privatgelehrter nach Berlin-Grunewald zurückgezogen. Harries hat seine Doktorarbeit unter Tiemann angefertigt, aber seine selbständigen Versuche begann er zu meiner Zeit, und bei dem dauernden persönlichen und wissenschaftlichen Verkehr, in dem wir jahrelang standen, hat er sicherlich so viel von mir gelernt, daß er wohl auch zu meinen Schülern gezählt werden darf. In der Tat bezog sich auch eine seiner ersten Arbeiten auf das von mir und Besthorn entdeckte sogen. Phenyl-Sulfocarbazin, von dem er nachwies, daß es nicht die von uns vermutete Struktur besitze, sondern den Schwefel im Benzolkern enthalte. Aber die endgültige Erkenntnis seiner Struktur ist ihm auch nicht gelungen. Harries hat im Berliner Institut seine Versuche über die Oxydation von ungesättigten Körpern mit Ozon begonnen, die er später auf den Kautschuk ausdehnte und damit seinen größten wissenschaftlichen Erfolg erzielte.

Die beiden von mir aus Würzburg mitgebrachten Unterrichtsassistenten Dr. Piloty und Dr. Fogh haben verschiedene Schicksale erlebt. Von dem Ersten ist früher schon ausführlich die Rede gewesen. Er hat im Berliner Institut die schönen Versuche über Dioxyaceton, über eine neue Synthese des Glycerins und eine besondere Klasse von Nitrosoverbindungen ausgeführt.

Seine Gattin Eugenie, Tochter von Baeyer, war schon von Würzburg her mit meiner Frau befreundet. Ihr erstes Kind, ein Sohn, wurde in Berlin Weihnachten 1892 geboren, und ich hatte die Ehre, als Pate gewählt zu werden. Er ist leider wie der Vater in dem unseligen Krieg im Westen gefallen.

Dr. Fogh, ein Däne, hat in Berlin wenig Erfolg gehabt. Die in Paris bei Berthelot begonnenen und in Würzburg fortgesetzten thermochemischen Versuche ließ er gänzlich liegen. Nach einigen Semestern wurde er auch noch krank, nahm zuerst Urlaub und da es nicht besser wurde, schied er freiwillig aus seiner Stelle. Er ist nach Kopenhagen zurückgekehrt und außer einer Verlobungsanzeige habe ich keine Nachricht mehr von ihm erhalten.

Viel besseren Erfolg hatte der von Würzburg mitgekommene Privatassistent Dr. Lorenz Ach, der mir bei der Untersuchung über Amidoacetaldehyd und die Alkoholglucoside wertvolle Hilfe leistete. Nach Aufgabe dieser Tätigkeit wurde er Unterrichtsassistent in der organischen Abteilung und ich schlug ihm eine gemeinsame Arbeit über die Verwandlung der 1.3-Dimethylpseudoharnsäure, die ich kurz zuvor von einem Studenten hatte darstellen lassen, in die entsprechende Harnsäure vor. In der Tat gelang die Reaktion zuerst durch Schmelzung mit Oxalsäure, die sich auch für die analoge Synthese der Harnsäure als brauchbar erwies. An diese Versuche hat sich später die erste Synthese des Coffeins angereiht. Zuvor aber war Dr. Ach in die Technik übergetreten und zwar in das wissenschaftliche Laboratorium der Firma C. F. Böhringer und Söhne zu Mannheim-Waldhof, dessen Leitung seinem Bruder, dem früher erwähnten Dr. Fritz Ach, anvertraut war. Nach dem frühzeitigen Tode des Bruders ist Lorenz an die Spitze dieses Laboratoriums getreten und hat hier hübsche Erfolge erzielt.

Durch diese persönlichen Beziehungen bin ich auch veranlaßt worden, derselben Firma die technische Ausarbeitung der Coffein-Synthese anzuvertrauen. Die daraus hervorgegangenen Arbeiten werde ich erst später im Zusammenhang mit den allgemeinen Untersuchungen über Purine besprechen.

Nachfolger von Dr. Ach im Privatlaboratorium wurde Dr. Rehländer, ein ebenfalls sehr tüchtiger, gewissenhafter und fleißiger Chemiker. Er hat nicht allein an der Ausarbeitung der Glucosidsynthese, sondern auch an den umfangreichen Enzymversuchen, welche mit dem charakteristischen Verhalten der α- und β-Glucoside gegen Emulsin begannen, regen Anteil gehabt. Er ist später in die Fabrik von Schering eingetreten. Gleichzeitig mit ihm fungierte Dr. Beensch als Privatassistent und ging dann wie die Brüder Ach zu der Firma Böhringer & Söhne. Er wurde im Privatlaboratorium durch P. Hunsalz abgelöst, der länger als gewöhnlich bei mir blieb und namentlich an den Purinsynthesen in hervorragendem Maße beteiligt gewesen ist. Er war in Memel zuhause, stammte aus bescheidenen Verhältnissen und war körperlich ein unansehnlicher Mann, aber klug und außerordentlich fleißig. Seine Neigung, zu hasten und die Versuche zu überstürzen, habe ich schon bei seiner Doktorarbeit über den Hydrazinoaldehyd energisch und auch erfolgreich bekämpft, denn er war später ein recht geschickter Experimentator, der besonders die Kunst sehr schnellen Arbeitens beherrschte. Er war der erste Privatassistent, dem ich mit Rücksicht auf seine Leistungen einen recht erheblichen Zuschuß zum staatlichen Gehalt gewährte. Schließlich kam er auch in das Laboratorium von Böhringer & Söhne, hat sich aber hier anscheinend wegen des Übergewichts der beiden Brüder Ach nicht wohl gefühlt und ist deshalb nach einigen Jahren zur Firma Schering in Berlin gegangen. Hinterher ist mir dieser Entschluß pathologisch vorgekommen; denn er war in Mannheim-Waldhof gut gestellt. In der Tat hat sich auch nicht lange nachher eine Geisteskrankheit entwickelt, die mit körperlichem Zerfall Hand in Hand ging und ihn schließlich in einen freiwilligen Tod geführt hat. Sein trauriges Schicksal ist mir sehr nahe gegangen, weil ich ihn gern hatte. Er selbst war aber durch die Krankheit so scheu geworden, daß er mich während seiner Anstellung bei Schering niemals mehr besuchte, sondern im Gegenteil jeder Begegnung auswich. Alles das war offenbar der Vorbote der schweren Krankheit.

Ein ähnliches, wenn auch nicht ganz so schweres Schicksal hatte Dr. Giebe, der mehrere Jahre Assistent in der Vorlesung war und mit dem ich eine ziemlich ausgedehnte Untersuchung über die Bildung der

Acetale aus Aldehyd und Alkohol bei Gegenwart von Salzsäure ausgeführt habe. Er war damals schon herzkrank, hätte aber mit dem Leiden sicherlich noch sehr lange leben können. Unglücklicherweise besaß er den Ehrgeiz, Soldat zu werden und trat deshalb gegen den Willen der Ärzte als Freiwilliger in ein Jägerbataillon ein. Infolge des anstrengenden Dienstes mußte er aber schon nach mehreren Monaten entlassen werden, und er ist etwa $^1/_2$ Jahr später fast gleichzeitig mit seinem Vater gestorben. Er war ein sehr tüchtiger, strebsamer und hoffnungsvoller junger Mann, dessen frühzeitiges Ende mir auch sehr leid getan hat.

Viel besser hat sich der Privatassistent Dr. Pinkus gehalten, der anfangs Hilfsassistent, aber von 1897/98 ab ordentlicher Assistent im Privatlaboratorium war. Er hat die Versuche über die verschiedenen isomeren Formen des Acetaldehyd-Phenylhydrazons ausgeführt und war mir eine besonders treffliche Hilfe in der Verwaltung des Instituts. Als ich ihm diese anbot, erklärte er mir, daß er nicht die geringste kaufmännische Erfahrung besitze und niemals ein Buch geführt habe. Ich vertraute aber doch seiner Veranlagung und es zeigte sich in der Tat, daß er all diese Dinge, man könnte sagen, instinktiv vernünftig und zweckmäßig besorgte. Er hat das Laboratorium geradezu vor manchen unnützen Ausgaben geschützt. Später ist er zu Nölting nach Mühlhausen gegangen, um sich in der Farbenchemie umzusehen und dann als Chemiker in die Actiengesellschaft für Anilinfabrikation zu Berlin eingetreten. Er hat diese später wieder verlassen, um sich selbständig zu machen.

Einen sehr tüchtigen Nachfolger für Hunsalz erhielt ich 1897 in Dr. Hübner, gebürtig aus Kreuznach, ein sehr verständiger ruhiger und gewissenhafter Chemiker. Er hat bei mir die schwierigen Versuche über das freie Purin und seine Methylderivate durchgeführt und mir auch bei der Abfassung der Biographie von A. W. von Hofmann, von der ich den wissenschaftlichen Teil übernommen hatte, in den Herbstferien 1898 während 3 Wochen stenographische Hilfe geleistet. Er bekleidet jetzt bei den Farbwerken zu Höchst eine angesehene Stellung.

Außer den Assistenten fanden immer einige Studierende, die ihre Doktorarbeit ausführten, im Privatlaboratorium Platz. Ich erwähne davon den späteren Direktor einer Schwefelsäurefabrik zu Duisburg Dr. Haenisch, den späteren technischen Direktor der Zellstoffabrik zu Waldhof Dr. Hans Clemm, den späteren Landwirt Dr. Kopisch, den jetzigen Professor der Pharmazie zu London Dr. Crossley.

Als besondere Gäste des Privatlaboratoriums sind zu verzeichnen Dr. Robert Pschorr und Fräulein Hertha von Siemens. Der erste hatte sich schon im Jahre 1892 bei mir in Würzburg als Student angemeldet, änderte aber seinen Entschluß, als er hörte, daß ich im Oktober nach Berlin übersiedle, weil München und Berlin diejenigen Städte seien, die er während der Studienzeit vermeiden müsse. Statt dessen ging er zu Knorr nach Jena und promovierte auch dort. Im Herbst 1895 kam er aber zu mir nach Berlin, und ich konnte ihm einen gerade frei gewordenen Platz im Privatlaboratorium überlassen. Er hat hier nach eigenen Plänen über Synthese von Phenanthrenderivaten gearbeitet, aber von mir doch manchen guten Rat empfangen. Nach $1^1/_2$jähriger eifriger und erfolgreicher Arbeit kündigte er mir an, daß er jetzt zusammen mit dem jungen Dr. Meister eine Weltreise unternehmen wolle. Ich riet davon ab mit der Begründung, daß man solche langen Reisen wohl zu ernsten Zwecken, aber nicht nur zur Unterhaltung als sogen. Globetrotter unternehmen solle. Da aber alle Vorbereitungen schon getroffen waren, so ließ er sich nicht abhalten und kehrte auch glücklicherweise unbeschädigt an Körper, Geist und Arbeitslust nach etwa $^2/_3$ Jahren zurück, um seine chemischen Untersuchungen zunächst wieder im Privatlaboratorium fortzusetzen. Bald nachher heiratete er ein Fräulein aus Frankfurt a. M. Durch den täglichen und vertrauten Verkehr im Laboratorium habe ich ihn rasch lieb gewonnen; denn er ist ein feiner, gut gebildeter Mann, musikalisch, auch zu hübschen Gelegenheitsgedichten in bayerischer Mundart befähigt und sehr bereit, anderen Leuten Hilfe zu gewähren oder eine Freude zu machen. Bei der Übersiedlung ins neue Institut konnte ich ihm eine Assistentenstelle anbieten, und als Harries nach Kiel berufen wurde, übernahm er die Stelle eines Abteilungsvorstehers, bis er im Frühjahr 1914 als Nachfolger von Liebermann an die technische Hochschule zu Charlottenburg kam. Seit Ausbruch des Krieges steht er im Felde, z. Zt. als Major im bayerischen Heere, und bei der Jubiläumsfeier der chemischen Gesellschaft erhielt er als früherer Redakteur der Berichte auf meinen Antrag den Titel Geheimer Regierungsrat.

Fräulein von Siemens war die erste Frau, die als Praktikantin in das chemische Institut aufgenommen wurde. Sie war mir von Anton Dohrn, dem Schöpfer der zoologischen Station zu Neapel, mit dem sie befreundet war, warm empfohlen. Sie hatte schon verschiedene naturwissenschaftliche Fächer studiert und wollte sich nun in Chemie, besonders dem organischen Teil, unterrichten, um später in Neapel biologische Studien anstellen zu können. Sie war damals schon 29 Jahre alt, und da sie wohl keine Zeit gehabt hatte, die Chemie von Grund aus zu betreiben, so nahm ich sie für ein Wintersemester ins Privatlaboratorium, wo sie organische Präparate und einige Analysen ausführte und der speziellen Fürsorge von Dr. Hübner anvertraut war. Sie nahm die Sache recht ernst, aber natürlich war es auch ihr nicht möglich, im

Fluge eine so schwierige Wissenschaft wie die Chemie zu bewältigen, zumal ich sie wegen Platzmangel nicht länger wie ein Semester im Privatlaboratorium behalten konnte. Ihren Dank für den genossenen Unterricht brachte sie dadurch zum Ausdruck, daß sie mehrere jüngere, unverheiratete Assistenten wiederholt in die von ihr und ihrer Mutter bewohnte prächtige Villa zu Charlottenburg einlud. Auf diese Weise machte sie auch die Bekanntschaft von Dr. Harries, der sie im Herbst 1899 heiratete. So ist sie der Chemie dauernd treu geblieben und hat ihr Interesse an unserer Wissenschaft später durch reichliche Gaben insbesondere auch zugunsten des Kaiser Wilhelm-Instituts für Chemie bekundet. Sie ist eine vornehme Frau, die vielleicht Gutes geleistet hätte, wenn sie früher in die Wissenschaft gekommen und dauernd dabei geblieben wäre.

Seitdem sind viele Frauen in unser Institut gekommen und während des Krieges haben sie sogar die Mehrheit unter den Praktikanten erreicht. Genau so wie bei den jungen Männern sind ihre Leistungen außerordentlich verschieden. Es gibt darunter oberflächliche, auch leichtfertige Elemente, die sich mehr zur Schau oder zur Unterhaltung in die Hörsäle und Institute drängen. Aber die Mehrzahl, besonders von den deutschen Frauen denken doch ernster, und unter den Mädchen, die in den letzten Jahren das Institut besuchten, habe ich 2 oder 3 kennen gelernt, die guten Chemikern an Leistungen ganz gleich zu stellen waren. Trotzdem habe ich mich von der Zweckmäßigkeit des Frauenstudiums in der Chemie nicht überzeugen können, weil die Erfahrung lehrt, daß die Mehrzahl der studierten Frauen und gerade die besten später heiraten und dann gewöhnlich nicht mehr imstande sind, ihren Beruf auszuüben. Sobald das eintritt, sind Geld und Arbeit, die für das Studium aufgewandt wurden, verloren, und dasselbe gilt für die große Mühe, welche die Dozenten der Chemie in den Laboratorien für den Unterricht aufwenden müssen. Für die Zeit des Krieges und vielleicht auch noch für einige Jahre des Friedens mag ja wohl die Hilfe der Frau in der Chemie unentbehrlich sein, weil die Männer fehlen. Auch die Gefahr der Verheiratung ist augenblicklich geringer geworden, aber sobald wieder normale Verhältnisse eintreten, wird voraussichtlich meine obige Ansicht von neuem zu Recht kommen. Ich hoffe deshalb, daß in Zukunft das Studium der Frau sich mehr auf die Fächer beschränkt, wo weibliche Gelehrte ein wirkliches Bedürfnis decken können. Das ist der Fall in gewissen Zweigen der Medizin und vor allen Dingen in dem Lehrberuf; denn ich erwarte, daß auch bei uns die Zeit kommen wird, wo man den Unterricht in der Volksschule und Mittelschule, auch für die Knaben bis etwa zum Alter von 11—12 Jahren, vorzugsweise der Frau anvertraut, ohne dabei Gefahr zu laufen, eine zu feminine männliche Jugend heranzuziehen.

Gegen Ende des 19. Jahrhunderts traten zwei Männer in die Reihe der Assistenten ein, die sich zu meiner Freude der anorganischen Chemie widmen wollten, und die auch später auf diesem Gebiete die schönsten Erfolge erzielt haben. Sie waren beide in unserem Institut herangebildet. Der eine war der Schwabe Otto Ruff, absolvierter Apotheker, aber im Besitz des Abituriums und bei uns vollständig als Chemiker ausgebildet. Seine Dissertation führte er als Organiker unter Piloty aus, fand dann selbständig einen Abbau der Zucker, der einfacher als das ältere Wohlsche Verfahren ist. Etwas später habe ich mit ihm noch eine kleine Arbeit über die Konfiguration der Xylose und den synthetischen Übergang von der Mannit- zur Dulcitreihe publiziert. Dann hat er sich fast ausschließlich mit anorganischen Aufgaben beschäftigt, wozu ich ihn auch gewissermaßen verpflichtet hatte, als er im Oktober 1897 eine Unterrichtsstelle in der analytischen Abteilung übernahm. Er ist hier 1900 Oberassistent und im Frühjahr 1903 Abteilungsvorsteher geworden. $1^1/_2$ Jahre später wurde er ordentlicher Professor der anorganischen Chemie an der neugegründeten Hochschule zu Danzig und ist jetzt in gleicher Eigenschaft an der technischen Hochschule zu Breslau tätig. Ruff ist ein begabter, sehr fleißiger und energischer Chemiker, der keine Mühe scheut, schwierige experimentelle Arbeiten zu unternehmen und durchzuführen. Auch als Lehrer genießt er guten Ruf. Von seiner Neigung, für seine Überzeugung zu kämpfen, habe ich einmal im Interesse des Instituts Nutzen gezogen. Bei dem Neubau war nämlich bei Bestellung von kleinen Rollwagen zum Transport von Chemikalien innerhalb des Instituts versäumt worden, einen Kostenanschlag einzufordern. Infolgedessen lieferte die betreffende Firma zwei für uns unbrauchbare Wagen und stellte dafür eine übertrieben hohe Rechnung aus. Dadurch entstand ein Prozeß, den Ruff mit Vergnügen und komischer Tatkraft für das Institut führte und auch glänzend gewann. Seine Freude am Disput hat ihn auch hier und da in Streit mit den Altersgenossen gebracht. Ich selbst habe in ihm aber stets einen verständnisvollen und zu jeder nützlichen Tätigkeit bereiten Mitarbeiter gefunden. In Anerkennung seiner eifrigen experimentellen Forschung hat es mir Freude gemacht, ihm wiederholt Geldunterstützungen für seine Arbeiten von Seiten der Akademie der Wissenschaften oder aus anderen Quellen zu verschaffen.

Der zweite Anorganiker war Alfred Stock, der im Oktober 1898 zunächst Vorlesungsassistent wurde. Auch er hat als Organiker promoviert, was seiner experimentellen Ausbildung nicht schädlich gewesen ist. Stock ist ein Berliner Kind und hatte sich schon im Friedrich Werderschen Gymnasium sehr ausgezeichnet, so daß ihm von dort ein dreijähriges Stipendium für die Studienzeit gewährt wurde. Er war auch als Vorlesungsassistent sehr brauchbar und darauf bedacht, neue Experimente zu finden oder die alten zu verbessern, bezw. zu verschönern. Als seine Neigung zur anorganischen Chemie sicher war, riet ich ihm im Herbst 99, zu Moissan nach Paris zu gehen und verschaffte ihm für diesen Zweck vom Kultusministerium ein Reisestipendium. Er hat auch Moissan so gut gefallen, daß er mich im Frühjahr 1900 ersuchte, Stock ein weiteres halbes Jahr bei ihm zu lassen. Im Spätsommer 1900 kam dann Stock von Paris zurück, dankerfüllt gegen Moissan, der ihn so freundlich aufgenommen und belehrt hatte, auch als Kenner von einzelnen Methoden und Apparaten, die in Frankreich üblich und in Deutschland kaum bekannt waren. Z. B. sind wir so in den Besitz der so bequemen Quecksilberwanne gekommen, die von Berthelot, Moissan und anderen französischen Gelehrten benutzt wurde und wegen ihrer Größe ein recht bequemes Arbeiten mit Gasen gestattet. Leider verlangt sie auch eine ziemlich kostspielige Quecksilberfüllung.

Nach seiner Rückkehr hat sich Stock ausschließlich der anorganischen Chemie gewidmet, durchlief die übliche Stufenleiter im Institut als Unterrichtsassistent und Abteilungsvorsteher und wurde 1909 als ordentlicher Professor der anorganischen Chemie an die technische Hochschule zu Breslau versetzt. Einige Jahre später erhielt er die Ernennung zum ordentlichen Professor und Direktor des chemischen Instituts an der Universität zu Münster. Damit wurde nach langen Jahren zum ersten Mal wieder eine ordentliche Professur der Chemie einem Anorganiker anvertraut und zwar ohne die Verpflichtung, die Elementarvorlesung über organische Chemie mitzulesen. Bevor aber Stock das neue Amt antrat, gab Willstätter seine Stellung am Kaiser Wilhelm-Institut für Chemie auf, um als Nachfolger von Baeyer nach München zu gehen. Stock bewarb sich nun um diese Stelle und wurde von dem Verwaltungsrat unter denselben Bedingungen wie Willstätter ernannt. Inzwischen war der Krieg ausgebrochen und Stock, der noch im wehrpflichtigen Alter stand, übernahm alsbald kriegswissenschaftliche Arbeiten. Auch die Räume im Kaiser Wilhelm-Institut mußte er abtreten, da sie für den Gaskampf in Anspruch genommen wurden, und so kehrte er denn für die Dauer des Krieges in unser Institut zurück. Stock ist ein sehr geschickter Experimentator und im Bau von wissenschaftlichen Apparaten dürfte er die erste Stelle unter den jüngeren Chemikern Deutschlands einnehmen. Den Beweis dafür liefert die glänzende Untersuchung über die Verbindungen des Wasserstoffes mit Silizium und Bor. Zudem ist er ein ausgezeichneter Redner und ich glaube nicht, das irgend ein anderer deutscher Chemiker ihm in der Gestaltung der anorganischen Experimentalvorlesung gleichkommt. Es ist mir deshalb eine große Beruhigung, daß er mich jetzt in dieser Vorlesung vertritt und dadurch den chemischen Unterricht in Berlin vor weiterem Niedergang schützt. Wenn Stock im zukünftigen Frieden so weiter arbeitet wie bisher, so wird er vermutlich einer der hervorragendsten Vertreter der anorganischen Chemie in der Welt werden.

Im Wintersemester 1899/1900, dem letzten im alten Institut, erhielt noch Otto Diels eine Assistentenstelle. Er hatte zuvor seine Doktorarbeit unter meiner Leitung ausgeführt und wurde Nachfolger von Stock, zunächst als Vorlesungsassistent. Von ihm stammt die neue Fassung des Vorlesungsbuches, die er als gewandter Zeichner mit vielen hübschen Illustrationen geschmückt hat. Er ist der Sohn des klassischen Philologen und jetzigen Sekretärs der Akademie der Wissenschaften, und war ebenso wie seine Brüder von vornherein entschlossen, wenn irgend möglich, die akademische Laufbahn einzuschlagen. Er ist später Unterrichtsassistent in der organischen Abteilung und nach dem Weggang von Pschorr dessen Nachfolger als Abteilungsvorsteher geworden. Inzwischen hat er sich in der Wissenschaft einen geschätzten Namen durch die Entdeckung und systematische Untersuchung des Kohlensuboxyds gemacht. Während des Krieges wurde er Nachfolger von Harries in Kiel, nachdem W. Traube den zuvor an ihn ergangenen Ruf abgelehnt hatte. Diels ist ein guter Experimentator und auch als Lehrer recht beliebt. Er hat mehrere Jahre für mich die Vorlesung über anorganische Chemie gehalten und auch einen in Berlin viel gebrauchten Grundriß der organischen Chemie geschrieben. Mir war er alle Zeit ein lieber Schüler und Kollege.

Mit der Eröffnung des neuen Instituts in der Hessischen Straße, das ungefähr dreimal so viel Arbeitsplätze hatte, wie das alte Haus und in technischer Beziehung viel vollkommener ausgestattet ist, wuchs natürlich die Zahl der Assistenten und gleichzeitig wurden drei Abteilungsvorsteherstellen geschaffen. Zwei davon fielen den beiden Leitern der analytischen Abteilung zu und die dritte wurde für die organische Chemie bestimmt. Das entsprach der räumlichen Einteilung des Instituts in vier gleich große Unterrichtssäle, von denen einer gleichsam als Abteilung von mir selbst besorgt worden ist.

Die ersten Abteilungsvorsteher waren Gabriel und Harries. Die dritte Stelle wurde provisorisch von Ruff verwaltet, der zu dem Zweck die Stelle eines Oberassistenten erhielt und 1903 zum Abteilungsvorsteher aufstieg. Unter den neuen Assistenten befanden sich auch der früher schon erwähnte R. Pschorr, ferner Dr. Lehmann, der sich namentlich um den Neubau Verdienste erwarb und später in die Farbenfabriken vorm. F. Bayer & Co. zu Elberfeld eintrat, endlich Dr. A. Wolfes und Dr. Poppenberg. Wolfes hat als mein Privatassistent bei den Arbeiten über Aminosäuren und Polypeptide vortreffliche Dienste geleistet. Er nahm auch teil an den Versuchen über Veronal, und das war wohl der Grund, weshalb er im Jahre 1903 von der Firma E. Merck in Darmstadt für das wissenschaftliche Laboratorium angeworben wurde, nachdem er zuvor sein militärisches Dienstjahr abgeleistet hatte. In dieser Stellung hat er sich sehr bewährt und wird von der Firma als treuer und tüchtiger Mitarbeiter ebenso sehr geschätzt, wie das von mir geschehen.

Während des Krieges ist er Leutnant in einem Infanterieregiment geworden und trotz seines anscheinend nicht starken Körpers hat er die großen Strapazen des Feldzuges an der Westfront gut vertragen.

Poppenberg war Unterrichtsassistent in der analytischen Abteilung. Er ist später Lehrer und Professor an der Artillerieschule zu Charlottenburg geworden, hat sich auch wissenschaftlich mit Erfolg auf artilleristischem Gebiet betätigt und während des Krieges mehrere in sein Fach einschlagende Erfindungen gemacht. Leider verlor er in der jetzigen Stellung bei der Ausführung einer Stickstoffanalyse durch einen Tropfen Kalilauge ein Auge.

Im Winter 1900/01 trat auch Dr. Rohmer in die Reihe der Assistenten ein und blieb einige Jahre in der analytischen Abteilung, wo er verschiedene kleine Erfindungen, z. B. Verbesserung der Arsenbestimmung durch Destillation als Arsen-Trichlorid machte. Er ist dann in den Dienst der Höchster Farbwerke getreten und hat hier als technischer Erfinder schöne Erfolge gehabt.

Zur selben Zeit war Dr. Franz, Sohn des Wirten vom Siechenbierhause, mein Vorlesungsassistent.

Ein Mann von besonderem Typ war Dr. Wolf von Loeben, Sprosse einer adeligen Offiziersfamilie in Sachsen. Er hatte sein Doktorexamen bei Behrend in Hannover über die γ-Methylharnsäure ausgeführt und war dann kurze Zeit Mitarbeiter bei den thermo-chemischen Untersuchungen von Stohmann in Leipzig gewesen. Er hat bei mir zunächst eine kleine Arbeit in der Harnsäuregruppe ausgeführt und wurde dann zunächst Hilfsassistent in der thermo-chemischen Abteilung, die ich im neuen Institut eingerichtet hatte. Es war mir sehr bequem, daß er hierfür die Erfahrungen aus dem Stohmann'schen Laboratorium mitbrachte. Die Resultate unserer gemeinsamen Versuche sind in einigen Abhandlungen in den Sitzungsberichten der Berliner Akademie niedergelegt. Loeben war ein behaglicher Sachse, der sich persönlich allgemeiner Beliebtheit erfreute, aber übermäßige Arbeit nicht liebte und deshalb auch vor der chemischen Industrie zurückscheute. Statt dessen bekam er die Stelle eines Assistenten in einer wissenschaftlichen Versuchsanstalt des Reichsschatzamtes, die unter Leitung des Professors K. von Buchka stand. Hier ist er kurz vor dem Kriege infolge einer Pyämie an einer kleinen Wunde gestorben. Trotz seiner persönlichen Gutmütigkeit war er politisch ein ausgesprochener Chauvinist und stellte sich als solcher an die Spitze einer Opposition in der chemischen Gesellschaft, welche die Wahl von Sabatier zum Ehrenmitglied aus politischen Gründen bekämpfte. Ich mußte ihm damals ziemlich scharf entgegentreten, was aber ohne Einfluß auf unser freundschaftliches persönliches Verhältnis geblieben ist.

Neben den ordentlichen Assistenten haben im neuen Hause immer einige Hilfsassistenten, teils besoldet, teils unbesoldet an meinen Arbeiten im Privatlaboratorium teilgenommen. Dahin gehören die Herren Dr. Bethmann und Dr. Hagenbach, ein Sohn des bekannten Professors der Physik an der Universität Basel. Sie sind beide später in die Höchster Farbwerke eingetreten und Dr. Hagenbach hat sich hier durch gute Leistungen eine recht geachtete Stellung geschaffen.

In derselben Eigenschaft war Dr. E. Frankland Armstrong, Sohn des Professors Henry Armstrong in London, mehrere Jahre bei mir tätig, nachdem er zuvor eine Untersuchung in der Puringruppe ausgeführt hatte, und daraufhin von der Berliner Fakultät zum Dr. phil. promoviert worden war. Er ist ein spekulativer und auch experimentell gut veranlagter Chemiker, der bei mir ziemlich schwere Versuche über die Synthese von Disacchariden und die Bereitung von Acetohalogen-Glucose durchgeführt hat. Leider ist die Existenz der isomeren als α- und β-Verbindung bezeichneten Acetochlor-Glucosen durch meine späteren Beobachtungen sehr zweifelhaft geworden und bei der Überführung unserer ursprünglichen Präparate in α-Methylglucosid, die als Beweis für die Struktur des vermeintlichen α-Halogenkörpers gedient hatte, muß entweder ein Irrtum passiert oder eine zufällige Änderung der Konfiguration eingetreten sein.

Armstrong hat später ein hübsches Büchlein über die einfachen Kohlenhydrate geschrieben, das von Dr. Eugen Unna ins Deutsche übersetzt wurde. Er hat ferner in London interessante Versuche über die Spaltung der α- und β-Glucoside mit Enzymen angestellt, mußte aber, weil er früh heiratete, wahrscheinlich aus materiellen Gründen eine Stellung in der Industrie annehmen, was seinen rein wissenschaftlichen Arbeiten natürlich Abbruch tat.

Von den neuen Assistenten des Wintersemesters 1901/02 sind Dr. Georg Röder und Dr. Alfred Dilthey besonders zu nennen. Ersterer hat bei mir die Synthese des Uracils, Thymins und ähnlicher Verbindungen nach neuen Methoden ausgeführt. Er ist ein begabter und auch theoretisch gut unterrichteter Chemiker, der sicherlich sehr hübsche Sachen hätte machen können, wenn er die nötige Ausdauer besessen hätte. Er hat es aber vorgezogen, nach einigen Semestern das Laboratorium zu verlassen und auf Reisen zu gehen. Von Zeit zu Zeit tauchte er wieder in Berlin auf und wußte dann Interessantes über seine Erlebnisse zu berichten. Wenn ich nicht irre, hat er mehrere fremde Kontinente kennen gelernt und zuletzt war er im Laboratorium von Piutty in Neapel als Unterrichtsassistent tätig. Hier wurde er durch den Krieg verscheucht, kehrte nach Berlin zurück und wurde dann bald Soldat. Wie ich höre, hat er durch Vermittlung seiner chemischen Freunde bald bei einem A. O. K. eine Anstellung gefunden, wo er trotz seines niedrigen militärischen Ranges seine vielfachen chemischen und technischen Kenntnisse glücklich verwerten konnte.

Alfred Dilthey war der jüngste Sohn meiner verstorbenen Schwester Mathilde, schon als Knabe bildhübsch und aufgeweckt. Nachdem er ein Semester in Genf studiert hatte, kam er im Herbst 1895 nach Berlin und blieb hier mehrere Jahre. Da er nur das Abiturium von einer Oberrealschule hatte und die Berliner Fakultät in solchen Fällen geneigt war, Schwierigkeiten zu machen, so ging er auf meinen Rat zur Promotion zu Hantzsch nach Würzburg. Dann diente er als Einjährig-Freiwilliger in Düsseldorf in einem Ulanenregiment und kehrte 1901 nach Berlin zurück. Hier nahm er im Privatlaboratorium teil an meinen Arbeiten über die Amidbildung bei alkylierten Malonestern und an den Synthesen von alkylierten Barbitursäuren, von denen das Veronal ein bekanntes Schlafmittel geworden ist. Da er zur wissenschaftlichen Laufbahn keine Lust hatte, so machte er 1902 eine Reise nach den Vereinigten Staaten von Nordamerika und kehrte von dort $^{3}/_{4}$ Jahre später, leider ziemlich heftig an Malaria erkrankt, zurück.

Er hat lange gebraucht, um sich zu erholen und ließ sich später in Berlin nieder, um ein eigenes Geschäft zu gründen. Meinen Rat, in eine der bestehenden chemischen Fabriken einzutreten oder sich an der blühenden Baumwollspinnerei seines Vaters und Bruders in Rheydt zu beteiligen, lehnte er ab, und bei den eigenen Geschäftsunternehmungen blieb er ohne Erfolg. So ist es gekommen, daß dem talentvollen, klugen und auch fleißigen jungen Mann eine richtige Lebensstellung, die seinen Fähigkeiten und seinen materiellen Mitteln entsprochen hätte, unerreichbar blieb. Er ist in dieser Beziehung, wenn man will, ein Opfer seiner Vorliebe für die Großstadt geworden. Beim Ausbruch des Krieges zog er ins Feld. Hier ist er als Offizierstellvertreter mit seiner Kolonne im Sommer 1915 von Kosaken überfallen und niedergemacht worden. Er ruht in polnischer Erde.

Solange er in Berlin weilte, war er regelmäßig an Sonn- und Festtagen mein Gast und bemühte sich, die jüngeren Vettern, meine Söhne, in allen Künsten männlicher Jugend, wie Kartenspiel, Weintrinken, Sportgeschichten, studentische Angelegenheiten, Tanzfragen, Verkehr mit Damen und dergl. heranzubilden. Die Jungen lauschten, besonders solange sie noch auf der Schule waren, mit Staunen und Hochachtung seinen Lehren und die vier jungen Leute bildeten ein durch Lustigkeit und Eintracht ausgezeichnetes Quartett. Es kam mir manchmal so vor, als wenn Alfred Dilthey mit zu meinen Söhnen gehörte. Von diesen vier prächtigen Menschen ist infolge des unseligen Krieges nur einer übrig geblieben.

Gleichzeitig mit Dilthey waren im Privatlaboratorium drei Herren, die trotz der Verschiedenheit ihres Wesens gute Freundschaft miteinander hielten und auch zu ausgelassenen Streichen sich öfters vereinten. Der unternehmendste davon war Dr. Theodor Dörpinghaus aus Elberfeld, ein schöner, durch körperliche Kraft ausgezeichneter Mann, in mancherlei Sport geübt und zu abenteuerlichem Leben geneigt. In der Chemie zeichnete er sich weniger durch Feinheit der Beobachtung, als durch schnelles, energisches Anfassen der experimentellen Aufgaben aus. Er hat die erste ziemlich mühsame Hydrolyse von Horn und ähnlichen Proteinen durchgeführt. Später ist er auf Reisen gegangen und hat sich nicht allein in Amerika und Asien, sondern auch ziemlich lange in Zentral-Afrika und Marokko aufgehalten. Die Frucht der letzten Reisen war eine Anklageschrift gegen die Verwaltung des belgischen Kongostaates, die einiges Aufsehen erregte. Seit Ausbruch des Krieges habe ich nichts mehr von ihm gehört, vermute aber, daß er tätigen Anteil genommen hat und ohne Schaden davon gekommen ist; denn in Überwindung von Gefahren hat er immer Geschick gezeigt und Glück gehabt. In der letzten Zeit seines Berliner Aufenthaltes war er zusammen mit seinen Freunden Dr. Ernst Königs und Eduard Andreae Besitzer eines alten Segelbootes, auf dem er alle freien Tage des Sommers zubrachte und uns auch zuweilen auf dem Wannsee seine Wasserkünste vorführte.

Dr. Peter Bergell, der Sohn eines mecklenburgischen Landwirten, ist Mediziner. Bevor er zu mir kam, war er Assistent an der Klinik der inneren Medizin zu Breslau. Infolge guter Vorstudien fand er sich rasch in unsere Arbeitsweise und hat bei mir die Isolierung der Aminosäuren als Derivate der β-Naphtalinsulfosäure bearbeitet. Mit Hilfe dieses Reagens gelang uns zum ersten Mal schon 1902 der Nachweis, daß bei gemäßigter Hydrolyse des Seidenfibroins ein Dipeptid, das Glycylalanin, entsteht. Nach dem Verlassen unseres Instituts wurde Bergell zuerst der medizinische Berater einer Fabrik für pharmazeutisch-chemische Präparate in Berlin. Später ist er zur reinen Medizin zurückgekehrt und hat augenblicklich eine umfangreiche Praxis als Spezialist für Stoffwechselkrankheiten. Seine Freunde rühmten ihm große Gewandtheit in geschäftlichen Dingen nach.

Neuerdings ist er unter die Schriftsteller gegangen und hat ein viel gelesenes Buch „Die linke Landgräfin" herausgegeben, in dem er das Problem der Bigamie behandelt.

Ganz anders geartet war Dr. Hermann Leuchs, eine stille Gelehrtennatur, ausgezeichnet durch Schweigsamkeit, Ruhe und Ernst. Er stammt aus einer Fabrikantenfamilie zu Nürnberg, kam anfangs des Jahrhunderts nach Berlin und führte unter meiner Leitung eine Doktorarbeit über die Synthese von Oxyaminosäuren aus. Ihre schönste Frucht war die künstliche Bereitung des Serins und Glucosamins. Die Geschicklichkeit und Sorgfalt, die er dabei bewies, war für mich die Veranlassung, ihn als Privatassistenten bei den schwierigen Arbeiten über Polypeptide zu wählen. Nachdem er hier zwei Jahre lang vortreffliche Dienste geleistet hatte, wurde er Unterrichtsassistent in der organischen Abteilung und schließlich nach dem Weggang von Diels dessen Nachfolger als Abteilungsvorsteher.

Leuchs ist ein grundgescheiter Chemiker und sehr geschickter Experimentator. Das beweisen seine späteren Untersuchungen besonders über den Abbau der Strychnosalkaloide. Ich hoffe, daß sie ihm in nicht allzu langer Zeit eine selbständige Stellung bringen werden.

Zusammen mit Dr. Dilthey haben diese drei jungen Männer im Privatlaboratorium trotz eifriger und auch erfolgreicher Arbeit ein vergnügtes Leben geführt, das sich zuweilen bis zu ausgelassenen Streichen steigerte. Besonders war das der Fall, als ich im Winter 1903/04 wegen Schlaflosigkeit Berlin für mehrere Monate verließ und mich in den Alpen herumtrieb. Da erwachte bei der Jugend ein starkes Selbständigkeitsgefühl, das sich zuerst in Erfindungen und Patentanmeldungen äußerte, die aber hinterher alle als wertlos erkannt wurden. Daneben kam es zu allem möglichen Schabernack. So wurde dem Dr. Bergell ein ganzer Waschkorb frischer Eier, die er für einen wissenschaftlichen Versuch benutzen wollte, heimlich gekocht und dadurch natürlich ihrem ursprünglichen Zweck entzogen. Ferner gab es eine freundschaftliche Rauferei unter den großen, meist sehr starken Herren, bei der Dr. Bergell einen Teil seines Bartes verlor und Dörpinghaus durch einen Fußtritt ins Gesicht beinahe ein Auge verloren hätte. Man wird daraus den Schluß ziehen, daß es im Privatlaboratorium auch andere Dinge als reine Wissenschaft gab, besonders wenn meine Abwesenheit von Berlin die Gefahr der Überraschung ausschloß. Glücklicherweise war damals das weibliche Element noch nicht bei uns eingedrungen. Sonst hätten schwierige Komplikationen entstehen können.

Gleichzeitig mit den eben genannten Herren waren in der anorganischen Abteilung zwei neue Assistenten ernannt worden, Dr. Blix, ein Schwede, und Dr. Arthur Stähler, ein Berliner. Blix hatte zuvor mit mäßigem Erfolg in Berlin promoviert, aber es wurde ihm experimentelle Geschicklichkeit von dem Leiter der Abteilung nachgerühmt. Dazu kam seine stattliche Persönlichkeit und sein gewinnendes Wesen. Aber spätere Schwierigkeiten mit anderen Assistenten zeigten doch, daß er nicht aus lauter Sanftmut zusammengesetzt war, und er hat die Assistentenstelle vor Ablauf der üblichen zwei Jahre wieder aufgegeben. Er ist dann nach den Vereinigten Staaten von Nordamerika gegangen und hat dort, wie ich indirekt erfuhr, in der Zuckerindustrie eine glänzende Stellung erworben.

Dr. Stähler ist als Anorganiker bei der Wissenschaft geblieben. Auf meinen Vorschlag wurde er vom Kultusminister 1906 zu Th. W. Richards an die Havard-University in Cambridge (Mass.) geschickt und hat dort ungefähr ein Jahr an den bekannten Atomgewichtsbestimmungen teilgenommen. Infolgedessen wurde er der vertraute Helfer von Richards, als dieser im Sommer 1907 als Austauschprofessor nach Berlin kam. Aus diesem Verkehr sind verschiedene gemeinsame Publikationen von Richards und Stähler über Atomgewichte hervorgegangen, und Stähler hat die Versuche später allein fortgesetzt, sich außerdem aber mit präparativen Aufgaben der Mineralchemie beschäftigt. Nach Ausbruch des Krieges ist er in die Dienste der Kriegsmetall-Gesellschaft getreten und leitet seit mehreren Jahren ein analytisches Laboratorium in Brüssel und zuletzt in Cöln a. Rh.

Epilogue

"There are two things which I am confident I can do very well: one is an introduction to any literary work, stating what it is to contain, and how it should be executed in the most perfect manner; the other is an epilog showing from various causes, why the execution has (not) been equal to what the author promised to himself and to the public (after Samuel Johnson: Boswell's Life modified from (1755))."

The "History of Organic Chemistry in the United States, 1875 - 1955", has been the subject of masterly essays by Dean and Ann Tarbell.[1] At the beginning of this period the laboratories of Liebig, von Baeyer, Emil Fischer were meccas for scientific pilgrims. As Professor Fruton points out, the characteristics and uniqueness of Fischer and his scientific progeny have been the subject of an article by L. Läsker.[2] The number of British and American visitors, pre- and post-doctoral, to Fischer's laboratory constitute a sizable part (27 American, 17 British) of the entire list of 310 Fischer-associates (Table 1 and 2). Appendix 1 is the result of years of diligent investigations by Professor Joseph Fruton[13] who kindly agreed to their publication side by side with Fischer's re-edited autobiography.

The seeds that Fischer sowed grew to large trees of knowledge with even larger ramifications, especially in the hands of disciples whose work was later recognized by the award of the Nobel Prize. Otto Diels (1876 - 1954), Adolf Windaus (1876 - 1959), Fritz Pregl (1869 - 1930) and Hans Fischer (1881 - 1945) represent the more chemically oriented group with strong

Table 1 (from Fruton, ref. 3)
Composition of Fischer's Group 1879 - 1919

	Dr. phil		Dr. med.		
	Predoct.	Postdoct.	Predoct.	Postdoct.	Total
Fischer					
German-speaking	172	46*		14	232
Other nations	46	23		9	78
Total	218	67		23	310

* Includes assistants (34) and guests (12); does not include 32 research assistants who had received their Ph. D. for work with Fischer.

Table 2 (from Fruton, ref. 3)
Principal Subsequent Activity of Junior Members of the Fischer's Group*

	Fischer	
	German-speaking nations	Other nations
Univ. full (ord.) Professor	31	18
Other university ranks	10	10
Scientific research institute	5	9
Hospital research and admin.	5	1
Private medical practice	2	
Industrial res. and admin.	107	17
Government techn. depts.	18	5
Chemical consultant	15	11
Other+	5	2
Total	198	73

* Not included in this table are persons known to have died soon after their doctoral work.

+ Includes publishing, pharmacy, farming, etc.

leanings toward natural products of key importance, such as steroids, vitamin D, hemin, bile pigments and chlorophyll, all investigations made possible by Pregl's elaboration of microanalytical methods.

Fischer's student Otto Warburg (1833 - 1870) left peptide synthesis and Berlin in 1906 and turned his attention to the respiration of fertilized sea urchin eggs at the Stazione Zoologica in Naples. He became a director of one of the newly created Kaiser Wilhelm Institutes and one of the founders of modern biochemistry. Part of his success lies in the old aperçu "The Style is the Man", in this case Warburg carried on Emil Fischer's style.

Emil Abderhalden (1877 - 1950), Carl Freudenberg (1886 - 1983), Geza Zemplen (1883 - 1956) and Burckhardt Helferich (1887 - 1982) continued Fischer's work in the field of proteins, lignins, glycosidases, general stereochemistry and Walden inversions. The dividends of Fischer's work far exceeded the original capital.

When we come to the next three names, Landsteiner, Bergmann and Levene, an anecdote by Willstätter illuminates Fischer's emancipated attitude, which was much more liberal than was customary in the Berlin of Adolf (sic!) Stoecker (1835 - 1909) and Heinrich von Treitschke (1834 -

1896) remembered as the author of the slogan: "Die Juden sind unser Unglück" (The Jews are our misfortune). When Fischer who had many collaborators with a Jewish background was asked: "How is it possible that Your Excellency does not harbor antisemitic feelings?" Fischer replied: "We Rhenish people are not so dumb that we would have to be afraid of the Jews."

Max Bergmann (1886 - 1944) two years after Fischer's death became director of the newly established Kaiser Wilhelm Institute for Leather Research in Dresden whence he was forced to move to the Rockefeller Institute in 1934. In 1932 he and Leonidas Zervas introduced the carbobenzoxy method of peptide synthesis, a milestone that made possible advances in solid phase synthesis, culminating in the research of Bruce Merrifield (Nobel Prize 1984). With Joseph S. Fruton Bergmann explored proteolytic enzymes. Fruton's own account of these historical aspects of biochemistry is recorded in his book "Molecules and Life" (New York 1972).[4] Bergmann's associates William Howard Stein (1911 - 1980) and Stanford Moore (1913 - 1982) received Nobel Prizes for their amino acid analyzer as applied to many enzymes, notably ribonuclease.

Karl Landsteiner (1868 - 1843) received his MD in Vienna in 1891, worked with Eugen Bamberger in Munich, Arthur Hantzsch in Zurich and Fischer (1892). With his discovery of specificity of serological reactions he founded modern immunology. He also continued at the Rockefeller Institute in New York where he received the Paul Ehrlich Prize in Frankfurt in 1929 and the Nobel Award in 1930. His book "The Specificity of Serological Reactions (New York 1936)" is a classic meriting reading and rereading.

Phoebus Aaron Theodore Levene (1869 - 1940) was the first American to work in Fischer's laboratory where, by the ester method, he analyzed hydrolyzates of gelatin from which Fischer had isolated the new amino acid 4-hydroxyproline in 1902. Levene is remembered for his work on (ribo)nucleotides in which he was supported by Walter Abraham Jacobs (1883 - 1967), the fourth Fischer student to end up at the Rockefeller Institute. Walter Jacobs was one of the few American chemists who established a school for natural products, cardiac steroids, such as digitoxigenin and strophantidin as well as ergot alkaloids. Levine's early work related some natural products to simple compounds, such as tartaric acid to which Fischer had assigned an absolute configuration that was confirmed by a pace-setting modification of Roentgen Ray crystallographic analysis (B. M. Bijvoet, Nature 168, 271 (1951)). Another milestone was the identification of D-ribose as the five-carbon sugar from yeast nucleic acid

(RNA) and of D-2-deoxyribose from thymonucleic acid (DNA). He, with Stuart Tipson, showed that adenine combined with D-ribose in the furanose form in the 7-nitrogen (later to be corrected to the 9-nitrogen) and that the sequence sugar-phosphate-base formed the repeatable "nucleotide" unit of nucleic acids. Here Fischer's legacy extends into modern genetic events. The fifth is Donald Dexter van Slyke (1883 - 1971) who stayed with Fischer in 1911 after having received his Ph. D. with Moses Gomberg (1866 - 1947), himself a student of Victor Meyer (1848 - 1897). Van Slyke has become a household word for physiologists and protein chemists because of his gasometric method for the determination of free amino nitrogen in enzymatic and non-enzymatic protein hydrolyzates.

The reader may like to continue this search for the origin of ideas and progress in organic and bio-chemistry as well as in "molecular" biology that are treaceable through the list of his 310 associates backward and forward in time and he may come to the conclusion that Fischer's autobiographical notes do not read like last year's newspaper but that they encourage the exploration of the birth and rebirth of guiding concepts in chemistry and biochemistry.

1. Dean Stanley and Ann Tarbell, History of Organic Chemistry in the United States, 1875 - 1955, Folio Publishers, Nashville, Tennessee 1986.

2. L. Läsker, "Charakterische Besonderheiten der Schule von Emil Fischer", in S. R. Mikulinsky, M. G. Jarosevskij, G. Kröber and H. Steiner, Wissenschaftliche Schulen (Berlin 1977, 1979, especially Vol. 2, pp. 99 - 112).

3. Contrasts in Scientific Style. Emil Fischer and Franz Hofmeister: Their Research Groups and Their Theory of Protein Structure, by Joseph S. Fruton, Proceedings of the American Philosophical Society, 129, 313 - 370 (1985).

4. Cf. Joseph S. Fruton, The Education of a Biochemist, in "Of Oxygen, fuels and Living Matter", Part 2, Edited by G. Semenza, Wiley & Sons, 1982, pp. 315 - 360.

Appendix 1

The Fischer Research Group

Name, Dates and Nationality*	Association with Fischer	Principal Subsequent Activity
ABDERHALDEN, EMIL (1877-1950)	Post-MD, asst 1902-07 (30)	Prof physiology Halle
Ach, Friedrich (1864-1902)	PhD, asst 1889-98 (7,3a)	Chemical industry (Boehringer)
Ach, Lorenz (1868-1948)	PhD, asst 1892-95 (5,10a)	Chemical industry (Schering)
Alexander, Walter (1871-) *USA*	PhD 1894 (2)	Chemical consultant
Andreae, Edward Philip (1879-1975) *Britain*	PhD 1905 (2)	Chemical consultant
Anger, Gerda	1915-19 (1,2a)	
Antrick, Otto (1862-1942)	PhD 1884 (3)	Chemical industry (Schering)
Arheidt, Richard (?-1924)	PhD 1887 (2)	Chemical industry (BASF)
Armstrong, Edward Frankland (1878-1945) *Britain*	Post-PhD 1901-02 (6)	Chemical industry (ICI)
Axhausen, Walter (1882-)	PhD, asst 1904-09 (2,10a)	
Baerwind, Heinrich (1892-1968)	PhD, asst 1915-19 (3)	Chemical industry (Freudenberg Co)
Beensch, Leo (1869-)	PhD, asst 1893-96 (4)	Chemical industry (Boehringer)
Behringer, Carl	PhD 1891 (2)	Chemical industry (Wilke-Dörfurt)
Bergell, Peter (1875-)	Post-MD 1902-04 (3)	Private medical practice
BERGMANN, MAX (1886-1944)	Post-PhD, asst 1912-19 (12,9a)	Res inst K. W. Dresden, Rockefeller‡
Besthorn, Emil (1858-1921)	PhD 1882 (2)	Chemical industry (Höchst)
Bethmann, Fritz (1874-)	PhD 1899 (1,3a)	Chemistry industry (Höchst)
Blank, Paul (1882-)	PhD 1906 (2)	
Blix, Martin (1877-1933) *Sweden*	PhD 1902 (1)	Chemical industry (US Sugar Co)
Blochmann, Richard Hermann (1877-)	1901 (1); PhD Rostock 1902	Chemical consultant (photography)
Blumenthal, Herbert (1885-1952)	PhD 1907 (2)	Govt testing bureau Berlin

Boeckler, August (1871-)	PhD 1892 (1)	Chemical industry (BASF)
Böhner, Reginald (1880-1945) *Canada*	Post-PhD 1910-11 (2)	Prof chemistry Syracuse
Boelsing, Friedrich	PhD 1892 (1)	Chemical industry (Neef & Co)
Bösler, Magnus (1852-1924)	PhD, asst 1880-82 (3,5a)	Chemical consultant
Bosart, Louis William (1874-1947)	PhD 1898 (1)	Chemical industry (Procter & Gamble)
Brauns, Friedrich Emil (1890-1982)	PhD 1913-15 (4,1a)	Chemical industry (Appleton, Wis)
Brieger, Walter (1891-)	PhD 1914 (3)	Chemical consultant, history of chem
Bromberg, Otto (1870-)	PhD 1895 (5)	Chemical consultant
Brüning, Gustav von (1864-1913)	PhD 1889 (3)	Chemical industry (Höchst)
Brunck, Rudolph (1867-1942)	PhD 1892 (2)	Chemical industry (IG Farben)
Brunner, Arnold (1880-1940)	PhD 1905 (2)	Chemical industry (Höchst)
Bülow, Karl (1857-1933)	Post-PhD asst 1884-86 (3)	Prof chemistry Tübingen
Bunte, Karl (1878-1944)	PhD 1903 (1)	Prof chem technol Karlsruhe TH
Cahn, Heinz (1892-)	PhD 1913, 19 (1)	Chemical consultant
Carl, Hans (1880-1966)	PhD 1906 (2)	Publisher
Chaplin, Edward Mitchell (1868-1948) *Britain*	PhD 1892 (2)	Goft chemical analyst
Clarke, Hans Thacher (1887-1972) *Britain*	Post-DSc 1912 (0)	Prof biological chem Columbia
Clemm, Hans (1872-1927)	Post-PhD asst 1897-98 (3)	Chemical industry (Zellstoff Fabrik)
Colman, Harold Govett (1866-1954) *Britain*	PhD 1888 (2)	Consultant gas industry
Cone, Lee Holt (1880-1957) *USA*	Post-PhD 1908 (1)	Chemical industry (National Aniline)
Crossley, Arthur William (1869-1927) *Britain*	PhD 1892 (3)	Chemical industry (British Cotton)
Culmann, Julius (?-1930) *USA*	PhD 1889 (3)	Chemical industry (Seigle Corp)
Curme, George Oliver (1888-1976) *USA*	Post-PhD 1914 (1)	Chemical industry (Union Carbide)
Curtiss, Richard Sydney (1863-1944) *USA*	PhD 1892 (2)	Various univ appointments in chem
Dangschat, Paul (1893-)	1919 (1)	Chemical consultant
Degen, Joseph (1861-1942)	PhD 1886 (2)	Chemical industry (Degen & Kuth)
Delbrück, Konrad 1884-1915	Post-PhD asst 1908-09 (2)	Chemical industry (Bayer)

DIELS, OTTO (1876-1954)	PhD, asst 1899-1904 (12)	Prof chemistry Kiel
Dilthey, Alfred (1877-1915)	Post-PhD asst 1902-03 (2)	Private chemical company
Dörpinghaus, Wilhelm Theodor (1878-)	PhD 1902 (2,1a)	Chemical consultant (mining)
Dyckerhoff, August	PhD 1894 (1)	Family cement factory
Easterfild, Thomas Hill (1866-1949) *Britain*	PhD 1894 (1)	Dir Cawthorn Inst Sci Res New Zealand
Ehrensberger, Emil (1858-1940)	1881 (1a)	Director Krupp works
Ehrhardt, Wilhelm	Post-PhD asst 1878-79 (2)	
Einbeck, Hans (1873-)	PhD 1905 (1)	Private chemical laboratory
Elbers, Alfred (1861-1936)	PhD 1884 (3)	Family textile factory
Elsinghorst, Gerhard (1858-1929)	PhD 1884 (2)	Chemical consultant
Evans, Thomas Brown (1863-) *USA*	PhD 1886 (1)	Chemical consultant
Fay, Irving Wetherbee (1861-1936) *USA*	PhD 1896 (2)	Prof chemistry Brooklyn Poly Inst
Felser, Heinrich (1881-)	PhD 1908 (1)	Staff K. W. Inst Textile Res.
Fiedler, Albert (1886-1961) *USA*	PhD 1910 (2)	Chemical consultant
FISCHER, FRANZ (1877-1947)	Post-PhD asst 1909-11 (0)	Dir K. W. Inst Coal Research
FISCHER, HANS (1881-1945)	Post-PhD, MD asst 1909 (1)	Prof chemistry Munich TH
FISCHER, HERMANN OTTO LAURENZ (1888-1960)	Post-PhD asst 1912-14 (4)	Prof biochemistry Toronto, Berkeley‡
Fischer, Wilhelm (1868-)	PhD 1892 (2)	
Flatau, Erich	1909 (2)	
Fodor, Kalman von (1881-1931)	Post-PhD 1914-15 (2,1a)	Govt exp station Budapest
Fogh, Johan Bertil (1865-1925) *Denmark*	Post-PhD asst 1895 (0)	Staff agric college Copenhagen
Forster, Martin Onslow (1872-1945) *Britain*	PhD 1892 (1)	Dir Indian Inst Sci Bangalore
FOURNEAU, ERNEST (1872-1949) *France*	Post-PhD 1901 (2)	Pasteur Institute Paris
Frank, Fritz (1868-1949)	PhD 1896 (2)	Govt testing bureau Berlin
FREUDENBERG, KARL (1886-1983)	PhD, asst 1910-14 (7)	Prof chemistry Heidelberg
Freytag, Paul (1876-)	PhD 1912 (1)	
Fritz, Victor (1872-1926)	PhD 1896 (2)	Chemical industry (Boehringer)

Geiger, Walter (1866-)	PhD 1908 (1)	
Gerlach, Ferdinand (1885-)	Post-PhD asst 1912 (1)	Chemical industry (Schuckerts)
Gerngross, Otto (1882-1966)	Post-PhD asst 1909 (1)	Prof chem technol Ankara‡
Geserick, Arthur (1860-)	PhD 1908 (2)	Private chemical factory
Gevekoht, Heinrich (1858-)	PhD 1888 (3)	
Giebe, Georg (1874-1899)	PhD, asst 1896-98 (4,1a)	
Gluud, Wilhelm (1887-1936)	Post-PhD asst 1909-12 (1,4a)	Dir Inst Coal Technol Dortmund
Göddertz, Bernhard Albert (1888-1916)	PhD 1911 (1,3a)	Chemical industry (Bayer)
Grävenitz, Richard von (1890-1918)	PhD 1914 (2)	
Groh, Reinhart	1910 (2)	
Guggenheim, Markus (1885-1970)	Post-PhD 1908 (1)	Chemical industry (Hoffmann-LaRoche)
Gwinner, Hans von (1887-1959)	PhD 1912 (2)	Chemical industry (Nitrit Fabrik)
Haberland, Hermann (1865-)	PhD 1888 (1)	Chemical industry (Neu-Stassfurt)
Hähnel, Otto (1884-)	PhD 1909 (1)	Chief chemist govt post office
Haenisch, Victor (1871-1940)	PhD 1894 (1)	Chemical industry (sulfuric acid fact)
Haga, Tamemasa (1856-1914) *Japan*	1897 (1)	Prof chemistry Tokyo
Hagenbach, Rudolf (1875-1927)	Post-PhD 1901 (1)	Chemical industry (Höchst)
Harper, Charles Athiel (1868-) *USA*	PhD 1896 (1)	Chemical consultant
HARRIES, CARL (1866-1923)	Post-PhD asst 1892-1900 (1)	Prof chemistry Kiel
Hartmann, Gerhard (1868-1945)	PhD 1891 (2)	Chemical industry (Griesheim)
Hauff, Friedrich (1863-1935)	PhD 1888 (2)	Family photochemical factory
Hegel, Sigmund (1863-)	PhD 1884 (2)	Govt patent office Berlin
HELFERICH, BURCKHARDT (1887-1982)	PhD, asst 1911-14 (3,1a)	Prof chemistry Leipzig
Heller, Gustav (1866-1946)	PhD, asst 1890-94 (1,4a)	Prof chemistry Leipzig
Heller, Hans (1870-)	PhD 1894 (2)	Chemical industry (Harburger Ölwerken)
Herborn (Avenarius-Herborn) Heinrich (1873-1955)	PhD 1896 (2)	Chemical industry (Avenarius Co)

Herrick, James Bryan (1861-1954) *USA*	Post-MD 1905 (1)	Prof medicine Rush Med Coll
Hertz, Johann Nicolaus (1869-1908)	PhD 1892 (2)	Chemical industry (Voelper)
Hess, Kurt (1888-1961)	Post-PhD asst 1912 (2)	K. W. Institute Chemistry Berlin
Hess, Otto Paul (1860-)	PhD 1884 (4)	Chemical industry (Höchst)
Hilpert, Siegfried (1883-1951)	PhD, asst 1905-07 (1,2a)	Prof chemistry Braunschweig TH
Hirschberger, Josef (1866-1954)	PhD, asst 1888-90 (5,4a)	Various chemical companies USA
Hirszowski, Alfred (1882-1943) *Poland*	PhD 1908 (2)	Chemical factories Lodz, Warsaw‡
Hoesch, Kurt (1882-1932)	PhD 1911 (2,1a)	Owner family paper factory
Hoffa, Erwin Friedrich (1875-1967)	PhD 1897 (2)	Chemical industry (Höchst)
Holzapfel, Julius (1883-1918) *Britain*	Post-PhD 1912 (1)	Chemical industry (Intern Paints)
Hübner, Friedrich (1876-1953)	PhD, asst 1897-99 (1,7a)	Chemical industry (Höchst etc)
Hütz, Hugo (1871-)	PhD 1895 (2)	Chemical industry (Höchst)
Hunsalz, Paul (1871-)	PhD, asst 1894-98 (3,13a)	Chemical industry (Schering)
Hutchinson, Arthur (1866-1937) *Britain*	PhD 1882 (2)	Prof mineralogy Cambridge
Ince, Walter Holinshed (1865-1907) *Britain*	PhD 1889 (2)	Various chemical companies
Jacobi, Friedrich Wilhelm (1863-)	Post-PhD 1895 (0)	
Jacobi, Hermann	PhD 1891 (3,1a)	
JACOBS, WALTER ABRAHAM (1883-1967) *USA*	PhD 1907 (3)	Rockefeller Inst Med Res
Jänecke, Ernst (1875-1957)	PhD 1898 (1)	Chemical industry (BASF)
Jennings, Walter Louis (1866-1944) *USA*	Post-PhD 1893-94 (2)	Prof chemistry Worcester Poly
Jessel, Henry Rose (1867-1933) *USA*	PhD 1900 (1)	Chemical industry (Scaife)
Joachimoglu, Georg (1887-1979) *Greece*	Post-MD 1912 (0)	Prof pharmacology Athens
Joedicke, Friedrich (1864-)	PhD 1887 (1)	Chemist asphalt industry
Johlin, Jacob Martin (1884-1954) *USA*	PhD 1910 (1)	Assoc prof biochem Vanderbilt
Johnson, Theodore (1884-) *USA*	PhD 1906 (1)	

Jourdan, Friedrich	Post-PhD 1883 (1)	Chemical industry Rome
Kaas, Carl (?-1915)	1906 (1)	Asst chemistry Graz
Kadisade, A. Refik (1892-) *Turkey*	PhD 1915, 18 (2)	
Kalantharianz, Anuschawan (1867-) *Russia*	PhD 1898 (2)	
Kall, Henry von der	PhD 1890 (2)	
Kametaka, Tokuhei (1872-1935) *Japan*	Post-PhD 1909 (2)	Prof chemistry Tokyo
Kauffmann, Hugo (1890-1916)	PhD 1914 (1)	
Kautzsch, Karl (1879-1920)	Post-PhD 1905 (1)	Chemical industry (Höchst)
Kay, Francis William (1883-1967) *Britain*	PhD 1908 (3)	Senior lecturer chemistry Liverpool
Kempe, Martin (1884-)	PhD 1907 (2)	
Kempf, Richard (1879-1935)	Post-PhD 1904 (0)	Govt chem techn inst Berlin
Kleberg, Werner	PhD 1891 (3)	
Klemperer, Georg (1865-1946)	Post-MD 1913 (1)	Hospital chief Berlin ‡
Klotz, Carl (1862-1938)	Post-PhD asst 1886 (3a)	Chemical industry (Höchst)
Knoevenagel, Oscar (1862-1944)	PhD 1887 (2)	Chemical industry (BASF)
KNORR, LUDWIG (1859-1921)	PhD, asst (3,1a)	Prof chemistry Jena
Koch, Hermann (1858-1939)	PhD 1883-85 (5)	Chemical industry (Kali-Chemie)
Koelker, Arthur Heinrich (1883-1911) *USA*	PhD 1908 (2)	Staff Roosevelt Hosp New York
Koelker, Wilhelm Friedrich (1880-1911) *USA*	PhD 1905 (2)	Asst prof chemistry Wisconsin
Köllisch, Anton (1888-1916)	PhD 1911 (1)	Chemical industry (Merck)
Königs, Ernst (1878-1945)	PhD 1903 (3)	Prof chemistry Breslau
Kötzle, Arthur (1865-1930)	PhD 1889 (1)	Chemist leather industry
Kohlrausch, Karl (1862-)	PhD 1889 (2)	
Kopisch, Friedrich (1867-)	PhD 1894 (2)	Landowner
Krämer, Adolf (1883-1914)	PhD, asst 1906-09 (3,1a)	Chemical industry (Bayer)
Kraus, Johann (1869-)	PhD 1895 (2)	
Kropp, Walther (1885-1939)	PhD 1908 (2)	Chemical industry (IG Farben)
Kühlewein, Malte von (1887-)	Post-MD, PhD 1915 (1)	Pharmacologist Utrecht
Kuzel, Hans (1859-1921)	PhD 1883 (6)	Chemical industry (Höchst)

LANDSTEINER, KARL (1868-1943)	Post-MD 1892 (1)	Rockefeller Inst Med Res
Langenwalter, Jacob (1868-1943)	PhD 1892 (2)	Chemical industry (BASF)
Laubmann, Heinrich (1865-1951)	PhD 1888 (2)	Chemical industry (Höchst)
Lawrence, William Trevor (1870-1934) *Britain*	PhD 1896 (2)	Lecturer chemistry Manchester
Lawson, Thomas Atkinson (1862-1903) *Britain*	Post-PhD 1890 (0)	Various chemical companies
Laycock, William Frederick (1866-1912) *Britain*	PhD 1889 (2)	Chemical consultant (dyes)
Le Count, Edwin Raymond (1868-1935) *USA*	Post-MD 1905 (1)	Prof pathology Rush Med Coll
Lehmann, Fritz Julius (1874-)	PhD, asst 1897-99 (1,5a)	Chemical industry (Bayer)
Lehnert, Hermann	PhD 1880 (1)	Govt patent office Berlin
Lemme, Georg (1865-1925)	Post-PhD 1894 (1)	Chemical industry (Haarmann & Reimer)
Lepsius, Richard (1885-1969)	PhD 1911 (2,1a)	Various govt agencies
LEUCHS, HERMANN (1879-1945)	PhD asst 1901-04 (3,4a)	Prof chemistry Berlin
LEVENE, PHOEBUS AARON (1869-1940) *USA*	Post-MD 1902 (1)	Rockefeller Inst Med Res
Lipschitz, Werner (1892-1948)	PhD 1915 (2)	Prof pharmacology Frankfurt‡
Loeb, Walther (1872-1916)	PhD 1894 (2)	PvDz chemistry Berlin
Loeben, Wolf von (1869-1913)	Post-PhD asst 1900-01 (2)	Govt testing bureau Berlin
London, Efim Semenovich (1868-1939) *Russia*	Post-MD 1911 (1)	Prof pathol anatomy Leningrad
Lorenz, Henry William Frederick (1871-) *USA*	PhD 1897 (1)	Chemical consultant (photography)
Lüppo-Cramer, Hinricus (1871-1943)	PhD 1894 (2)	Chemical industry (Gelatin Fabrik)
Luniak, Andrei Ivanovich (1881-1957) *Russia*	Post-PhD 1909 (1)	Prof chemistry Kazan
Macnair, Duncan Scott (1861-1937) *Britain*	PhD 1889 (2)	Govt educational inspector
Marko, Dmitri Militiadov (1878-) *Russia*	Post-PhD 1908 (1)	Chem-Techn Institute Perm
Marx, Max (1868-)	PhD 1891 (2)	Chemical consultant
Matsubara, Koichi (1872-1955) *Japan*	Post-PhD 1906 (1)	Prof chemistry Tokyo
Mauritz, Alfred (1867-1938)	PhD 1891 (1)	Brewing industry Dortmund

Max, Jules (1882-) *France*	PhD 1906 (2)	Private chemical company
Mechel, Lucas von (1893-1956)	PhD 1916 (2)	Chemical industry (CIBA)
Meyer, Jacob (1863-)	PhD 1889 (3)	Business manager Fischer institute
Meyer, Paul (?-1914)	PhD 1892 (2)	Chemical industry (Kalle)
Michaelis, Ludwig (1869-)	PhD 1893 (2)	Various chemical companies
Minovici, Stefan (1867-1935) *Rumania*	PhD 1897 (2)	Prof chemistry Bucharest
Moreschi, Annibale (1886-1931) *Italy*	Post-PhD 1912 (1)	Chemical industry (Stacchini)
Morrell, Robert Selby (1867-1946) *Britain*	PhD 1893 (2)	Chemical industry (Manders)
Mouneyrat, Antoine (1870-1952) *France*	Post-PhD 1900 (2); MD Paris 1902	Assoc prof medicine Paris
Müller, Franz (1874-)	1894 (1); PhD Erlangen 1898	Chemical industry (Geigy)
Müller, Hermann (1867-)	PhD 1894 (2)	Chemical consultant
Mylo, Bruno (1884-1915)	PhD 1906 (1)	Asst chemistry Danzig
Nastvogel, Oscar (1863-1910)	PhD 1887 (2)	Chemical industry (Bayer)
Neufeld, Karl Albert (1866-1914)	PhD 1887 (2)	Govt nutrition institute
Neymann, Hans von Splawa (1885-1916)	Post-PhD asst 1913-14 (2,1a)	Chemical industry (Wolfener Farbenfabr)
Niebel, Wilhelm (?-1901)	1896 (1)	
Noth, Hartmut (1892-)	PhD 1917 (2,2a)	
Nouri, Osman (1885-) *Turkey*	PhD 1917 (2)	
Oetker, Rudolf (1889-1916)	PhD 1914 (2)	
Osa, Adolphe de (1873-) *France*	PhD 1904 (1)	Chemist steel industry
Ostmann, Paul (1893-1965)	PhD 1919 (2)	
Otto, Erich (?-1902)	1902 (1)	
Ou, Ching Ko (1885-) *China*	PhD 1911 (2)	
Passmore, Francis William (1867-1921) *Britain*	PhD 1889-90 (3)	Chemical consultant
Pfähler, Ernst (1890-1981)	PhD, asst 1912-14, 19 (2,5a)	Chemical consultant Argentinia
Pfeffer, Otto (1885-1918)	Post-PhD asst 1912-1913 (4)	Govt testing bureau Berlin
Pfülf, August (1862-1921)	PhD 1887 (2)	Chemical industry (Buckau)
Pickel, Max (1861-1933)	PhD 1886 (2)	Chemical industry (Nürnberg)
Piloty, Oskar (1866-1915)	PhD, asst 1889-96 (4,3a)	aoProf chemistry Munich

Pinkus, Georg (1870-1943)	PhD, asst 1893-98 (1,13a)	Chemical consultant‡
Plimmer, Robert Henry Aders (1877-1955) *Britain*	Post-PhD 1902 (1)	Prof biochem St Thomas Hosp Med Sch
Poppenberg, Otto (1876-1956)	PhD, asst 1900-02 (1)	Prof chemistry Charlottenburg TH
PREGL, FRITZ (1869-1930)	Post-MD 1905 (1)	Prof medical chemistry Graz
Rabe, Artur (1889-)	PhD 1919 (1)	Chemical industry (Bayer)
Rahnenführer, Carl (1859-1921)	Post-PhD asst 1885-88 (1,7a)	Chemical industry (Weiler-terMeer)
Ramsay, Henrik (1886-1951) *Finland*	Post- PhD 1908 (1)	Various govt commissions
Rapaport, Max (1889-1915)	PhD, asst 1912-14 (2,2a)	
Raschen, Julius (1863-) *Britain*	PhD 1887 (2)	Chemical industry (United Alkali)
Raske, Karl August Heinrich (1863-)	Post-MD, PhD, asst 1903-09 (8)	
Reese, Ludwig	1883 (1)	
Rehländer, Paul (1869-)	PhD, asst 1892-95 (1,10a)	Various chemical companies
Reif, Johann Georg (1880-1964)	Post-PhD asst 1907-1908 (1)	Govt health institute
Reinbold, Bela (1875-1927)	Post-MD 1905 (1)	ao Prof physiol chem Szeged
Reinbrecht, Otto	PhD 1891 (2)	
Reisenegger, Hermann (1861-1930)	PhD, asst 1882-84 (4,3a)	Chemical industry (Höchst)
Reitzenstein, Friedrich (1868-1940)	PhD 1892 (1)	aoProf chemistry Würzburg
Renouf, Edward (1848-1934) *USA*	PhD 1881 (2,1a)	Prof chemistry Johns Hopkins
Reuter, Ferdinand (1879-1942)	Post-PhD asst 1904-06 (1,4a)	Chemical industry (Didier Werke)
Richter, Ernst (1861-1917)	Post-PhD 1893 (0)	Chemical consultant
Ristenpart, Eugen (1873-1953)	PhD 1896 (1)	Govt textile institute
Roder, Anton (1864-1931)	PhD 1887 (2)	Private chemical company
Roeder, Georg (1874-)	Post-PhD 1900-01 (2)	Chemical industry (Gerber Milk Co)
Roesner, Hans (1886-)	PhD 1911 (2)	Chemical industry (Merck)
Rhode (Davis), Alice (1882-1933) *USA*	Post-MD 1914 (1)	Asst prof several medical schools

Rohmer, Martin (1878-1941)	Post-PhD asst 1900-04 (0)	Chemical industry (Höchst)
Rona, Peter (1871-1945)	Post-MD 1904 (1)	Hospital chief Berlin‡
Rose, Frederick (1867-1932) *Britain*	PhD 1894 (1)	Govt education commission
Rostoski, Otto (1872-1962)	Post-MD 1905 (2)	Hospital chief Dresden
Rudolph, Otto (1864-)	PhD 1888 (1)	Chemical consultant
RUFF, OTTO (1871-1939)	Post-PhD asst 1900 (1)	Prof chemistry Danzig, Breslau
Rund (Schmidt-Rund), Charlotte (1890-1970)	PhD 1916 (1); MD Berlin 1918	Private medical pracitce
Samuely, Franz (1879-1913)	Post-MD 1905 (1)	aoProf medicine Freiburg
Sasaki, Takaoki (1878-1966) *Japan*	Post-MD 1910 (0)	Hospital chief Tokyo
Sattler, Wilhelm (1859-)	PhD 1890 (1)	Owner dye factory
SCHEIBLER, HELMUTH (1882-1966)	PhD, asst 1908-11 (7)	Prof chemistry Charlottenburg TH
Schenkel, Julius Wilhelm (1878-1967)	PhD 1906 (2)	PvDz Munich TH
Schittenhelm, Alfred (1874-1954)	Post-MD 1904 (1)	Prof medicine Kiel
Schlieper, Adolf (1865-1945)	PhD 1887 (3)	Chemical industry (Baum & Co)
Schlotterbeck, Fritz (1876-1940)	Post-PhD 1903-04 (1)	Chemical industry (Stockhausen)
Schmidlin, Julius (1880-1962)	Post-PhD 1905 (1)	Prof chemistry ETH Zurich
Schmidmer, Eduard (1861-1933)	PhD 1892 (2)	Owner factory Nürnberg
Schmidtmann, Hermann (1868-1919)	PhD 1896 (1)	Owner Kaliwerke Aschersleben
Schmitt, Theodor Friedrich (1859-)	PhD 1889 (2)	Govt agricult exp station Vienna
Schmitz, Wilhelm (1879-1939)	Post-PhD asst 1905-06 (2)	Chemical industry (BASF)
Schneider, Wilhelm (1882-1939)	PhD 1909 (2)	Prof chemistry Jena
Schoeller, Walter (1880-1965)	PhD 1906 (2)	Chemical industry (Schering)
Schotte, Herbert (1897-1950)	PhD 1919 (1)	Chemical industry (Schering)
Schrader, Hans (1887-1982)	PhD 1910 (2)	Chemical industry (Goldschmidt Co)
Schrauth, Walther (1881-1939)	PhD 1906 (2)	Chemical industry (Deutsche Hydrierwerke)
Schuler, Josef (1883-1963)	PhD 1910 (1)	Chemical industry (CIBA)

Schulze (Schulze-Forster), Arnold (1882-1946)	PhD 1907 (1)	Govt institute of hygiene
Sembritzki, Kurt (1878-1943)	PhD 1897 (2)	Chemical consultant
Seuffert, Otto (1875-1952)	Post-PhD 1901 (1)	Chemical industry (Merck)
Severin, Josef (1866-)	Post-MD 1912 (1)	Hospital chief Breslau
Skita, Aladar (1876-1953)	Post-PhD asst 1901-02 (2)	Prof chemistry Hannover TH
Slimmer, Max Darvin (1877-) *USA*	PhD 1902 (2)	Chemical consultant
Smith, William Stanley (1863-) *Britain*	PhD 1891 (2)	Chemical consultant (rubber)
Sonn, Adolf (1882-1957)	PhD, asst 1907-10 (1,7a)	Prof chemistry Königsberg
Speier, Arthur	1895 (2); PhD Rostock 1896	Chemical industry (BASF)
Sperling, Rudolf (1888-1914)	PhD 1913 (1)	
SPIRO, KARL (1867-1932)	PhD 1889 (1)	Prof physiological chem Basel
Stahel, Rudolf (1866-)	PhD 1891 (5)	Chemical industry (Sydowsawe)
Stahlschmidt, Axel (1882-1966)	PhD 1910 (2)	Chemical industry (Bayer)
Steche, Albert (1862-)	PhD 1887 (3)	Chemical industry (Heine & Co)
Steingroever, Joseph (1884-)	PhD 1907 (2)	Chemical industry (Deutsche Oxyhydrierges)
Stewart, Anthony *USA*	PhD 1892 (1)	Landowner
Strauss, Hermann (1884-)	Post-MD 1912 (3)	Hospital chief Berlin
Suzuki, Umetaro (1874-1943) *Japan*	Post-PhD 1904-05 (5)	Prof chemistry Tokyo
Täuber, Ernst (1861-1944)	Post-PhD asst 1884 (2a)	Govt patent office Berlin
Tafel, Julius (1862-1918)	PhD, asst 1884-89 (9)	Prof chemistry Würzburg
Tappen, Hans (1879-1969)	Post-PhD asst 1906-07 (4a)	Zeiss-Ikon AG
Techow, Walter (1870-)	PhD 1894 (1)	
Teru-uchi, Yukata (1873-1936) *Japan*	Post-MD 1905 (1)	Kitasato Inst Tokyo
Thieme, Bruno	PhD 1891 (2)	Pulverfabrik Troisdorf
Torrey, Henry Augustus (1871-1910) *USA*	Post-PhD 1898 (1)	Asst prof chem Harvard
Trenkler, Bruno	PhD 1887 (2)	
Troschke, Hermann Oswald (1851-)	Post-PhD asst 1879-80 (2)	
Tüllner, Hermann (1879-1948)	PhD 1902 (2)	
Unna, Eugen (1885-1958)	PhD 1911 (1)	Gen Sec German Pharmaceutical Soc

VAN SLYKE, DONALD DEXTER (1883-1971) *USA*	Post-PhD 1911 (1)	Rockefeller Inst Med Res
Voss, Arthur (1882-1940)	PhD 1908 (1,1a)	Chemical industry (Höchst)
Wagner, Philipp (1863-)	PhD 1887 (2)	Private chemical company
Wagner, Theodore Brentano (1867-1936) *USA*	1891 (1)	Various food companies
WARBURG, OTTO (1883-1970)	PhD 1904-05 (3)	K. W. Inst. Cell Physiology
Weichhold, Oskar (1883-1965)	PhD 1909 (1)	
Weigert, Fritz (1876-1947)	Post-PhD asst 1902 (1)	Prof photochemistry Leipzig
Weil, Albert (1867-)	Post-PhD 1892 (0)	Chemical industry Görlitz
Weil, Hugo (1863-)	PhD 1885 (2)	Chemical industry (Höchst)
Weller, Heinrich (1853-1923)	Post-PhD 1883 (0)	Govt food res inst
WELLS, HARRY GIDEON (1875-1943) *USA*	Post-MD 1905 (1)	Prof pathology Chicago
Wenzing, Max	PhD 1887 (1)	PvDz Chemistry Graz
Wheeler, Alvin Sawyer (1866-1940) *USA*	Post-PhD 1910 (1)	Prof chemistry North Carolina
WINDAUS, ADOLF (1876-1959)	Post-PhD 1900 (2)	Prof chemistry Göttingen
Wirthle, Ferdinand (1873-1936)	PhD 1890 (1)	Govt food res inst
Wislicenus, Wilhelm (1861-1922)	PhD, asst 1883-85 (1,2a)	Prof chemistry Tübingen
Wolfes, Otto (1877-1942)	Post-PhD asst 1900-01 (1,6a)	Chemical industry (Merck)
Wrede, Franz (1877-1946)	PhD, asst 1903-07 (3)	Chemical industry (Schering)
Zach, Karl (1888-1968)	PhD, asst 1910-13 (5)	Chemical industry (Schuster & Wilhelmy)
ZEMPLÉN, GÉZA (1883-1956)	Post-PhD 1908-1910 (8)	Prof chemistry Budapest
Zöllner, Clemens (1884-)	PhD 1908 (1)	Chemical industry (Schering)

* The names in capital letters denote those who achieved considerable distinction. The absence of a date of death for persons born prior to 1900 indicates only that I have not been able to ascertain the date, and is not bo be construed as an affirmation that the person is still alive, where this is not a reasonable assumption. The nationality is indicated only if other than German, Austrian or Swiss.

† PhD refers to a dissertation based on work in Fischer's institute, the year(s) to the approximate time of association, and the figures in parentheses to the number of publications (including dissertations) as author or co-author; the papers in which Fischer made acknowledgement are indicated separately, *e. g.* 3a.

‡ Emigrated 1933 - 1945.

GPSR Compliance

The European Union's (EU) General Product Safety Regulation (GPSR) is a set of rules that requires consumer products to be safe and our obligations to ensure this.

If you have any concerns about our products, you can contact us on ProductSafety@springernature.com

In case Publisher is established outside the EU, the EU authorized representative is:

Springer Nature Customer Service Center GmbH
Europaplatz 3
69115 Heidelberg, Germany

Batch number: 10620141

Printed by Printforce, the Netherlands